Frontispiece. This photograph of Archie McIntyre was taken by Professor G. H. Bell in Dundee while Archie was visiting the University of St Andrews as Praelector in 1974. We are grateful to Professor Bell for permission to reproduce the photograph.

Studies in neurophysiology

Presented to A. K. McIntyre

EDITED BY R. PORTER

PROFESSOR OF PHYSIOLOGY AND CHAIRMAN
DEPARTMENT OF PHYSIOLOGY, MONASH UNIVERSITY

CAMBRIDGE UNIVERSITY PRESS

CAMBRIDGE

LONDON · NEW YORK · MELBOURNE

Published by the Syndics of the Cambridge University Press
The Pitt Building, Trumpington Street, Cambridge CB2 1RP
Bentley House, 200 Euston Road, London NW1 2DB
32 East 57th Street, New York, NY 10022, USA
296 Beaconsfield Parade, Middle Park, Melbourne 3206, Australia

First published 1978

Printed in Great Britain
at the University Press, Cambridge

Library of Congress Cataloguing in Publication Data
Main entry under title:
Studies in neurophysiology.
Includes index.
1. Sense-organs. 2. Neurophysiology.
I. McIntyre, Archibald Keverall, 1913–   .
II. Porter, Robert, 1932–   . [DNLM:
1. Neurophysiology. WL102. 3 S933]
QP431. S76   599'. 01'88   78–1695
ISBN 0 521 22019 X

Contents

CONTENTS

List of contributors

PROFESSOR MAMORU AOKI, Department of Physiology, Asahikawa Medical College, Asahikawa 071-01, Japan

PROFESSOR C. B. B. DOWNMAN, Professor and Head, Department of Physiology, The Royal Free Hospital School of Medicine, 8 Hunter Street, London WC1N 1BP, England

SIR JOHN ECCLES, 'Ca' a La Gra', Contra (Locarno) 11, CH 66 11, Switzerland

PROFESSOR CARLOS EYZAGUIRRE, Chairman, Department of Physiology, College of Medicine, 410 Chipeta Way, Room 156, Research Park, University of Utah, Salt Lake City, Utah 84108, USA

DR S. J. FIDONE, Department of Physiology, College of Medicine, 410 Chipeta Way, Research Park, University of Utah, Salt Lake City, Utah 84108, USA

PROFESSOR M. E. HOLMAN, Department of Physiology, Monash University, Clayton, Victoria 3168, Australia

PROFESSOR JOHN I. HUBBARD, Professor of Neurophysiology, Department of Physiology, University of Otago, P.O. Box 913, Dunedin, New Zealand

PROFESSOR C. C. HUNT, Department of Physiology and Biophysics, Washington University School of Medicine, 660 S. Euclid Avenue, St Louis, Missouri 63110, USA

PROFESSOR A. IGGO, Department of Veterinary Physiology, Royal (Dick) School of Veterinary Studies, Summerhall, Edinburgh EH9 1QH, Scotland

DR J. JULIAN B. JACK, University Laboratory of Physiology, Parks Road, Oxford OX1 3PT, England

PROFESSOR SIR BERNARD KATZ, Department of Biophysics, Gower Street, London WC1E 6BT, England

PROFESSOR Y. LAPORTE, Laboratoire de Neurophysiologie, Collège de France, 11 Place Marcelin Berthelot, 75231 Paris Cedex 05, France

PROFESSOR DAVID LLOYD, New Cottage, Greatham, Pulborough, Sussex RH20 2ES, England

PROFESSOR JOHN LUDBROOK, Department of Surgery, Royal Adelaide Hospital, University of Adelaide, Adelaide, S.A. 5000, Australia

PROFESSOR W. I. MCDONALD, Institute of Neurology, University Department of Clinical Neurology, The National Hospital, Queen Square, London WC1N 3BG, England

PROFESSOR RICHARD F. MARK, Department of Behavioural Biology, Research School of Biological Sciences, ANU, P.O. Box 475, Canberra City, ACT 2601, Australia

PROFESSOR R. PORTER, Chairman, Department of Physiology, Monash University, Clayton, Victoria 3168, Australia

DR UWE PROSKE, Department of Physiology, Monash University, Clayton, Victoria 3168, Australia

PROFESSOR WILFRED RALL, Department of Health, Education and Welfare, Public Health Service, National Institutes of Health, Bldg 31, Rm 9A-17, Bethesda, Maryland 20014, USA

PROFESSOR J. R. ROBINSON, Department of Physiology, University of Otago, P.O. Box 913, Dunedin, New Zealand

PROFESSOR G. H. SATCHELL, Chairman, Department of Zoology, University of Otago, Box 56, Dunedin, New Zealand

PROFESSOR G. C. SCHOFIELD, Dean of Medicine, Monash University, Clayton, Victoria 3168, Australia

PROFESSOR J. D. SINCLAIR, Department of Physiology, School of Medicine, University of Auckland, Private Bag, Auckland, New Zealand

PROFESSOR G. SOMJEN, Department of Physiology and Pharmacology, Duke University Medical Center, Durham, North Carolina 27710, USA

DR D. TRACEY, Department of Physiology, Monash University, Clayton, Victoria 3168, Australia

DR COLWYN TREVARTHEN, Department of Psychology, University of Edinburgh, 1–7 Roxburgh Street, Edinburgh EH8 9TA, Scotland

PROFESSOR J. L. VEALE, Department of Physiology, University of Adelaide, P.O. Box 498, Adelaide, S.A. 5001, Australia

DR R. A. WESTERMAN, Department of Physiology, Monash University, Clayton, Victoria 3168, Australia

DR POH TECK YEO, 5 Lucknow Court, Wessex Estate, Singapore 5

PROFESSOR MANFRED ZIMMERMANN, Universität Heidelberg, II Physiologisches Institut, D 6900 Heidelberg 1, West Germany

A tribute to A. K. McIntyre

In the university world there can be few more exciting or more demanding situations than assuming responsibility for establishing a new department. When this exhilarating task is shared with a progressively increasing number of colleagues in the course of development of a new university, an occasion when planning strategies for material development and introduction of new programmes of research and instruction are as competitive as they will ever be, the resulting department is a fair reflection of both the talents of its initiator and the breadth of experience he has had to offer. In 1968 Sir Robert Blackwood, the first Chancellor of Monash University, reviewed events at the university during its first decade.* The Chancellor wrote of the difficulties met in acquiring a Professor of Physiology who could match the exacting requirements of the Council of the university. Subsequently 'Leading figures in the field were consulted and as a result Professor Archibald Keverall McIntyre, Professor of Physiology at the University of Otago, New Zealand, was invited to the Chair. When Professor McIntyre accepted the invitation the Interim Council felt that it had obtained a man whose research interests and whose previous experience as head of a department of physiology would ensure the rapid development of a substantial and first class department at Monash.' This expectation has been met in full measure, with inestimable advantage to biological science in Australia and to the great benefit of staff members in the Faculties of Medicine and Science at Monash University and their many graduate and undergraduate students.

Professor McIntyre's interest in physiology was established early in his career. After an outstanding undergraduate course and clinical experience within hospitals immediately after graduation from Sydney University, he entered the mainstream of neurophysiological research with experience in a variety of settings including the Rockefeller Institute. The School of

* Blackwood, R. (1968). *Monash University: the first ten years*. Melbourne: Hampden Hall.

Neurophysiology at Otago University was strengthened by the arrival of Professor McIntyre who remained to develop the school further after Sir John Eccles' departure to join the John Curtin School of Medical Research at Canberra. These periods were exciting ones for New Zealand students and the standards of excellence set by the two Australians have meant much for those who experienced them, whatever their subsequent careers.

Professor McIntyre's influence on physiological studies in Australasia has been quite remarkable. His achievements scientifically and the formal acknowledgement of these through various distinctions and awards will be well known to those who will have a special interest in the contents of this volume. Those who have had the privilege of association with him know that these achievements belong to a modest, unassuming and generous man whose continuing concerns are the welfare of those about him, the provision of resources, facilities and a milieu which best enables them to achieve the highest goals they can set for themselves and, perhaps most importantly that scientific and societarian movement shall be in the best interests of society itself. It is not surprising that he should have been joined both at Otago and Monash by men and women of outstanding quality who continue in scintillating fashion to promote the interests of the discipline for which he himself has worked so effectively.

In the following pages friends and colleagues have paid their own tributes to Archie McIntyre in a way that they know he would like best, and several particular themes encompassing areas of special interest to him have emerged. Many other themes of collateral rather than subsidiary interest could also have been presented for he has an extensive range of interests and an even wider range of friends who would have liked to acknowledge publicly his achievements and to express in the same way their gratitude for his warm friendship. It has been left for me to convey something of their understanding of the part he has played in issues of real significance to the world of scholarship and to express something of their gratitude to him for his humour, his companionship and his generosity of thought. But perhaps the best tribute to his energy and judgement and capacity for scholarship is the present high standing of the department he now leaves in the hands of others, all highly talented and superbly equipped to proceed, as each succeeding generation must, towards new goals. This department is indeed a towering tribute to the Foundation Professor of Physiology at Monash University.

G. C. Schofield
Dean, Faculty of Medicine, Monash University

Foreword

Archie McIntyre's contributions to neurophysiology, particularly in Australia and New Zealand, are recognized here by contributions from some of his ex-students, colleagues and collaborators. It is the practice in Australian universities for professors to retire from their departmental duties at the end of the year in which they reach 65 years of age. This book of collected works is presented to Archie McIntyre as a tribute by his associates on that occasion.

Archie's personal contributions to neurophysiological knowledge are well recognized. Even this book cannot indicate his important influence on the work of a whole generation of associates and on the development of neurophysiology as a respected discipline in the two universities to which he has contributed so greatly – Otago, where he succeeded Sir John Eccles in the Chair of Physiology and Monash, where, in 1961, he founded a thriving and vital department in a new university. But an attempt is made here to convey some of the enthusiasms for physiology which have been engendered by Archie in his associates during the last 30 years and the enduring influence which Archie has had on their work.

Not all those who would have regarded themselves as Archie's pupils in graduate school have been able to contribute to this volume. Some now work in fields remote from neurophysiology. Others felt that the things they might write would be repetitions of work presented here by others. But all those who could be contacted wished to be associated with this venture, with the tribute to past achievements and with the good wishes for future successes during Archie's retirement from active teaching but not, we hope, from collaborative research in neurophysiology.

Some of these contributions are from ex-students who began their careers in neurophysiology under Archie's guidance. From him they learned the skills of operative procedures, of dissection of nerve filaments and of controlled stimulation and recording techniques using a variety of electronic apparatus. They would all agree that they learned much more and that their present attitudes to their scientific work owe much to Archie's influence. Other contributions come from those with whom Archie worked when he travelled overseas (to the Rockefeller Institute, to Salt Lake City and to University College, London) or from those who sought out his laboratory in the Antipodes to spend their own sabbatical leave. Finally, there are tributes from Archie's contemporary colleagues in teaching and research.

Most of the presentations deal with the neurophysiology of sensory receptors and the influences produced by nerve impulses in their afferent fibres. But Archie's interests have never been limited to these matters and some of his associates have been encouraged to explore fields far removed from the physiology of the muscle spindle or the Pacinian corpuscle. An indication of the scope of these other interests is contained within these pages.

R. Porter
Monash University

September 1977

The release of the
neuromuscular transmitter and
the present state of the
vesicular hypothesis

B. KATZ

In discussing the transmission of brief nerve messages one can draw a line between two different processes. There is an important distinction between conduction of impulses in individual cells, and the transfer of signals from one cell to another, across synapses which are the points of functional contact between excitable cells.

The nerve impulse is a wave of electric excitation, involving the influx of Na^+, which is passed on electrically from one region of the cell to the next: its continuous and automatic propagation depends, as in a submarine cable, on the continuity of the internal core, that is, the cytoplasm, and on the continuity of a high-resistant, insulating membrane which forms a cylindrical sheath. This type of continuity, of a conducting core and an insulating surface membrane, is broken at most synaptic contacts, where the two cells are in close, but not in intimate, apposition, and in fact are separated from one another by a small extracellular fluid shunt.

We have known since the work of H. H. Dale and his colleagues 40 years ago that a special process of chemical mediation is interposed at such points (Dale, Feldberg & Vogt, 1936). Dale examined the particular case of the neuromuscular junction. He showed very clearly that, although each impulse in the motor nerve gives rise without fail to an immediate impulse and contraction in the muscle fibres, there is no direct continuity of electric conduction. On the contrary, the nerve impulse causes the secretion from the terminals of a specific substance, namely acetylcholine (ACh), and ACh diffuses across the small intercellular gap, reacts with specific receptive molecules in the muscle membrane, at the so-called endplate, and starts up a new wave of electrically propagated excitation. My colleagues and I have been working on these two peculiar intermediate steps: the neurosecretory process in the motor nerve terminals and the chemo-receptor mechanism in the muscle endplate. We have used an electro-physiological approach, by applying microscopic electrical probes to single nerve–muscle junctions and measuring localized changes in the membrane potential of the muscle fibre at its junctional region, and the local ionic

currents which are associated with the transmission of impulses and with
the local action of ACh. This type of technical approach has been very
rewarding. It is a fruitful technique because nature has provided us in the
endplate membrane of the muscle fibre with a powerful pre-amplifier,
perhaps I should say with a highly sensitive chemo-electric transducer,
which is adapted to receive a small quantity of ACh and to translate it into
a large electrical potential change, sufficient to start up a muscle impulse.

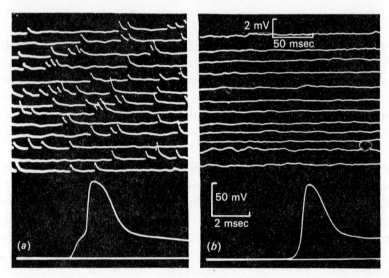

Fig. 1. Spontaneous 'miniature endplate potentials' (from Fatt & Katz, 1950).
(a) Intracellular recording at a frog endplate; (b) recorded 2 mm away in same
muscle fibre. Upper portions were recorded at low speed and high amplification;
they show the localized spontaneous activity at the junctional region. Lower records
show the electric response to a nerve impulse, taken at high speed and lower gain.
The stimulus was applied to the nerve at the beginning of the trace; response (a)
shows a step-like endplate potential leading to a propagating muscle spike; in (b),
the spike alone is recorded after additional delay due to conduction along 2 mm of
muscle fibre.

Of the various experiments which my colleagues, Paul Fatt, Jose del
Castillo, Ricardo Miledi, and I have been doing over the years, I want to
single out one or two phenomena which we have found of great interest.

The first one concerns the mechanism by which the transmitter substance
is secreted from the nerve terminals. When one inserts an electrical
microprobe into a muscle fibre at various points, one finds that the endplate
region, the point of the muscle fibre which is contacted by the nerve, is in a
state of spontaneous local activity, even when the nerve–muscle system is
otherwise in a state of rest, when there is no impulse traffic and no visible

movement among the muscle fibres. The endplate region shows a random sequence of minute local waves of depolarization, about 0.5 mV in amplitude and all of the same standard time course.

When a nerve impulse arrives, it produces a similar localized potential change, but a hundred times larger in size; this is the so-called endplate potential (EPP), and this large wave immediately exceeds the firing level of the muscle, and gives rise to a propagating electrical impulse in the muscle fibre which in turn causes the fibre to twitch. When Paul Fatt and I examined the nature of the little spontaneous blips, which we called the miniature EPPs, we were left in no doubt that they were due to the impact of small quantities of ACh, which were discharged spontaneously at random intervals by the adjacent nerve ending (Fatt & Katz, 1952). The local region of the muscle fibre is a very sensitive indicator for this substance, so that it registers the intermittent release of a droplet of ACh in the form of a 0.5-mV potential change. That it *is* ACh and not some other agent, became clear when we examined the effects of specific ACh antagonists like curare, and of specific cholinesterase inhibitors. Normally, any ACh which is discharged from the nerve is rapidly hydrolysed, and split into choline and acetic acid by a potent enzyme which is concentrated in the synaptic cleft between nerve and muscle membrane. When we stop the hydrolysis by esterase inhibitors like prostigmine or edrophonium, we potentiate the effects of any artificially applied ACh, and at the same time, we cause the amplitude and the duration of the spontaneous blips to increase, to the same extent.

It also became quite clear that these discrete all-or-none blips, the miniature EPPs, must be due to a synchronous discharge of a large packet, probably thousands of molecules at a time, because individual molecular effects are much too small to be resolved as discrete events by the ordinary intracellular recording methods. If we apply graded doses of ACh in solution, by discharging them ionophoretically from a nearby micropipette, the response appears to be continuously graded in size and duration: the individual molecular components are evidently below the resolving power of our instruments.

The spontaneous miniature potentials are of very great functional importance. They form the statistical unit, the basic coin of neuromuscular transmission (del Castillo & Katz, 1956; Katz, 1969; Ginsborg & Jenkinson, 1976). When a nerve impulse arrives at the junction, it produces a much larger effect; but it has been shown quite convincingly that this large, impulse-evoked EPP is made up statistically of an integral number of quantal units which are identical with the spontaneous miniature potentials. The nerve impulse merely accelerates the process of secretion of ACh, of the same standard packets, so that for a brief period of about 1 msec, after the arrival of an impulse, a few hundred packets are discharged in synchrony, while at rest the rate of secretion is a few hundred thousand times lower, with an average occurrence of about 1 packet per second at each endplate.

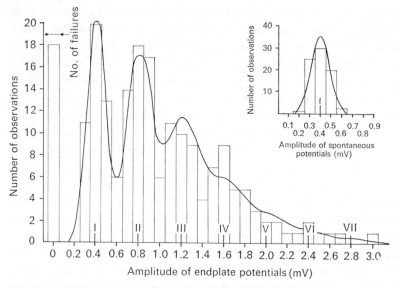

Fig. 2. Histograms of endplate potential (EPP) and (inset) spontaneous miniature potential amplitudes, from a mammalian endplate (from Boyd & Martin, 1956). Peaks of EPP amplitude distribution occur at one, two, three and four times the mean amplitude of the spontaneous potentials. A simple Gaussian curve has been fitted to the latter and used to calculate the theoretical distribution of EPP amplitudes (multi-peaked curve). Arrows indicate expected number of failures (zero amplitude).

There are several questions which arise from these observations, and to which we cannot give a complete answer, though a good deal of knowledge has come to light. We know that the rate of secretion depends on a local chemical reaction on the inside of the nerve membrane, in which Ca^{2+} play an essential part. The rate of secretion can be enormously increased above the background level, if one raises the concentration of free Ca^{2+} inside the nerve terminal (Miledi, 1973). We also know that depolarization of the terminal, such as occurs during the nerve impulse, opens up gates in the nerve membrane through which external Ca^{2+} can enter, flowing down a very steep concentration gradient from outside to inside (Katz & Miledi, 1967b, 1969a,b; Baker, Hodgkin & Ridgway, 1971; Blaustein, 1975). If one replaces external Ca^{2+} by Mg^{2+}, transmission fails because, in the absence of local Ca^{2+} influx, the impulse no longer evokes secretion of ACh packets.

The low spontaneous background frequency of miniature potentials also depends on the presence of Ca^{2+}, this time not on external calcium – because during periods of rest the calcium gates in the membrane are shut – but on the steady level of ionized calcium in the cytoplasm (Rahamimoff, 1976). This is normally kept very low by the calcium-absorbing power of the intracellular mitochondria and possibly other subcellular particles (Kendrick, Blaustein, Fried & Ratzlaff, 1977). If one disturbs the calcium-binding capacity of the mitochondria, by metabolic inhibitors or by certain heavy metal ions, the spontaneous discharge rate soon goes up together with the rising level of intracellular Ca^{2+}.

How do we picture this process of a calcium-dependent quantal secretion, always occurring in the same discrete steps, in minimum packets of very large multimolecular content? Is there a recognizable structural basis for this apparent pre-packaging of the transmitter substance in standard parcels, and for the vast acceleration of the rate of release during excitation?

When Paul Fatt and I saw the spontaneous miniature EPPs (see Fatt & Katz, 1950), we thought at first that they might possibly originate from a process of spontaneous self-excitation in exceedingly fine nerve terminals. The idea was that in sufficiently small axon branches, thermal agitation of ions might produce large enough voltage noise in the membrane to rise occasionally above threshold and so elicit local spikes at random intervals. It was conceivable that such spikes might remain localized through failure to invade large axon branches in the retrograde direction, but there would certainly ensue transmitter release from the locally activated terminal. However, it was not long before we abandoned this hypothesis, because it appeared, on any reasonable calculation, that the motor nerve terminals are in fact not small enough to produce the required voltage noise, and for the same reason, that if an action potential did arise in one of the terminals, it could hardly fail to spread into all the other axon branches. We then pursued a different line of speculation (cf. Fatt & Katz, 1953), and considered that ACh might be released from the terminal axoplasm by means of a membrane carrier, in exchange for the Na^+ which were thought to enter the axon via a similar carrier mechanism. The miniature EPP, we thought, might be the result of spontaneous activation of a single carrier molecule, while the nerve impulse was associated with synchronous activation of a large number of such carriers. In a modified form this idea still persists, though one generally prefers nowadays to invoke specific membrane 'gates' or 'channels' for the rapid transfer of substances;

carrier molecules are believed to be too sluggish to shuttle across the membrane at the required rate.

The concept of quantal ACh release being linked with normal nerve excitation or with the entry of Na^+ lost its attraction completely when del Castillo and I (1954, 1955) discovered that miniature EPPs still occur in sodium-free media, and after all traces of electric excitability have been removed by application of isotonic potassium sulphate. This was followed by the observation that abolition of electric excitation by the use of tetrodotoxin (TTX) has no influence whatever on the quantal secretion process and that, in TTX, not only do miniature EPPs persist, but full-size EPPs can be evoked if one locally depolarizes the paralysed nerve endings, provided only that Ca^{2+} are still present in the surrounding medium (Katz & Miledi, 1967a).

In 1954, we had reached a stage where it appeared that the transmitter ACh is always delivered in standard-sized, multimolecular packets, whose rate of secretion can be made to vary enormously from moment to moment, by altering the membrane potential of the axon terminal, and by altering the chemical composition of the environment, but whose size remained constant under these changing conditions. Since that time means have been found to alter the packet size, for instance as a result of exhausting stimulation (Elmqvist & Quastel, 1965; Kriebel & Gross, 1974), but the fact remains that over a very wide experimental range the size of the packet does not depend on the way in which it is set free, whether spontaneously, or by a nerve impulse, or by local electrical, osmotic or chemical stimulation. The amplitude of the miniature EPP does, of course, depend on other factors, such as the presence or inhibition of cholinesterase activity, and the availability or blockage of postsynaptic receptors, but if these conditions which concern the stages following the release are kept constant, *pre*synaptic changes which temporarily raise the frequency of miniature EPPs by several orders of magnitude do not alter their individual size.

This constancy of the transmitter packet, regardless apparently of changing electrical and chemical potential gradients, caused us to query the view that constant quanta of ACh might escape from the cytoplasm through a pre-set 'membrane shutter' or 'carrier' mechanism. The idea of a fixed 'ACh-gate' which opens for a pre-determined interval, seemed possible, but not very attractive to us, because we did not regard it as very likely that both the gating action and the rate of efflux of ACh from the axoplasm would remain uninfluenced by the physical and chemical changes which had been imposed on the axon membrane (Katz & Miledi, 1965). We preferred, therefore, to think in terms of a concentrated package of the transmitter substance built up within cytoplasmic organelles, separate from and independent of the axon membrane, and providing a discrete quantity of ACh for instant release into the synaptic cleft (del Castillo & Katz, 1955, 1956).

It was at that time that the first important clues came from examinations of the ultrastructure of the synapse in the electron microscope (de Robertis & Bennett, 1954; Palade, 1954; Palay, 1954; Robertson, 1956). Of the numerous subcellular features which are characteristic of synapses, one of the most striking is the accumulation of vesicles – about 50 nm in diameter – in the nerve terminal. These 'synaptic vesicles' are clustered together in large numbers at the very points where we think the neuromuscular transmitter is discharged into the synaptic cleft (Birks, Huxley & Katz, 1960; Akert, Moor, Pfenninger & Sandri, 1969; Couteaux & Pécot-Dechavassine, 1970, 1974; Heuser, Reese & Landis, 1974; Heuser, 1976). Biochemical analysis of subcellular fractions from cholinergically inner-vated tissue have shown that a large part of the ACh is concentrated inside these vesicles (Israël, Gautron & Lesbats, 1968; Zimmermann & Whittaker, 1974), and there is a widespread belief that the transmitter is normally concentrated and pre-packaged inside them, and that during certain critical collisions with the nerve membrane a vesicle fuses and breaks open, discharging its content into the synaptic interspace – a process of so-called 'exocytosis'. If one accepts this idea, it is clear that Ca^{2+} are needed for this to occur, and it has been suggested that Ca^{2+} react with molecules on the inside of the nerve membrane so as to provide reactive attachment sites for colliding vesicles.

There are now many examples of pictures of synapses, obtained with different methods of fixation and after varying experimental pre-treatment, which show vesicles situated at characteristic zones of the presynaptic membrane and opening into the synaptic cleft, thus seemingly demonstra-ting what could be regarded as a structural counterpart of the mechanism of quantal release and of the production of a miniature EPP (Nickel & Potter, 1971; Couteaux & Pécot-Dechavassine, 1970; Heuser, Reese & Landis, 1974). There is other converging ultrastructural evidence which supports the idea of exocytosis of vesicles (Ceccarelli, Hurlbut & Mauro, 1973), and although the interpretation of static pictures in terms of dynamic events will always remain hypothetical, few will deny that the examination of the synapse with the electron microscope has provided us with a very important clue and greatly advanced our thoughts about the mechanism of quantal transmitter release.

One of the serious criticisms which has been directed against the vesicular release hypothesis was that it seemed doubtful whether a 50-nm vesicle is capable of accommodating enough ACh to produce a miniature

EPP. For a long time, we were confronted with a wide range of uncertain estimates of the number of molecules required for a quantal effect. Values of 50000 or more were current in the literature, and such large estimates were indeed difficult to reconcile with the concept that a single vesicle might contain them (see Katz, 1969; Katz & Miledi, 1973b). More recently the estimates have been lowered, and their range has been narrowed down to somewhere between 10^3 and 10^4 molecules per quantal packet. These values were derived by three different, but converging, experimental approaches: (a) by an attempt to determine the depolarizing effect of single ACh molecules or, more precisely, by determining the depolarization due to the opening of a single postsynaptic ion channel by the transmitter molecules (Katz & Miledi, 1970, 1972, 1976). This will not tell us directly how many molecules make up the quantal packet, but at least will give us an idea of how many elementary or molecular 'gating' actions are needed to produce a miniature EPP. (b) The second approach was made by Kuffler & Yoshikami (1975) in an experiment designed to find the minimum pulse of ionophoretically applied ACh to imitate the miniature EPP. (c) The third way was to use a very sensitive assay technique to measure the amount of ACh liberated during a period of nerve stimulation and relate it to the total number of miniature EPPs produced (Fletcher & Forrester, 1975).

I propose to deal at some length with the first of these methods, which has wide applications in the field of synaptic transmission and, more generally, of drug–membrane interactions. It is based on the analysis of electric membrane fluctuations, or 'noise' phenomena, which were found to be invariably associated with steady-dose effects of ACh and other depolarizing agents on a motor endplate. While I shall be referring largely to experiments by Ricardo Miledi and myself, I want to emphasize that the subject has been greatly advanced by the successful application of the voltage clamp technique in the hands of Anderson & Stevens (1973), and quite recently by Neher & Sakmann (1976a) who developed an ingenious method of recording discrete current pulses flowing through single endplate channels.

When R. Miledi and I started the experiments, we had no direct way of recording the action of a single transmitter molecule as a discrete event. As I mentioned previously, when we applied a dose of ACh, the membrane response looked continuously graded, and we could not resolve its small molecular components with our recording instruments. It was clear that they are very much smaller than the 0.5-mV blip due to a single packet of ACh coming from the nerve. Nevertheless, miniature EPPs and the graded ACh potential must be made up of a statistical superposition of elementary

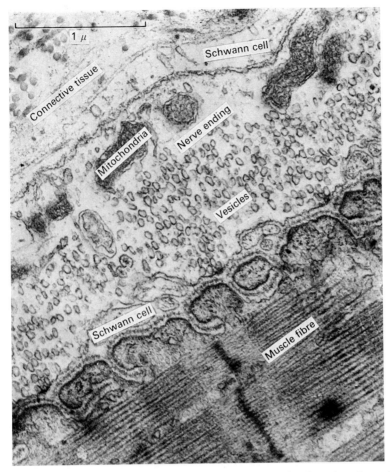

Fig. 3. Electron micrograph of portion of neuromuscular junction from frog sartorius (from Birks, Huxley & Katz, 1960). Longitudinal section of muscle. Note: presence of mitochondria and synaptic vesicles inside the nerve ending.

depolarizations, due to the transient opening and closing of ionic gates in the cell membrane by brief molecular interactions. During a steady dose of ACh, one observes a persistent depolarization which corresponds to a maintained average frequency of effective collisions between ACh and receptor molecules. Each of these collisions opens a channel in the membrane through which a pulse of current – consisting mainly of Na^+ ions – enters the fibre and displaces its membrane potential by a very small amount. Although we could not detect the individual pulses, we found that one can readily observe the statistical fluctuations due to the random variation in pulse rate from moment to moment. We decided to study this ACh-produced 'noise', and to analyse the minute fluctuations in membrane voltage which characteristically accompany the depolarization induced by the drug.

On simple Poisson statistics, it is not difficult to determine the approximate *size* of the elementary 'molecular' depolarization: all that one needs is to measure the mean and the variance and calculate the ratio of the variance to the mean. The important assumption involved here is that both variance and mean are the result of a statistical superposition of the *same* elementary quantities. The method for the noise analysis, apart from conventional computations of mean and variance, was the derivation of the so-called power spectrum, that is, a Fourier analysis of the spectral distribution of the variance. In this way, one can determine, at least to a first approximation, the average amplitude of the molecular shot effect, that is, the voltage pulse, the intensity of the underlying current pulse and the ionic conductance change, and finally the average lifespan, or relaxation time of the ionic channel.

To summarize the results: At an endplate where the miniature potential is 0.5 mV, the 'molecular' depolarization amounts to roughly 0.25–0.30 μV, that is, 1500–2000 times less. This suggests that the quantal action, produced by the standard packet of ACh, is made up of a superposition of a few thousand molecular effects, that is, the simultaneous opening of a few thousand ion channels. The actual number of ACh molecules released with each quantal packet is probably appreciably larger because (*a*) it is possible that not one, but two or three molecules are needed for each elementary action (see Colquhoun, 1975; Rang, 1975) and (*b*) not all the transmitter molecules which are discharged into the synaptic cleft are likely to be effective in opening ion gates.

Using the voltage clamp technique, Anderson & Stevens (1973) found that each of these ion channels passes an inward current of a few picoamperes and has a conductance of approximately 2 to 3×10^{-11} reciprocal ohms. Summarizing, the molecular effect of the ACh–receptor interaction

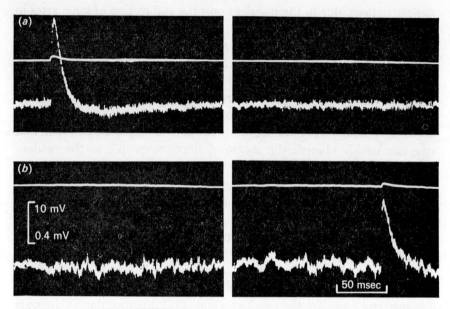

Fig. 4. 'Acetylcholine (ACh) noise' recorded from an endplate in frog sartorius (from Katz & Miledi, 1970). In each block, the upper trace was recorded on a low-gain d.c. channel (scale 10 mV); the lower was simultaneously recorded on a high-gain, condenser-coupled channel (scale 0.4 mV). (a) Controls (no ACh); (b) membrane noise during ACh application, by diffusion from a micropipette. In (b) the increased distance between the low- and high-gain traces is due to upward displacement of the d.c. trace because of ACh-induced depolarization. Two spontaneous miniature endplate potentials are also seen.

is some three orders of magnitude smaller than the effect of the minimal packet which is normally released by the nerve, and five to six orders of magnitude smaller than the transmitter action of a full-fledged nerve impulse at each endplate.

The elementary depolarization of 0.25 μV is too small to be detected directly. With the conventional micro-electrode set-up one cannot easily measure membrane potential changes which are less than 20 or 30 μV, that is, two orders of magnitude greater than the molecular action. By using the statistical approach, one is looking at continuous fluctuations which involve – not discrete single events – but hundreds of events at a time. In effect, one applies a dose of ACh which will keep a large average number of gates open at each endplate. Suppose the mean number is 10000, with a Poissonian standard deviation equal to the square root of the mean, that is, one hundred ion gates, one obtains fluctuations large enough to measure and analyse.

The other interesting property of the ACh noise is its temporal characteristic displayed by its Fourier spectrum. If one measures intracellular voltage noise only, the information is rather limited, because the main part of the spectrum is then determined by the large membrane capacity of the muscle fibre, and the time course of the analysed elementary voltage pulse simply reflects the slow passive decay of the brief displacement caused by each pulse effect. The result of the analysis of such a voltage noise recording shows a small blip, a fraction of a microvolt in amplitude, which resembles in time course the miniature EPP. It is more pertinent to record the fluctuations in membrane current. One can do this either by the voltage clamp technique, or by placing an electrode extracellularly, right at the focus where the current converges into the active region. One then finds noise spectra of a much higher cut-off frequency, and on analysis these spectra show a brief elementary current pulse of about 1 msec in duration at 20 °C. This presumably corresponds to the average lifespan, or relaxation time of the ionic channels.

It is interesting that the ionic current during the nerve-evoked EPP is only slightly longer; it has a relaxation time of about 1.5 msec, 50 % longer than the effect of the molecular action, which indicates that the thousands of molecules discharged in packets from the nerve terminal are very rapidly removed from their site of action, so that most of the molecules have no time to open more than one ionic channel. By the time the channels close, few ACh molecules remain to repeat their action. The rapid removal is brought about by the specific hydrolytic enzyme ACh-esterase, which is present at high concentration in the synaptic cleft. If one treats the preparation with a cholinesterase inhibitor, or removes the esterase by a proteolytic enzyme, the duration of the endplate currents can be lengthened severalfold, while there is no change in the time characteristics or the size of the ACh noise; in other words, when the rapid local hydrolysis of ACh has been stopped, the molecular effect – the opening and closing of the single channel – remains unaltered, but the ACh molecules in the cleft can now repeat their action several times before they are dispersed by thermal diffusion (see Katz & Miledi, 1973b).

The analysis of endplate noise has thrown interesting sidelights on the molecular characteristics of various other drug actions (Katz & Miledi, 1972, 1973a; Colquhoun, 1975; Neher & Sakmann, 1976a,b). To give a few examples: ACh is the natural transmitter, but it is not the only substance which is capable of reacting with the endplate receptor and of depolarizing the endplate. There are many other chemical agents which have a similar effect, though most of them are less potent. We have tried quite a number of them, for instance carbachol, which has a similar action, but has the practical advantage of not being attacked by the hydrolytic enzyme. When we compared the effects of ACh and carbachol, we were very surprised to find that carbachol is much less 'noisy' than ACh. That is to say, when we gave equivalent doses of the two agents which produced the same depolarization, the ACh effect was accompanied by much larger voltage fluctuations than carbachol. It was clear from an analysis of the records, that the elementary effect, that is, the depolarization produced by the carbachol molecule, is much smaller than the molecular effect of ACh, and the difference is due to a briefer lifetime of the channel opened by carbachol, only about 0.35 msec instead of 1 msec. It is very probable that the lifespan of the ion channel is related and possibly coincident with the lifetime of the drug–receptor complex. If this complex is less stable, and dissociates more rapidly in the case of carbachol than of ACh, then this would explain the shorter lifetime of the open channel, hence the smaller amount of total charge flowing through it, and thus the smaller depolarizing potency per molecule.

We have tried not only a variety of ACh-like agonists, but several antagonists – inhibitors of ACh action – among which curare is the best-known agent. It has for a long time been suggested that curare is a competitive inhibitor which can attach itself reversibly to the same primary binding site of the endplate receptors so that it blocks the normal access to ACh ions. Analysis of ACh noise strongly supports this interpretation: when one gives a large blocking dose of curare which prevents transmission and reduces the EPP to a small residue, one then finds that a much larger dose of ACh is needed to depolarize the endplate, but the accompanying noise fluctuations are unaltered. That is to say, the ratio of variance to mean depolarization and the time characteristics of the noise remained the same, which means that the elementary pulse is unchanged. The conclusion was, therefore, that curare blocked most of the receptors so that ACh molecules had no access, but those receptor molecules which happened to remain free, were reacting in the normal way and produced the same minute depolarization as before.

But there are many other types of ACh-antagonists of a non-competitive kind, among them a variety of anaesthetic membrane agents, which act quite differently. For instance, atropine, procaine, ether and several other substances reduce ACh sensitivity, not by preventing the access to individual receptor molecules, but by changing the kinetics of the ACh–receptor interaction and so reducing the depolarizing potency of ACh (Katz & Miledi, 1973a, 1976). We found, for instance, that atropine reduces the lifetime of the ACh-operated ion channel so that it becomes as short as, or even shorter than, the normal carbachol channel. It looks then as though atropine does not attach itself to the primary binding site, but to some

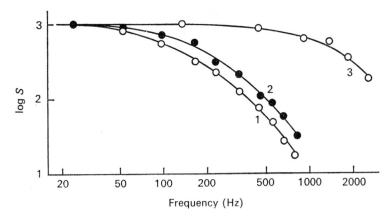

Fig. 5. Spectral distribution of endplate *current* noise produced by different de-polarizing agents (from Katz & Miledi, 1973a). Curve 1, suberyldicholine; curve 2, acetylcholine (ACh); curve 3, acetylthiocholine. The noise fluctuations were recorded with focal micro-electrodes applied externally to the endplate surface. Temperature approximately 22 °C. Ordinate: 'variance density' $S = \overline{dE^2}/df$, scaled to common maximum. Abscissa: frequency (f) in Hz. Double log co-ordinates. Note: the higher frequency characteristic of curve 3 (acetylthiocholine) implies that ion channels produced by this agonist have a briefer lifespan than those due to ACh (curve 2).

adjacent site of the receptor molecule or of its immediate environment, and so affects the kinetics and reduces the stability of the ACh–receptor complex.

The method of ACh noise analysis is relatively new, and there are many potentialities for improving it, and extending it to other cases of drug–receptor interactions, at synapses and membranes more generally. Never-theless, the statistical analysis and interpretation of the ACh noise measure-ments rests on a number of assumptions whose validity can only be checked by resolving the elementary pulses as discrete events. Until very recently this did not seem to be within practical reach, but in 1976, Neher & Sak-mann published a remarkable experiment in which they succeeded in recording the individual current pulses elicited by low doses of ACh and certain depolarizing analogues. They were able to accomplish this by a combination of methods which greatly reduced the background noise and improved the 'signal' produced by the inward current. Neher & Sakmann used an external pipette containing a low concentration of the agonist and applied it closely to the surface of the muscle fibre, effectively insulating a small membrane patch, with a leak resistance of many megohms to the surrounding medium. To do this, collagen and other particles had to be cleaned from the fibre surface by proteolytic enzymes. Furthermore, a long 'lifespan' of the ion channel was aimed at, in order to be able to reduce the band-width, and therefore the noise, of the recording system. This was achieved by working at low temperature, using suberyldicholine (a 'long-channel' agonist) and extrajunctional regions of denervated muscle whose ion channels are known to have long duration. The results of these experiments have in important respects extended the noise analysis and confirmed its main conclusions regarding the conductance of the channel and its average lifetime. Moreover, Neher & Sakmann have shown beyond doubt that the individual current pulses are normally of the square-pulse type, their durations being distributed in the random exponential manner, which had been foreseen but not proven by previous workers.

While this work indicated that the molecular content of a transmitter packet was at least one or a few thousand ACh molecules, the experiments of Kuffler & Yoshikami (1975) led them to an estimate of an upper limit amounting to 10000 molecules. They reached this conclusion by ingenious improvements of the ionophoretic technique, employing a special optical system for optimal placement of the ACh pipette and – even more important – using a special micro-assay method to measure the ionophoretically delivered amounts of ACh. This was done by comparing the depolarizing effects on the same endplate of two small droplets of measured volume, one containing a known concentration of ACh, the other having ACh injected by a series of ionophoretic pulses.

At about the same time, Fletcher & Forrester (1975) re-examined the amount of ACh released from muscles into the surrounding fluid during periods of rest and of nerve stimulation. Considering only the extra amount evoked by stimulation, Fletcher & Forrester arrived at an equivalent of approximately 6000 molecules per quantal packet. On the other hand, the release of ACh during periods of rest gives an entirely different picture: it is much too large to be equated with the small number of quantal packets which are discharged in the absence of stimulation. The discrepancy amounts to two orders of magnitude. This had been known for some time (Mitchell & Silver, 1963), and it was thought to reflect a release from pre-terminal portions of the nerve, too far away from the endplate receptors to produce miniature EPPs. It is, however, quite possible that most of the background release of ACh is due to steady, diffuse leakage from the terminal itself. The motor endplate is a sensitive detector for concentrated packets of ACh, if they are delivered instantaneously from a nearby point source. But if the same, or even a hundred times larger, amount becomes dissipated over the whole synaptic area and spread out over a period of a second instead of less than a millisecond, the result will be an incon-

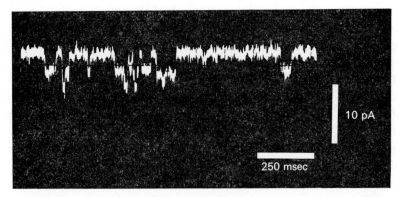

Fig. 6. 'Single channel' current pulses, evoked by suberyldicholine in a small patch of frog muscle membrane (from Neher & Sakmann, 1976a). Temperature 8 °C.

spicuous steady depolarization, less than 100 μV in amplitude, which is only just detectable even with special techniques (see Katz & Miledi, 1977). Thus, in resting muscle there is probably diffuse 'non-quantal' leakage, as well as spontaneous quantal secretion of ACh, while the nerve impulse operates effectively through the quantal mechanism alone.

To sum up after this lengthy diversion, present estimates place the content of the ACh packet above 1000 and below 10000 molecules. To accommodate some 5000 molecules in a vesicle with an outside diameter of 50 nm and an inside diameter of 40 nm is not at all impossible, though it would give a hypertonic ACh concentration of approximately 250 mM. Of course, this estimate is quite compatible also with alternative propositions, such as a discharge of the quantum from the cytoplasm through a membrane gate, or with the possibility that single vesicles may contain much less ACh, and the miniature EPP may be due to the combined discharge of several vesicles. This last suggestion has been put forward to explain the presence of a distinct population of spontaneous miniature EPPs of 'dwarf' size, side by side with the ordinary miniature EPPs whose amplitude is nearly 10 times larger. There are indeed 'giants' as well as 'dwarfs', well outside the normal distribution of miniature EPP amplitudes, and the interpretation of these curious potentials remains as intriguing as it is obscure. Under normal resting conditions, under- and outsize potentials are statistically very infrequent and make up no more than a few per cent of the total population of spontaneous potentials, but by special treatments one can bring them out in larger numbers (Katz & Miledi, 1969b; Pécot-Dechavassine & Couteaux, 1972; Kriebel & Gross, 1974). Neither the 'giants' nor the 'dwarfs' appear to contribute to the impulse-evoked EPP (Liley, 1957; Pécot-Dechavassine & Couteaux, 1972; Bevan, 1976). It has been suggested that many of the giant potentials arise from the discharge of large intracellular vacuoles which have been formed by fusion of normal-sized vesicles. Indeed, some support for this view has been provided by correlated ultrastructural findings (Pécot-Dechavassine & Couteaux, 1972).

The 'sub-standard' population has been studied in detail by Kriebel & Gross (1974) who made the very interesting observation that the normal histogram of miniature EPPs can be, to a large extent, reversibly trans-formed into 'dwarfs' by interposing a period of massive, exhausting transmitter release. In Kriebel & Gross' view the 'dwarfs' represent the basic unit, and the traditional miniature EPP is a composite, multi-quantal product. This is a possible interpretation, but it has not so far been sup-ported by other evidence (see Bevan, 1976), and in the absence of a structural correlate for multi-vesicular exocytosis, it lacks the attraction of the univesicular hypothesis.

It is, in fact, not difficult to think of alternative explanations for the 'sub-standard' miniature EPPs, though again there is no proof. For example, if the recycling and replenishing of discharged synaptic vesicles proceeds in two relatively rapid steps separated by a relatively long time interval, the phenomenon of the 'dwarfs' could be interpreted without abandoning the single vesicular scheme. Suppose the first step, after reconstitution of a vesicular envelope from the axon surface, consists of passive equilibration with the cytoplasm and with its low ACh concentration. The second step is active accumulation of ACh to a much higher intravesicular level. If we assume that the acquisition of this uptake process is delayed, because of the involvement of intermediate reactions, the occurrence of two populations of miniature EPPs with a substantial number of 'dwarfs', during a period of replenishing of quantal packets, would not be difficult to understand.

One of the chief embarrassments to the vesicular hypothesis is the fact that the biochemical studies of cholinergic nerve terminals and their sub-cellular fractions, after a promising start, have remained in a highly con-troversial state (MacIntosh & Collier, 1976). Although ACh-charged vesicular fractions have been isolated, especially from the electric organ of elasmobranchs, these particles show little or no biochemical activity, they appear to be unable to accumulate ACh and merely lose it through gradual leakage. Nor do they show any significant rapid depletion if taken from stimulated tissue. The changes in ACh content and turnover, which occur as a result of stimulation, are not reflected in the 'vesicular' fraction, but in what appears to be the 'free' cytoplasmic fraction of the tissue homogenate. This finding is taken by some authors to be contrary to the vesicular hypothesis and to give strong support for the view that ACh is liberated directly from the axoplasm. Others think that the issue will remain obscure until methods have been found to isolate viable and biochemically competent vesicles from cholinergic nerve endings. It is, after all, conceivable that the particulate residue of the homogenized tissue represents no more than a shadow of functional ACh-loaded vesicles.

The cholinergic system, and especially the vertebrate nerve–muscle junction has been a more fruitful object for electrophysiological than for biochemical exploration of the presynaptic mechanism. It has served to show up the quantal process of neurosecretion, but has not enabled us to decide between vesicular exocytosis and 'membrane shutter' mechanisms. The structures revealed in the electron microscope point to the former mechanism, but decisive ultrastructural and biochemical evidence for exocytosis has so far been confined to the adrenergic system. In the adrenal medulla (Grynspan-Winograd, 1971), and in certain invertebrate neurosecretory organs (Smith & Smith, 1966) very convincing pictures of exocytosis have been seen. This was possible owing to the presence in these cells of vesicles with electron-dense cores: not only can one see vesicles fused with the surface membrane and opened to the external medium, but the picture is completed by the observation of discharged dense cores outside the cell. Moreover, catecholamine-containing vesicles have been isolated which have retained much of their normal biochemical function and are able actively to accumulate and to release the transmitter. Finally, the existence of specific intravesicular proteins, which – in contrast with cytoplasmic proteins – are always released in close association with the catecholamine, has been a decisive argument for exocytosis and against cytoplasmic efflux through a membrane gate (Kirshner & Kirshner, 1971; Smith, 1971). ACh-containing structures have not lent themselves to such a powerful approach, and this remains indeed a continuing biochemical challenge.

Perhaps the most direct attack against the vesicular idea of ACh release has come from recent experiments of Tauc *et al.* (1974) on a molluscan cholinergic axon. These authors found that an intraneuronal injection of cholinesterase led to a progressive failure of transmitter release, apparently by destroying the ACh contained in the cytoplasm, but not that protected inside vesicles. This is an important experiment, and it will be interesting to see whether the authors' interpretation will stand up to further controls and checks: for example, whether the intracellular enzyme injection interferes specifically and exclusively with cholinergic neurones, and whether vesicular ACh content is really maintained even when the surrounding cytoplasmic concentration falls to a very low level.

In reviewing the present status of the vesicular release theory – and confining myself here to the cholinergic system – I think it is fair to say that it maintains its attraction as a working hypothesis, and that it is supported by a mass of converging, but still only circumstantial evidence. It is surrounded by serious problems, and earlier hopes of finding a decisive experimental answer for or against it have not yet been fulfilled. Nevertheless, if the vesicular hypothesis still lacks convincing confirmation, the evidence which has been cited as opposing it seems to me to rest on even less secure foundations.

In view of the technical difficulties which have been encountered in the past, it would be somewhat gratuitous to try to prescribe ways in which the exocytosis problem might be tackled in future. Important advances are at present being made in the isolation of macromolecular constituents of the postsynaptic receptor membrane. Presumably some of the techniques developed in this process will be found to be applicable to the presynaptic release sites in the not too distant future. But if I may try to speculate about much more remote possibilities, I wonder when it will become feasible to observe ultrastructural events with the directness and resolution which is now obtainable only on fixed material. I sometimes doubt whether our ideas about exocytosis in nerve terminals and other dynamic events of subcellular dimensions can really be put to a decisive test except by seeing them in action! This may seem quite fantastic, but perhaps no more so than present-day technical approaches to the study of single neuronal events would have appeared to the pioneers of neurophysiology in the nineteenth century.

To Archie McIntyre

Some 14 years ago you asked me to contribute an article to a 'Festschrift' for Jack Eccles, and I wrote a few pages on the same theme as now. At that time I had genuine hopes that the vesicular hypothesis might be securely established – or definitely thrown out – by now. Not so! The argument goes on. But if you look at the remarkable advances which have been provoked by the continuing controversy about vesicles in the field of ultrastructural, neurochemical and physiological studies, even its adversaries must admit that this is a 'working hypothesis' *par excellence*!

REFERENCES

Akert, K., Moor, H., Pfenninger, K. & Sandri, C. (1969). Contribution of new impregnation methods and freeze etching to the problems of synaptic fine structure. *Prog. Brain Res.* **31**, 223–40.

Anderson, C. R. & Stevens, C. F. (1973). Voltage clamp analysis of acetylcholine produced end-plate current fluctuations at frog neuromuscular junction. *J. Physiol., Lond.* **235**, 655–92.

Baker, P. F., Hodgkin, A. L. & Ridgway, E. B. (1971). Depolarization and calcium entry in squid giant axons. *J. Physiol., Lond.* **218**, 709–55.

Bevan, S. (1976). Sub-miniature end-plate potentials at untreated frog neuromuscular junctions. *J. Physiol., Lond.* **258**, 145–55.

Birks, R., Huxley, H. E. & Katz, B. (1960). The fine structure of the neuromuscular junction of the frog. *J. Physiol., Lond.* **150**, 134–44.

Blaustein, M. P. (1975). Effects of potassium, veratridine and scorpion venom on calcium accumulation and transmitter release by nerve terminals *in vitro*. *J. Physiol., Lond.* **247**, 617–55.

Boyd, I. A. & Martin, A. R. (1956). The end-plate potential in mammalian muscle. *J. Physiol., Lond.* **132**, 74–91.

Ceccarelli, B., Hurlbut, W. P. & Mauro, A. (1973). Turnover of transmitter and synaptic vesicles at the frog neuromuscular junction. *J. Cell Biol.* **57**, 499–524.

Colquhoun, D. (1975). Mechanisms of drug action at the voluntary muscle end-plate. *A. Rev. Pharmac.* **15**, 307–25.

Couteaux, R. & Pécot-Dechavassine, M. (1970). Vésicules synaptiques et poches au niveau des zones actives de la jonction neuromusculaire. *C.r. hebd. Séanc. Acad. Sci., Paris, D* **271**, 2346–9.

Couteaux, R. & Pécot-Dechavassine, M. (1974). Les zones spécialisées des membranes présynaptiques. *C.r. hebd. Séanc. Acad. Sci., Paris, D* **278**, 291–3.

Dale, H. H., Feldberg, W. & Vogt, M. (1936). Release of acetylcholine at voluntary motor nerve endings. *J. Physiol., Lond.* **86**, 353–80.

del Castillo, J. & Katz, B. (1954). Action, and spontaneous release, of acetylcholine at an 'inexcitable' nerve–muscle junction. *J. Physiol., Lond.* **126**, 27P.

del Castillo, J. & Katz, B. (1955). Local activity at a depolarized nerve–muscle junction. *J. Physiol., Lond.* **128**, 396–411.

del Castillo, J. & Katz, B. (1956). Biophysical aspects of neuromuscular transmission. *Prog. Biophy. biophys. Chem.*, **6**, 121–70.

de Robertis, E. D. P. & Bennett, H. S. (1954). Submicroscopic vesicular component in the synapse. *Fedn Proc. Fedn Am. Socs exp. Biol.* **13**, 35.

Elmqvist, D. & Quastel, D. M. J. (1965). Presynaptic action of hemicholinium at the neuromuscular junction. *J. Physiol., Lond.* **177**, 463–82.

Fatt, P. & Katz, B. (1950). Some observations on biological noise. *Nature, Lond.* **166**, 597–8.

Fatt, P. & Katz, B. (1952). Spontaneous subthreshold activity at motor nerve endings. *J. Physiol., Lond.* **117**, 109–28.

Fatt, P. & Katz, B. (1953). Chemo-receptor activity at the motor end-plate. *Acta physiol. scand.* **29**, 117–25.

Fletcher, P. & Forrester, T. (1975). The effect of curare on the release of acetylcholine from mammalian motor nerve terminals and an estimate of quantum content. *J. Physiol., Lond.* **251**, 131–44.

Ginsborg, B. L. & Jenkinson, D. H. (1976). Transmission of impulses from nerve to muscle. *Handb. exp. Pharmak.* **42**, 229–364.

Grynspan-Winograd, O. (1971). Morphological aspects of exocytosis in the adrenal medulla. *Phil. Trans. R. Soc., B* **261**, 291–2.

Heuser, J. (1976). Morphology and synaptic vesicle discharge and reformation at the frog neuromuscular junction. In *Motor innervation of muscle*, ed. S. Thesleff, pp. 51–115. London: Academic Press.

Heuser, J. E., Reese, T. S. & Landis, D. M. D. (1974). Functional changes in frog neuromuscular junctions studied with freeze-fracture. *J. Neurocytol.* **3**, 109–31.

Israël, M., Gautron, J. & Lesbats, B. (1968). Isolément des vesicules synaptiques de l'organe électrique de la torpille et localisation de l'acétylcholine à leur niveau. *C.r. hebd. Séanc. Acad. Sci., Paris, D* **266**, 273–5.

Katz, B. (1969). *The release of neural transmitter substances*. Liverpool: Liverpool Univ. Press.

Katz, B. & Miledi, R. (1965). The quantal release of transmitter substances. In *Studies in physiology*, pp.118–25. Berlin: Springer-Verlag.

Katz, B. & Miledi, R. (1967a). Tetrodotoxin and neuromuscular transmission. *Proc. R. Soc., B* **167**, 8–22.

Katz, B. & Miledi, R. (1967b). A study of synaptic transmission in the absence of nerve impulses. *J. Physiol., Lond.* **192**, 407–36.

Katz, B. & Miledi, R. (1969a). Tetrodotoxin-resistant electric activity in pre-synaptic terminals. *J. Physiol., Lond.* **203**, 459–87.

Katz, B. & Miledi, R. (1969b). Spontaneous and evoked activity of motor nerve endings in calcium Ringer. *J. Physiol., Lond.* **203**, 689–706.

Katz, B. & Miledi, R. (1970). Membrane noise produced by acetylcholine. *Nature, Lond.* **226**, 962–3.

Katz, B. & Miledi, R. (1972). The statistical nature of the acetylcholine potential and its molecular components. *J. Physiol., Lond.* **224**, 665–99.

Katz, B. & Miledi, R. (1973a). The characteristics of 'end-plate noise' produced by different depolarizing drugs. *J. Physiol., Lond.* **230**, 707–17.

Katz, B. & Miledi, R. (1973b). The binding of acetylcholine to receptors and its removal from the synaptic cleft. *J. Physiol., Lond.* **231**, 549–74.

Katz, B. & Miledi, R. (1973c). The effect of atropine on acetylcholine action at the neuromuscular junction. *Proc. R. Soc., B* **184**, 221–6.

Katz, B. & Miledi, R. (1976). The analysis of end-plate noise – a new approach to the study of acetylcholine/receptor interaction. In *Motor innervation of muscle*, ed. S. Thesleff. pp.31–50. London: Academic Press.

Katz, B. & Miledi, R. (1977). Transmitter leakage from motor nerve endings. *Proc. R. Soc., B* **196**, 59–72.

Kendrick, N. C., Blaustein, M. P., Fried, R. C. & Ratzlaff, R. W. (1977). ATP-dependent calcium storage in presynaptic nerve terminals. *Nature, Lond.* **265**, 246–8.

Kirshner, N. & Kirshner, A. G. (1971). Chromogranin A, dopamine β-hydro-xylase and secretion from the adrenal medulla. *Phil. Trans. R. Soc., B* **261**, 279–89.

Kriebel, M. E. & Gross, C. E. (1974). Multimodal distribution of frog miniature endplate potentials in adult, denervated and tadpole leg muscle. *J. gen. Physiol.* **64**, 85–103.

Kuffler, S. W. & Yoshikami, D. (1975). The number of transmitter molecules in a quantum: an estimate from iontophoretic application of acetylcholine at the neuromuscular synapse. *J. Physiol., Lond.* **251**, 465–82.

Liley, A. W. (1957). Spontaneous release of transmitter substances in multi-quantal units. *J. Physiol., Lond.* **136**, 595–605.

MacIntosh, F. C. & Collier, B. (1976). Neurochemistry of cholinergic terminals. *Handb. exp. Pharmak.* **42**, 99–228.

Miledi, R. (1973). Transmitter release induced by injection of calcium ions into nerve terminals. *Proc. R. Soc., B* **183**, 421–5.

Mitchell, J. F. & Silver, A. (1963). The spontaneous release of acetylcholine from the denervated hemidiaphragm of the rat. *J. Physiol., Lond.* **165**, 117–29.

Neher, E. & Sakmann, B. (1976a). Single-channel currents recorded from mem-brane of denervated frog muscle fibres. *Nature, Lond.* **260**, 799–802.

Neher, E. & Sakmann, B. (1976b). Noise analysis of drug induced voltage clamp currents in denervated frog muscle fibres. *J. Physiol., Lond.* **258**, 705–29.

Nickel, E. & Potter, L. T. (1971). Synaptic vesicles in freeze-etched electric tissue of *Torpedo*. *Phil. Trans. R. Soc., B* **261**, 383–5.

Palade, G. E. (1954). Electron microscope observations of interneuronal and neuro-muscular synapses. *Anat. Rec.* **118**, 335–6.

Palay, S. L. (1954). Electron microscope study of the cytoplasm of neurons. *Anat. Rec.* **118**, 336.

Pécot-Dechavassine, M. & Couteaux, R. (1972). Potentiels miniatures d'amplitude anormale obtenus dans des conditions expérimentales et changements concomitants des structures présynaptiques. *C.r. hebd. Séanc. Acad. Sci., Paris, D* **275**, 983–6.

Rahamimoff, R. (1976). The role of calcium in transmitter release at the neuromuscular junction. In *Motor innervation of muscle*,ed. S. Thesleff, pp. 117–49. London: Academic Press.

Rang, H. P. (1975). Acetylcholine receptors. *Q. Rev. Biophys.* **7**, 282–399.

Robertson, J. D. (1956). The ultrastructure of a reptilian myoneural junction. *J. biophys. biochem. Cytol.* **2**, 381–94.

Smith, A. D. (1971). Secretion of proteins (chromogranin A and dopamine β-hydroxylase) from a sympathetic neuron. *Phil. Trans. R. Soc., B* **261**, 363–70.

Smith, U. & Smith, D. S. (1966). Observations on the secretory processes in the corpus cardiacum of the stick insect, *Carausius morosus. J. Cell Sci.* **1**, 59–66.

Tauc, L., Hoffmann, A., Tsuji, S., Hinzen, D. H. & Faille, L. (1974). Transmission abolished on a cholinergic synapse after injection of acetylcholinesterase into the presynaptic neurone. *Nature, Lond.* **250**, 496–8.

Zimmermann, H. & Whittaker, V. P. (1974). Effect of electrical stimulation on the yield and composition of synaptic vesicles from the cholinergic synapses of the electric organ of *Torpedo*; a combined biochemical, electrophysiological and morphological study. *J. Neurochem.* **22**, 435–50.

The problem of identifying the neurotransmitters of smooth muscle

M. E. HOLMAN

The aim of this essay is to point out some of the difficulties involved in identifying the neurotransmitters which help to control the contractile activity of mammalian smooth muscles. Two examples are discussed:

(1) The nature of the excitatory transmitter to the longitudinal layer of the guinea pig vas deferens.

(2) The transmitter released from enteric neurones which causes inhibitory junction potentials in smooth muscles of the gastro-intestinal tract.

In 1904, Elliott foreshadowed our present ideas about chemical transmission as follows, 'And the facts suggest that the sympathetic axons cannot excite the peripheral tissue except in the presence, and perhaps through the agency of the adrenalin or its immediate precursor secreted by the sympathetic paraganglia . . .Adrenalin might then be the chemical stimulant liberated on each occasion when the impulse arrives at the periphery'. Much evidence accumulated during the first half of this century in support of Elliott's idea and today most neurophysiologists have come to accept the proposition that electrical transmission is the exception rather than the rule, at least for vertebrate synapses.

Transmission at many peripheral synapses in mammals can be blocked quite readily and specifically with drugs such as curare, atropine and propranolol. Other characteristics of these synapses are also indicative of chemical transmission; for example, the postsynaptic response is variable and may show facilitation or depression, the amplitude of the response varies with the membrane potential of the postsynaptic cell and transmission is affected by the ratio of the concentration of Ca^{2+} and Mg^{2+} in the environment of the synapse. There are only two situations where electrical transmission is known to occur in the vertebrate autonomic nervous system – the chick ciliary ganglion (Martin & Pilar, 1963), and mudpuppy cardiac ganglia (Roper, 1976). So far no-one has found a peripheral synapse in mammals with similar properties.

Much of the classical work which supported the hypothesis of chemical transmission came from work on transmission from autonomic nerves to various effector organs (see Eccles, 1964). However, the synapse which has proved to be by far the most amenable for studies at the cellular level is the frog skeletal neuromuscular junction. Our understanding of the operation of all other chemical synapses is based on this model. Unlike most mammalian autonomic synapses the electrical properties of the postsynaptic structure, the skeletal muscle fibre, are well understood. Often, one skeletal muscle fibre receives one synapse and it is possible to visualize this synapse. Recently it has become feasible to study the molecular events which underly the postsynaptic action of acetylcholine (see the preceding chapter by B. Katz).

Kuffler and his colleagues have done much to close the gap between our knowledge of the frog skeletal neuromuscular junction and other synapses by seeking out preparations with relatively simple anatomy which can be visualized *in vitro* by means of a compound microscope. These include the atrial septa of frogs and mudpuppies, and skeletal neuromuscular junctions of frogs and snakes (Dennis, Harris & Kuffler, 1971; Kuffler & Yoshikami, 1975; Roper, 1976). It is difficult to see how it is going to be possible to make equivalent preparations of mammalian smooth muscles, although recent studies by Hirst & Neild (1977) on small intestinal arteries and arterioles are promising.

Almost all vertebrate synapses that have provided useful data at the cellular level are cholinergic. There are great deficits in our knowledge of the functioning of synapses where noradrenaline is generally considered to be the neurotransmitter and also at synapses where acetylcholine acts on muscarinic receptors. One of the most important transmitters in vertebrates, that which causes inhibition of the smooth muscle of the gastro-intestinal tract, remains to be positively identified.

The technical problems involved in making proper records of the membrane potentials of smooth muscles are not generally recognised and a great many published data must be regarded as unhelpful in solving problems concerned with the innervation of this tissue. The interpretation of the most accurate of records is also difficult. Nevertheless it seems that an electrophysiological approach may still be useful.

THE TRANSMISSION OF EXCITATION IN THE
GUINEA PIG VAS DEFERENS

In 1961, Huković described a new 'nerve–muscle' preparation for studying the pharmacology of the action of sympathetic nerves on smooth muscle. This consisted of the vas deferens of guinea pigs in continuity with the hypogastric nerve which is very prominent in this species. Geoffrey Burnstock and I had decided to study this tissue as an example of a smooth muscle which was not 'spontaneously' active but which could be stimulated electrically to give reproducible action potentials. When we moved our stimulating electrodes from the vas deferens to the hypogastric nerve it was apparent that action potentials (and contractions) were initiated as a consequence of what we called excitatory junction potentials (EJPs). Unlike endplate potentials (EPPs), EJPs were graded in amplitude according to the strength of stimulation, and their time course was relatively slow. The time to reach peak amplitude of an EJP in the guinea pig vas deferens was variable (50–100 msec) and EJPs decayed exponentially with a time constant of about 300 msec. With hindsight, it is clear that similar EJPs had been recorded with extracellular electrodes from the nictitating membrane of the cat some 20 years earlier by Eccles & Magladery (1937). In the guinea pig vas deferens there was no evidence that contraction occurred unless an action potential was initiated. This required depolarization to a threshold of about -35 mV.

When the guinea pig vas deferens was examined, using the fluorescence histochemical technique for the localization of catecholamines, developed in Scandinavia during the early 1960s, it was apparent that both the outer longitudinal layer and the inner circular layer were densely innervated by adrenergic nerves (Falck, 1962). This abundance of nerve terminals appeared to provide an explanation for the gradation of the amplitude of the EJP with the strength of stimulation. The concentration of noradrenaline (NA) in this tissue was exceedingly high (10 μg g^{-1} tissue; Sjöstrand, 1965). Since 1961 the vas deferens of small laboratory animals (guinea pigs, rabbits, rats and mice) has been widely used by biochemists, pharmacologists and physiologists as a model for studies on transmission at sympathetic nerve terminals. Details of the ultrastructure of the innervation of this smooth muscle and the evidence that NA is released upon nerve stimulation have been discussed in numerous reviews (see, for example, Bennett, 1972; Burnstock & Costa, 1975; Gabella, 1975) and these topics are only referred to briefly here.

The vas deferens is innervated by ganglion cells with relatively short axons which are found within the hypogastric plexus (paravisceral ganglia, Gabella, 1975). When these axons reach the wall of the vas deferens they become varicose in appearance and branch extensively to form a typical autonomic ground plexus (Hillarp, 1960). Small nerve trunks, consisting of many axons within the same Schwann cell sheath, are found in the spaces between groups of smooth muscle cells which are tightly packed into 'bundles' (Bennett, 1972). There is evidence that some of the axons passing through the longitudinal layer of the guinea pig vas deferens are cholinergic but these axons are mainly to be found within larger bundles of axons. According to Furness & Iwayama (1972), 90% of smaller bundles (containing up to six axons) are exclusively adrenergic. The longitudinal layer of the guinea pig vas deferens appears to be innervated by single adrenergic axons, some of which may be buried deeply within the cytoplasm of the smooth muscle cells. These terminal axons have usually lost their Schwann cell sheath and the membranes of axon and smooth muscle cell may be separated by less than 20 nm. About one third of the axon varicosities observed in both rat and guinea pig vas deferens formed such close contacts with smooth muscle cells (Goto, Millecchia, Westfall & Fleming, 1977).

Axon varicosities contain synaptic vesicles which are characteristic of chemical synapses. If the tissue is prepared for electronmicroscopy in the appropriate way many of these vesicles show an electron-dense core; when the vas deferens was incubated with 5-hydroxydopamine before fixation all the vesicles within adrenergic varicosities had this appearance (Furness & Iwayama, 1972). This substance is regarded as a specific marker for adrenergic vesicles (see Burnstock & Costa, 1975, p. 25). Other histochemical techniques have confirmed that the longitudinal layer of guinea pig vas deferens is innervated, almost exclusively, by nerves containing NA.

The overflow of NA and its metabolites into the medium bathing the vas deferens of small laboratory animals has also been studied (see, for example, Langer, 1970; Farnebo & Malmfors, 1971; Hughes, 1972). The interpretation of the results of these studies is complicated. It is now apparent that adrenergic nerve terminals possess a membrane 'pump' which transports certain amines from extracellular solution to within the nerve terminal by a process comparable with the uptake of choline by cholinergic nerve terminals. Further, NA also appears to be taken up by smooth muscle cells (Gillespie & Muir, 1970). However, there is no evidence that pharmacological interference with these processes has any effect with the time course of the EJP (see below).

'Overflow' studies have also suggested that many of the traditional blockers of the excitatory receptors for NA in smooth muscle (α-receptors) may have significant effects on the evoked release of NA. There is indirect evidence that NA, and especially dopamine, may combine with receptors on the membrane of terminal axons to reduce the release of NA (see Burnstock & Costa, 1975, p. 59). Attempts to test this hypothesis using electrophysiological techniques have been reported by Bennett & Middleton (1975) who used the mouse vas deferens. The duration of the EJP in this preparation is much briefer than that of the guinea pig for reasons that are not understood (Holman, Taylor & Tomita, 1977). It was therefore possible to stimulate nerves at a high frequency (10 Hz) and to distinguish individual EJPs. The initial EJPs during such a train showed facilitation but this rapidly reversed to depression. No such depression was observed in the presence of drugs considered to block presynaptic α-receptors.

There appears to be much circumstantial evidence that NA is the excitatory transmitter in the vas deferens. This proposition has been questioned by Ambache and his colleagues on the basis of a series of carefully controlled pharmacological studies. Ambache & Zar (1971) drew attention to the lack of sensitivity of the longitudinal muscle of the guinea pig vas deferens to NA: adrenaline and dopamine failed to elicit contractions. NA caused a powerful inhibition of contractions elicited by brief trains of stimuli; dopamine had a similar effect but was less active. In guinea pigs, phentolamine failed completely to block contractions in response to nerve stimulation but did block contractions in response to added NA. The rat vas deferens was more sensitive to NA than the guinea pig preparation (Ambache, Dunk, Verney & Zar, 1972) but phenoxybenzamine failed to block responses to nerve stimulation.

It may be argued that some of the results obtained by Ambache and his colleagues could be explained in terms of presynaptic α-receptors which depress the release of NA. Alternatively, the addition of NA to the fluid bathing the vas deferens may not necessarily mimic the effect of NA when it is released from nerve terminals. There is some evidence for the latter proposition from experiments on vascular smooth muscle. The addition of NA to a solution bathing the rabbit ear artery caused contraction in the absence of any change in membrane potential (Casteels & Droogmans, 1976). In contrast, stimulation of perivascular adrenergic nerves caused EJPs similar to those of the vas deferens (Speden, 1967). In order to find answers to these questions it is necessary to have a better understanding of the processes which occur during transmission at the cellular level.

Studies on nicotinic receptors suggest one possible explanation for a difference between the response to transmitter when it is released from nerve terminals and when it is added to the bathing solution. The nicotinic receptors at skeletal neuromuscular junctions undergo desensitization when they are exposed to doses of acetylcholine which produce a sub-threshold depolarization of 10–20 mV for a few seconds (Katz & Thesleff, 1957). In the desensitized state, these receptors are not blocked by α-bungaro toxin but become more susceptible to blockade by irreversible antagonists; it has also been suggested that desensitized receptors have an increased affinity for acetylcholine (see Potter, 1977). Normally, acetylcholine is present in the vicinity of its receptors for a period of about 1 msec and desensitization has never been observed to occur during nerve stimulation. It is tempting to suggest that receptors for NA in the vas deferens and possibly other tissues may also change their properties when they are exposed to sub-threshold doses of catecholamine for prolonged periods (seconds or minutes). This hypothesis would only be feasible if it could be shown that the time course of action of NA, after its release from sympathetic nerve terminals, was relatively brief.

As already indicated, it is necessary to know about the passive and active membrane properties of the postjunctional element in order to describe transmission at the cellular level, including the duration of action of NA (Fatt & Katz, 1951). This has proved to be a difficult problem for smooth muscles although it now seems clear that they are functional syncytia like cardiac muscle.

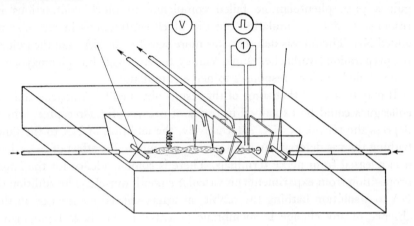

Fig. 1. The method of Abe & Tomita (1968) for stimulating and recording from smooth muscle. The muscle chamber was divided by one of the stimulating electrodes into stimulating (right) and recording (left) compartments which were irrigated separately (arrows). The stimulating electrodes were silver plates with a small hole (about 1 mm in diameter) through which the preparation could be passed; inter-electrode distance was about 10 mm. The surface of the electrode facing the recording chamber was covered with Araldite. A constant current stimulator (⊓) was used and relative values of current intensity were obtained from the voltage gradient between two silver electrodes 2 mm apart (1). An intracellular electrode was used to record membrane potentials.

The spread of electrotonus in certain preparations of cardiac muscle (e.g. Purkinjĕ fibres) suggested that these preparations may have cable-like properties (Noble, 1966) and the same appears to be true for smooth muscle. One method that has frequently been used to estimate passive membrane properties of smooth muscle was developed by Abe & Tomita (1968) and this is shown in Fig. 1. In order to use this technique it is necessary to make a sufficiently long strip from a smooth muscle which is cut in the direction of orientation of the cells. This can be done easily for the taenia coli of guinea pigs, uterine smooth muscle, the walls of large blood vessels, etc. In all preparations tested so far (except the rat and mouse vas deferens) when a voltage step is applied to the plate electrodes, changes in membrane potential are observed to occur some millimetres beyond the insulated face of one of the electrodes. If voltage pulses of long duration are used (at least 1 sec) the change in membrane potential reaches a steady level ($\overline{V}$). It has been found that $\overline{V}$ decays exponentially with distance away from the plate and hence it was possible to define a length constant (λ) whose value for various smooth muscles varied from 0.5 to 3 mm. When the vas deferens of guinea pigs was investigated in this way, the electrotonic potentials recorded from the outer longitudinal layer behaved like other smooth muscles ($\lambda = 2.1$ mm; Tomita, 1967). No electrotonic potentials could be recorded beyond the polarizing plate from rat or mouse vas deferens (Goto *et al.*, 1977; Holman *et al.*, 1977).

The first analysis of the time course of electrotonic potentials when these were recorded intracellularly was that of Tomita (1966). During the last 10 years, numerous workers have confirmed his observation that the strips of smooth muscles behaved in accordance with the predictions of cable theory (Hodgkin & Rushton, 1946). Values of the time constant (τ) derived from cable analysis varied from 55 to 470 msec, that is, over almost one order of magnitude (Tomita, 1975, Table 1).

Various models have been put forward to explain why bundles of smooth muscle cells should behave in this way. It is generally assumed that when a large number of smooth muscle cells are polarized in a uniform way in the radial direction by large extracellular electrodes, current spreads in a longitudinal direction across electrical couplings between neighbouring cells. There should be little or no radial current flow between neighbouring cells provided the length of individual cells is small compared with λ. Smooth muscle cells of the guinea pig taenia coli are about 200 μm in length and λ was found to vary from 1.4 to 1.9 mm. In the vas deferens, λ was longer (1.5 to 2.6 mm) but so was the length of the smooth muscle cells (about 400 μm). When electrotonic potentials were measured in radial and circumferential directions in the vas deferens there was no consistent difference in their amplitude at any given distance from the polarizing electrode (Bennett, 1972, Fig. 15).

The aim of these studies was to arrive at values for R_m and C_m (the resistance and capacitance of a unit area of smooth muscle cell membrane, respectively), R_i (the resistivity of the sarcoplasm) and the value of the coupling resistance between cells. Tomita's model proposed that, because there was no radial flow across the preparation, the parameters which were determined from cable analysis of strips of smooth muscle could also be applied to individual smooth muscle cells coupled to each other longitudinally. He made independent measurements of the longitudinal impedance of taenia coli, using alternating currents and found that this decreased from a value of about 250 Ωcm at low frequencies, to a steady level of 190 Ωcm at frequencies above 100 Hz (Tomita, 1969). He proposed that this was due to capacitative coupling across the junctions between cells and that the coupling resistance was 180 Ωcm. More recent experiments (Tomita, 1975) suggested that these values were too low and that internal myoplasmic resistance was 218 Ωcm, junctional resistance 267 Ωcm and the capacity component of the junctions was 1.4 μF cm^{-1}.

From further predictions of the cable theory derived by Tasaki & Hagiwara (1957) it was possible to make independent estimates of R_m, C_m and the diameter of the smooth muscle cell, a, (see Tomita, 1970; Bennett, 1972). For a smooth muscle with a time constant of about 100 msec, C_m was about 1 μF cm^{-2} and R_m about 100 kΩ cm^2. As pointed out by Tomita (1975), $\lambda = aR_m/2R_i$; since λ has a similar value in smooth and skeletal muscle one might expect the value of R_m to be at least 10 times higher in smooth muscle, since the diameter of these fibres is at least one tenth that of skeletal muscle. It has been argued that the value of C_m (1 μF cm^{-2}) is unlikely to differ greatly from one type of excitable membrane to another (Cole, 1968). If one assumes that the value of τ obtained from cable analysis also applies to the smooth muscle cell membrane, then it is unreasonable to consider that the membranes of smooth muscle cells are leaky. Indeed the input resistance of a single smooth muscle cell, if it were uncoupled from its neighbours, would be very high indeed (at least 2000 MΩ).

Apart from the observation that smooth muscles show cable-like properties there are other reasons for believing that individual cells must be coupled to each other. When an intracellular electrode is used to pass current across the cell membrane ('point polarization') no changes in membrane potential could be detected within one cell length of the stimulating electrode. Likewise if an action potential was initiated by the intracellular polarization, it failed to cause a detectable contraction since such an action potential was probably not able to excite neighbouring cells (Bennett, 1972; Tomita, 1975; Holman & Hirst, 1977). Similar observations had been made on cardiac muscle. As pointed out by Woodbury & Crill (1961), if cells were coupled electrically changes in membrane potential induced by point polarization would be rapidly attenuated since the area of membrane available for the movement of ions would increase very rapidly with distance from the source of current. The threshold current for the initiation of an action potential would be high in view of the large membrane capacitance which would need to be discharged in order to lower the membrane potential. It is clear that the characteristics of the electrotonic potentials measured at the point of polarization, by a Wheatstone bridge or an equivalent circuit, are not likely to provide information about the time constant of the smooth muscle cell membrane (τ_m). As pointed out by Jack, Noble & Tsien (1975) the time course of such changes in membrane potential would be expected to be at least an order of magnitude faster than τ_m.

The input resistance for a variety of smooth muscles was found to range from 10 to 50 MΩ, with the exception of the mouse vas deferens (Bennett, 1972, Table 2; Holman et al., 1977). This is several orders of magnitude lower than that estimated for a single uncoupled cell, from the results of cable analysis. In 1972, Bennett described a model of a syncytium which predicted the way in which input impedance would vary with the frequency of sinusoidal input currents. Individual cells of this syncytium consisted of a resistance and capacitance in parallel and no account was taken of myoplasmic resistance. Any one cell was coupled to n others by electrical bridges with a time constant which was short compared with the membrane time constant. This model predicted that the locus of the impedance curve (reactance versus resistance) for different frequencies should be a semicircle. Data obtained for frequencies of up to 600 Hz fitted the predicted curve when τ_m was 100 msec, the bridge time constant (τ_b) 2 msec and n was 6. This model enabled Bennett to derive values for R_m, C_m and the specific resistance of a 1-cm length of muscle which were in agreement with values derived from cable analysis. The impedance curve predicted by this model depended on the values taken for τ_m, τ_b and the measured value of input resistance. Given the more recent estimate of τ_b by Tomita (1975) it would be interesting to know what range of values of τ_m could be tolerated by this model. Bennett's syncytial model also predicted the differences in spatial spread and time course of electrotonic potentials in response to extracellular versus intracellular polarization.

The wide variation in the estimates of τ_m from the analysis of electrotonic potentials in response to extracellular polarization has been noted. Some of this variation may arise from the uncertainties involved in attempting to fit data to exponential and error functions. This is especially difficult if the stimulating current spreads into the recording area. Some variation may be due to non-uniform radial polarization. It is possible that some of the preparations studied by the method of Abe & Tomita (1968) may not have been long enough to represent infinite cables.

Recently there have been at least two attempts to measure τ_m by more direct methods. Walsh & Singer (1977) were able to impale single, isolated smooth muscle cells from the stomach of the toad (*Bufo marinus*) with two micro-electrodes; one was used for passing current and the other to record changes in membrane potential. Their preliminary studies suggested that τ_m varied from 40 to 90 msec. Hirst & Neild (1977) have also used two intracellular electrodes to study the passive electrical properties of isolated arterioles. They found values for τ_m of about 300 msec. If it is true that τ_m varies so widely for smooth muscle cells and C_m is constant (Cole, 1968), the range of values of R_m might provide some explanation for the diverse properties of different smooth muscles.

It is perhaps more important to establish the value of τ_m for smooth muscles in order to know about the duration of action of neurotransmitters. The time course of the EPP is determined by the passive electrical properties of the skeletal muscle fibre and it has been shown that the duration of action of acetylcholine is brief compared with τ_m (Fatt & Katz, 1951). The EJPs (and inhibitory junction potentials) recorded from many smooth muscles decay exponentially, with time constants ranging from 250 to 400 msec (with the exception of mouse and rat vas deferens and rat seminal vesicle – see Bennett 1972, Table 10). My colleague David Hirst has drawn attention to the observation that many junction potentials recorded from different smooth muscles decay with a time constant of about 300 msec, regardless of the nature of the transmitter or whether they are excitatory or inhibitory. It remains to be shown whether this is coincidental, or due to some property which is common to many different smooth muscles.

As already indicated, the vas deferens of rats and mice appear to have different passive electrical properties from those of all other mammalian smooth muscles investigated so far. When these tissues were studied, using extracellular polarization according to the method of Abe & Tomita (1968), it was not possible to detect any electrotonic potentials beyond the region of the polarizing plate (Goto *et al.*, 1977; Holman *et al.*, 1977). Thus it has not been possible so far to estimate the value of τ_m for these preparations.

In the guinea pig vas deferens there is consistent evidence that τ_m is about 100 msec. If it is assumed that NA acts in the same way on all the cells of a syncytium so that there is no current flow between cells, the syncytium may be treated as a single 'giant' cell and the time course of the conductance change can be estimated from

$$G = \frac{((E-V)-\tau_m \mathrm{d}V/\mathrm{d}t)}{VR_m} \tag{1}$$

where G is the conductance change, V is the membrane potential and E is the resting membrane potential (see Martin & Pilar, 1963). This equation implies that the reversal potential is 0 mV. When the duration of action of NA in the guinea pig vas deferens was calculated in this way this appeared to be prolonged (about 400 msec, time constant for decay, 120 msec, see Bennett, 1972, Fig. 73); the time course of the calculated conductance change was considerably longer than the duration of spontaneous EJPs (see below). Clearly, if τ_m had been underestimated the calculated time course of G would have been shorter, and vice versa.

In the absence of nerve stimulation, spontaneous EJPs (SEJPs) can be recorded in the vas deferens. Their time course is about one tenth that of the EJPs. In the guinea pig vas deferens, SEJPs reach peak amplitude in about 25 msec and decay exponentially with a time constant of about 30 msec. In the rat and mouse vas deferens, SEJPs are even faster (time to peak, about 8 msec, time constant of decay about 20 msec, Holman, 1970; Bennett, 1972; Goto et al., 1977). It has been argued that SEJPs are due to the release of quanta of NA from nerve terminal varicosities which make close contacts with smooth muscle cell membranes. Since it is likely that the concentration of NA, following the release of such a quantum, would decay very rapidly, and that only a small region of membrane would be affected, SEJPs may be compared with point polarization (Bennett, 1972; Holman & Hirst, 1977). Indeed Tomita (1967) found that the SEJPs recorded simultaneously from two cells 100 μm apart were completely independent. When the recording electrodes were closer (less than 50 μm apart) occasional SEJPs were seen which were synchronous (Burnstock & Holman, 1966, Fig. 2). Since the time course of a change in membrane potential caused by point polarization is virtually independent of τ_m, it has been argued that the time course of the SEJP may approximate that of the conductance change associated with a quantum of NA. In the guinea pig vas deferens the conductance change assessed in this way would last for about 100 msec, that is, for less than 25 % of the time estimated for the duration of action of NA during an EJP.

Bennett (1972) has put forward an explanation for this difference. Using data available from electron microscopy he found solutions to the diffusion equation which suggested that the time constant of decay of the concentration of NA within the middle of a 50 μm diameter bundle of smooth muscle cells was about 160 msec. This was very similar to the time constant for the decrease in conductance referred to above (120 msec). Since Bennett (1972) has already developed a syncytial model for a smooth muscle bundle, he was able to compute the changes in membrane potential which would occur if conductance changes were proportional to the concentration of NA. The EJPs computed in this way decayed with a time constant of 400 msec.

Many of the EJPs recorded from guinea pig vas deferens have faster components with a time course resembling that of the SEJP. Bennett was also able to compute the changes in membrane potential which might occur in a smooth muscle bundle if NA was released only from close-contact varicosities which were assumed to occur on 20 % of the smooth muscle cells. In this case the EJP was much faster and decayed at a rate which approximated the membrane time constant (100 msec). The computed EJP was still somewhat slower than those recorded from rat and mouse vas deferens where it is likely that many more cells are innervated by close-contact varicosities. However, the value of τ_m has yet to be determined for these muscles and their SEJPs are briefer than those of the guinea pig.

Bennett's computations suggested that NA may be released from varicosities within a muscle bundle which are situated at considerable distances from the smooth muscle membrane. These distances would need to be large enough to ensure that the release of a single quantum of NA failed to cause any detectable change in conductance because the concentration of NA was rapidly reduced, in time and space, by diffusion. On the other hand, the release of NA from many such varicosities, in response to nerve stimulation, might lead to a build-up and subsequent decline of the concentration of NA within the bundle. This would be associated with the exposure of receptors to NA for a period of more than 500 msec. In contrast, if NA was released only from close-contact varicosites the effective interaction of NA with its receptor might last for 50 msec or less. (Note that if a quantum of NA caused a conductance change over an area that was significantly larger than that associated with point polarization the role of the passive properties of the muscle membrane in determining the time course of the SEJP would become increasingly apparent.)

It is clear that the relation between the concentration of NA and the magnitude of the conductance change must be known before Bennett's model can be developed any further. The problem is to obtain such data from a syncytial preparation. If the receptors for NA are uniformly distributed throughout the syncytium, it may be argued that this could be done by simply adding NA to the solution bathing the tissue. As pointed out by Ambache et al. (1972) 'there is a wide variation, not only between species but also within a species, in the sensitivity of vas deferens longitudinal muscle to the motor action of noradrenaline'. Since motor responses to nerve stimulation vary very little within a species, it must be concluded that NA is unlikely to be the excitatory transmitter or, that the release of NA from nerve terminals cannot be mimicked by bathing the preparation in a solution containing NA.

Theoretically it should be possible to mimic an SEJP by the iontophoretic application of NA. When this was attempted in vas deferens (in the presence of tetrodotoxin) no changes in membrane potential were detected (M. E. Holman, unpublished). But this could have been due to poor visibility and failure to position the extracellular iontophoretic electrode in the optimal position with respect to the recording electrode. Once again, the rapid spatial attenuation of changes in membrane potential which occur when current flow occurs 'at a point' would tend to make this a difficult experiment to achieve. It appears to be impossible to mimic experimentally the conditions discussed by Bennett (1972) in which the concentration of NA within a smooth muscle bundle rises and falls with a relatively slow time course. The present approach to this problem in this laboratory is to study the EJPs of small blood vessels where NA is released from varicosities which do not penetrate the media (see Hirst & Neild, 1977).

Not only is knowledge of the dose–response curve for NA important in relation to Bennett's interpretation of the EJP, but if this were known it might be possible to explain why drugs such as cocaine and desipramine, which increase the overflow of NA by blocking its uptake into nerves, have little effect on the time course of the EJP and none whatsoever on SEJPs (Holman, 1967; Bennett & Middleton, 1975). It appears that the slight prolongation of the EJP observed in these experiments could be due to a local anaesthetic action of these drugs when they are present in high concentrations.

As mentioned above, the time course of the SEJP may be comparable with the time course of the conductance change caused by a quantum of NA. If the receptors for NA were activated, more or less synchronously, by such a quantum then it would seem that the average time course of the NA-activated ionophore is very long compared with those activated by nicotinic receptors at skeletal neuromuscular junctions (some 30 msec compared with about 1 msec). However, there is no evidence for an enzyme with properties resembling acetylcholine esterase at adrenergic junctions and it is possible that individual molecules of NA may activate more than one receptor. This is implied by Bennett's model.

An explanation must also be found for the resistance of EJPs and SEJPs of the vas deferens, to the action of α-blocking drugs. When it became apparent from the electron microscopic studies of Richardson (1962), Merrillees (1968) and Furness & Iwayama (1972) that many varicose nerve terminals made close (20 nm) contacts with the smooth muscle membrane, the suggestion was made that the high local concentrations of NA which might be present in these gaps could overcome the action of antagonists. However, this seems unlikely on two counts. Firstly, the vas deferens is resistant to blockade by antagonist drugs considered to act in a non-competitive way and secondly, it is well known that atropine blocks salivary secretion even though parasympathetic nerve terminals are buried deeply between the cells of acini, where nerve and gland cell membranes are separated by less than 20 nm (Emmelin, 1967; Creed & Wilson, 1969).

It seems more likely that this problem is associated with the differences in the interaction of NA with smooth muscle receptors when it is released in a discontinuous fashion from nerve terminals and when it is applied to muscle in an isolated organ bath continuously and for relatively long periods. These controversies might be resolved if more was known about the receptor for NA. Does it change its properties in the presence of catecholamines and if so, how do its properties change? Could the apparently abnormal pharmacology of the vas deferens be explained by the density of its innervation which might 'protect' its receptors from circulating catecholamines? The skeletal neuromuscular junction has proved to be an accurate model so far for other chemical synapses and it will be interesting to discover if its predictive value extends to adrenergic junctions with smooth muscle.

THE TRANSMISSION OF INHIBITION BY NON-ADRENERGIC, NON-CHOLINERGIC NERVES

A second autonomic nerve–smooth muscle junction where the nature of the transmitter is still debatable is the inhibitory synapse between certain enteric neurones and smooth muscles of the gastro-intestinal tract. Again, many of the data concerning the action of these nerves are based on recordings of contractile activity, and few electrophysiological data are available. There is evidence that this inhibitory junction potential (IJP) is caused by an increase in potassium conductance (Den Hertog & Jager, 1975). The IJP can be reversed to depolarization when the membrane is hyperpolarized to about − 100 mV (see Fig. 2) and the IJP is unaffected by replacing chloride with an impermeant anion (Bennett, Burnstock & Holman, 1963).

In the experiment of Fig. 2, a strip of circular muscle (guinea pig small intestine) was mounted in a bath with polarizing plate electrodes, according to the method of Abe & Tomita (1968). As the current between the plates was progressively increased the IJP became smaller in amplitude and eventually reversed to a depolarization. It is of interest that the time course of the 'reversed' IJP was faster than that of the control. In this experiment, only a small segment of tissue (about 1 mm) protruded into the recording chamber, in order to ensure that changes in membrane potential were more or less uniform in the recording area. It is possible that the faster rate of

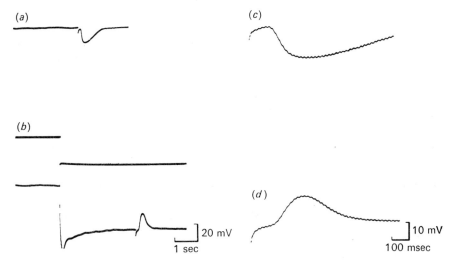

Fig. 2. Intracellular records from the circular muscle of the small intestine of a guinea pig. (a) Inhibitory junction potential (IJP) recorded at the resting membrane potential; (b) the membrane was hyperpolarized and the IJP is reversed in direction; (c) control IJP and (d) reversed IJP recorded at higher amplification and faster time base. Calibrations: (a) and (b) 20 mV and 1 sec, current monitor in (b) 0.4 V cm^{-1}; (c) and (d) 10 mV and 100 msec. (R. A. Bywater and G. S. Taylor, unpublished observations.)

decay of the response to the inhibitory transmitter may reflect a decrease in the value of τ_m when the membrane potential exceeds the potassium equilibrium potential.

The IJP has a long latency of 90 msec at 35 °C, (guinea pig taenia coli, Bennett, Burnstock & Holman, 1966). The latency of the IJP also has a high temperature coefficient of 2.7 (25–35 °C, guinea pig small intestine, R. Lang, unpublished data). It is now widely recognized that junctions where acetylcholine acts on muscarinic receptors have latencies of the order of 100 msec. As was discussed by Koketsu (1969) it is difficult to explain the slow time course of the muscarinic action of acetylcholine in terms of diffusion barriers. Koketsu studied the response of bull frog sympathetic ganglion cells to the iontophoretic application of acetylcholine and observed a depolarization whose initial rapid phase could be blocked by curare and nicotine. This was accompanied by a much slower phase of depolarization which was readily blocked by atropine. More recently the long latency of action of muscarinic receptors has been confirmed, for example in salivary gland cells (Creed & McDonald, 1970) and smooth muscle (see Bolton, 1976).

The latency of IJPs is similar to that of muscarinic EJPs. It has been suggested that the action of muscarinic receptors might involve a bio-chemical step associated with an increase in guanosine $3':5'$-cyclic phos-phate (Kanof, Ueda, Uno & Greengard, 1977). One might suppose that such a process could involve a lag time of this order of magnitude. Perhaps the IJP involves a comparable biochemical step which leads to an increase in potassium conductance caused by an initial increase in Ca^{2+}. But such speculation does not help to identify the nature of the transmitter!

Since it is difficult to block the effects of sympathetic nerve stimulation to the vas deferens, using many of the classical antagonist drugs which are considered to block the postsynaptic actions of catecholamines, it may not be surprizing that no drug has emerged which selectively blocks the IJP. This has proved to be a somewhat inhibiting observation. Numerous antagonists were tested against the IJP by ourselves and others during the 1960s. For example, Hidaka & Kuriyama (1969) reported that the following substances had no specific effect on the inhibitory potential; picrotoxin, strychnine, morphine, lysergic acid, ouabain, P-chloromecuribenzoic acid and 2-4 dinitrophenol.

In 1969, Geoffrey Burnstock and his colleagues put forward the sugges-tion that the inhibitory transmitter might be ATP or a related compound (Satchell, Burnstock & Campbell, 1969). Subsequently the term 'puriner-gic' was introduced to describe these nerves (Burnstock, 1972).

All the problems discussed above in relation to mimicking the effects of the release of NA from nerve terminals in the vas deferens with exogenous NA will need to be solved once again in relation to the ATP theory. The cellular actions of ATP, when this was applied to the bathing solution in an isolated organ bath, were reported by Tomita & Watanabe (1973). Low concentrations of ATP had an action on the taenia coli which closely resembled the inhibitory action of the β class of adrenergic agonists. The 'spontaneous' initiation of action potentials was abolished with little or no change in membrane potential or membrane properties. If it is true that inhibition of smooth muscle by β agonists depends on a β-activated adenyl cyclase and a subsequent biochemical reaction leading to an increase in cyclic AMP it might be possible to explain this action of ATP. When higher concentrations of ATP were used (10^{-3} M) the membrane hyperpolarized, and this hyperpolarization was found to be due to an increase in potassium conductance, thus mimicking the IJP.

When IJPs are observed in tissues like the taenia coli (guinea pigs) which usually have low resting membrane potentials and tend to be spontaneously active, IJPs are often followed by a myogenic 'rebound' excitation. If ATP were the inhibitory transmitter and if its concentration decayed relatively slowly, as predicted by Bennett's model, it would be difficult to understand how rebound could occur. As its concentration decreased, ATP should begin to exert its β-like inhibition of myogenic activity. However, if ATP were to be released in a slug at a high concentration and then be rapidly broken down or dispersed by diffusion, the observations of Tomita & Watanabe (1973) could be taken as evidence that ATP has the appropriate action in mimicking the IJP. Clearly more work needs to be done at the cellular level to resolve this question.

It was pointed out by Rall *et al.* (1967) that the criteria that have been established for the identification of transmitters are simple in concept but difficult to establish experimentally. In conclusion, it may be of interest to consider these criteria in relation to the theory that ATP is the neurotransmitter which mediates the IJPs observed in the gastro-intestinal tract.

(1) It must be shown that the proposed transmitter is synthesized (or taken up and concentrated) by the nerve terminals. It is certain that all nerve terminals synthesize ATP. In view of the observation that some afferent nerve terminals contain an especially large number of mitochondria it will be necessary to ensure that any nerve terminals which apparently contain a high concentration of ATP are efferent (see Burnstock, 1975).

(2) The transmitter must be released in appropriate amounts upon nerve stimulation. There is evidence that ATP is released from sympathetic nerve terminals together with NA (see Smith & Winkler, 1972) and at the skeletal neuromuscular junction, with acetylcholine (Silinsky, 1975). It is possible that ATP may also be released together with the transmitter which causes the IJP.

(3) It should be possible to mimic the effects of nerve stimulation by the appropriate application of the transmitter with respect to concentration, space and time, to its postsynaptic receptors. As pointed out above, this will be a difficult but not impossible task for smooth muscle.

(4) There should be a mechanism for the rapid inactivation of the transmitter. This presents no problem in relation to ATP. However, it seems that this criterion for the identification of transmitters needs further examination. For example, it is not clear how relevant this might be in relation to the immediate, short term (msec) action of transmitters such as NA, gamma aminobutyric acid, glycine, glutamate or the muscarinic action of acetylcholine.

(5) Drugs which potentiate or depress the action of nerve stimulation should have similar effects on the action of exogenous transmitter. In my opinion, no drugs have emerged which cause a significant change in the action of purinergic nerves (other than tetrodotoxin). However, I do not believe that this criterion should be used to rule out the ATP theory, for the reasons discussed in this essay.

For many years it has been considered that smooth muscle played an important role in pharmacology because it was easy to mimic the effects of nerve stimulation by the application of exogenous transmitters. The time has come to ask if this is really true and if so, why it should be true.

REFERENCES

Abe, Y. & Tomita, T. (1968). Cable properties of smooth muscle. *J. Physiol., Lond.* **196**, 87–100.

Ambache, N., Dunk, L. P., Verney, J. & Zar, M. A. (1972). Inhibition of post-ganglionic motor transmission in vas deferens by indirectly acting sympatho-mimetic drugs. *J. Physiol., Lond.* **227**, 433–56.

Ambache, N. & Zar, M. A. (1971). Evidence against adrenergic motor transmission in the guinea-pig vas deferens. *J. Physiol., Lond.* **216**, 359–89.

Bennett, M. R. (1972). *Autonomic neuromuscular transmission.* Cambridge: Cambridge Univ. Press.

Bennett, M. R., Burnstock, G. & Holman, M. E. (1963). The effect of potassium and chloride ions on the inhibitory potential recorded from guinea-pig taenia coli. *J. Physiol., Lond.* **169**, 33P.

Bennett, M. R., Burnstock, G. & Holman, M. E. (1966). Transmission from intra-mural inhibitory nerves to the smooth muscle of the guinea-pig taenia coli. *J. Physiol., Lond.* **182**, 541–58.

Bennett, M. R. & Middleton, J. (1975). An electrophysiological analysis of the effects of amine-uptake blockers and α-adrenoceptor blockers on adrenergic neuromuscular transmission. *Br. J. Pharmac.* **55**, 87–95.

Bolton, T. B. (1976). On the latency and form of the membrane responses of smooth muscle to the iontophoretic application of acetycholine or carbachol. *Proc. R. Soc. B* **194**, 99–119.

Burnstock, G. (1972). Purinergic nerves. *Pharmac. Rev.* **24**, 509–81.

Burnstock, G. (1975). Purinergic transmission. In *Handbook of psychopharmacology,* ed. L. L. Iverson, S. D. Iverson & S. H. Snyder, vol. 5, pp.131–94. New York: Plenum.

Burnstock, G. & Costa, M. (1975). *Adrenergic neurones.* London: Chapman & Hall.

Burnstock, G. & Holman, M. E. (1966). Junction potentials at adrenergic synapses. *Pharmac. Rev.* **18**, 481–93.

Casteels, R. & Droogmans, G. (1976). Membrane potential and excitation–contraction coupling in the smooth muscle cells of the rabbit ear artery. *J. Physiol., Lond.* **263**, 163P–4P.

Cole, K. S. (1968). *Membranes ions and impulses.* Los Angeles: Univ. Calif. Press.

Creed, K. E. & McDonald, I. R. (1970). Salivary secretory potentials in the marsupial *Trichosurus Vulpecula. Comp. gen. Pharmac.* **1**, 285–92.

Creed, K. E. & Wilson, J. A. F. (1969). The latency of response of secretory acinar cells to nerve stimulation in the submandibular gland of the cat. *Aust. J. exp. Biol. med. Sci.* **47**, 135–44.

Den Hertog, A. & Jager, L. P. (1975). Ion fluxes during the inhibitory junction potential in the guinea-pig taenia coli. *J. Physiol., Lond.* **250**, 681–91.

Dennis, M. J., Harris, A. J. & Knuffler, S. W. (1971). Synaptic transmission and its duplication by focally applied acetylcholine in parasympathetic neurons in the heart of the frog. *Proc. R. Soc., B* **177**, 509–39.

Eccles, J. C. (1964). *The physiology of synapses.* Berlin: Springer-Verlag.

Eccles, J. C. & Magladery, J. W. (1937). The excitation and response of smooth muscle. *J. Physiol., Lond.* **90**, 31–7.

Elliott, T. R. (1904). On the action of adrenalin. *J. Physiol., Lond.* **31**, xx–xxi.

Emmelin, N. (1967). Nervous control of salivary glands. In *Handbook of physiology alimentary canal*, ed. C. F. Code, vol. 2, pp. 595–632. Washington: Am. Physiol. Soc.

Falck, B. (1962). Observations on the possibilities of the cellular localization of monoamines by a fluorescence method. *Acta physiol. scand.* **56** (Suppl. 197), 1–25.

Farnebo, L-O. & Malmfors, T. (1971). ^{3}H-Noradrenaline release and mechanical response in the field stimulated mouse vas deferens. *Acta physiol. scand.* Suppl. 371, 1–18.

Fatt, P. & Katz, B. (1951). An analysis of the end-plate potential recorded with an intracellular electrode. *J. Physiol., Lond.* **115**, 320–70.

Furness, J. B. & Iwayama, T. (1972). The arrangement and identification of axons innervating the vas deferens of the guinea-pig. *J. Anat.* **113**, 179–96.

Gabella, G. (1975). *Structure of the autonomic nervous system.* London: Chapman & Hall.

Gillespie, J. S. & Muir, T. C. (1970). Species and tissue variation in extraneuronal and neuronal accumulation of noradrenaline. *J. Physiol., Lond.* **206**, 591–604.

Goto, K., Millecchia, L. L., Westfall, D. P. & Fleming, W. W. (1977). A comparison of the electrical properties and morphological characteristics of the smooth muscle of the rat and guinea-pig vas deferens. *Pflügers Arch. ges. Physiol.* **368**, 253–61.

Hidaka, T. & Kuriyama, H. (1969). Responses of the smooth muscle membrane of guinea pig jejunum elicited by field stimulation. *J. gen. Physiol.* **53**, 471–86.

Hillarp, N-Å. (1960). Peripheral autonomic mechanisms. In *Handbook of Physiology*, vol. II, Neurophysiology, ed. J. Field, pp. 979–1006. Washington: Am. Physiol. Soc.

Hirst, G. D. S. & Neild, T. O. (1977). Spread of electrical current in arterioles. *Proc. Aust. physiol. pharmac. Soc.* **8**, 165P.

Hodgkin, A. L. & Rushton, W. A. H. (1946). The electrical constants of a crustacean nerve fibre. *Proc. R. Soc.*, B **133**, 444–79.

Holman, M. E. (1967). Some electrophysiological aspects of transmission from noradrenergic nerves to smooth muscle. *Circulation Res.* **21** (Suppl. III), 71–81.

Holman, M. E. (1970). Junction potentials in smooth muscle. In *Smooth muscle*, ed. E. Bulbring, A. F. Brading, A. W. Jones & T. Tomita, pp. 244–88. London: Edward Arnold.

Holman, M. E. & Hirst, G. D. S. (1977). Junctional transmission in smooth muscle and the autonomic nervous system. In *Handbook of physiology*, The cellular biology of neurones, ed. E. R. Kandel, pp. 417–61. Washington: Am. Physiol. Soc.

Holman, M. E., Taylor, G. S. & Tomita, T. (1977). Some properties of the smooth muscle of mouse vas deferens. *J. Physiol., Lond.* **266**, 751–64.

Hughes, J. (1972). Evaluation of mechanisms controlling the release and activation of the adrenergic transmitter in the rabbit portal vein and vas deferens. *Br. J. Pharmac.* **44**, 472–91.

Huković, S. (1961). Responses of the isolated sympathetic nerve-ductus deferens preparation of the guinea pig. *Br. J. Pharmac.* **16**, 188–94.

Jack, J. J. B., Noble, D. & Tsien, R. W. (1975). *Electric current flow in excitable cells.* Oxford: Clarendon Press.

Kanof, P., Ueda, T., Uao, I. & Greengard, P. (1977). Cyclic nucleotides and phosphorylated proteins in neuronal function. In *Approaches to the cell biology of neurones*, ed. W. M. Cowan & J. A. Ferrendelli, pp. 399–434. Bethesda: Soc. Neurosci.

Katz, B. & Thesleff, S. (1957). A study of the 'desensitization' produced by acetyl-choline at the motor endplate. *J. Physiol., Lond.* **138**, 63–80.

Koketsu, K. (1969). Cholinergic synaptic potentials and the underlying ionic mechanisms. *Fedn Proc. Fedn Am. Socs exp. Biol.* **28**, 101–12.

Kuffler, S. W. & Yoshikami, D. (1975). The number of transmitter molecules in a quantum: an estimate from iontophoretic application of acetylcholine at the neuromuscular synapse. *J. Physiol., Lond.* **251**, 462–82.

Langer, S. Z. (1970). The metabolism of [^{3}H] noradrenaline released by electrical stimulation from the isolated nictitating membrane of the cat and vas deferens of the rat. *J. Physiol., Lond.* **208**, 515–46.

Martin, A. R. & Pilar, G. (1963). Dual mode of synaptic transmission in the avian ciliary ganglion. *J. Physiol., Lond.* **168**, 443–63.

Merrillees, N. C. R. (1968). The nervous environment of individual smooth muscle cells of the guinea-pig vas deferens. *J. Cell Biol.* **37**, 794–817.

Noble, D. (1966). Applications of Hodgkin–Huxley equations to excitable tissues. *Physiol. Rev.* **46**, 1–50.

Potter, L. T. (1977). Molecular properties of acetylcholine receptor-channel molecules, and an oligomeric model for their activation, inactivation and de-sensitization. *Proc. Aust. physiol. pharmac. Soc.* **8**, 55–74.

Rall, W., Burke, R. E., Smith, T. G., Nelson, P. G. & Frank, K. (1967). Dendritic location of synapses and possible mechanisms for the monosynaptic EPSP in motoneurones. *J. Neurophysiol.* **30**, 1169–93.

Richardson, K. C. (1962). The fine structure of autonomic nerve endings in smooth muscle of the rat vas deferens. *J. Anat.* **96**, 427–42.

Roper, S. (1976). An electrophysiological study of chemical and electrical synapses on neurones in the parasympathetic cardiac ganglion of the mudpuppy *Necturus Maculosus*: evidence for intrinsic ganglion innervation. *J. Physiol., Lond.* **254**, 427–54.

Satchell, D. G., Burnstock, G. & Campbell, G. (1969). Evidence for a purine com-pound as the transmitter in non-adrenergic inhibitory neurones in the gut. *Aust. J. exp. Biol. med. Sci.* **47**, 24.

Silinsky, E. M. (1975). On the association between transmitter secretion and the release of adenine nucleotides from mammalian motor nerve terminals. *J. Physiol., Lond.* **247**, 145–62.

Sjostrand, N. O. (1965). The adrenergic innervation of the vas deferens and the accessory male genital glands. *Acta physiol. scand.* **65** (Suppl. 257), 1–82.

Smith, A. D. & Winkler, H. (1972). Fundamental mechanisms in the release of catecholamines. In *Catecholamines*, ed. H. Blaschko & E. Muscholl, pp. 538–617. New York: Springer-Verlag.

Speden, R. N. (1967). Adrenergic transmission in small arteries. *Nature, Lond.* **216**, 289–90.

Tasaki, I. & Hagiwara, S. (1957). Capacity of muscle fibre membrane. *Am. J. Physiol.* **188**, 423–29.

Tomita, T. (1966). Electrical responses of smooth muscle to external stimulation in hypertonic solution. *J. Physiol., Lond.* **183**, 450–68.

Tomita, T. (1967). Current spread in the smooth muscle of the guinea-pig vas deferens. *J. Physiol., Lond.* **189**, 163–76.

Tomita, T. (1969). The longitudinal impedence of the guinea-pig taenia coli. *J. Physiol., Lond.* **191**, 517–27.

Tomita, T. (1970). Electrical properties of mammalian smooth muscle. In *Smooth muscle*, ed. E. Bülbring, A. F. Brading, A. W. Jones & T. Tomita, pp. 197–243. London: Edward Arnold.

Tomita, T. (1975). Electrophysiology of mammalian smooth muscle. *Prog. Biophys molec. Biol.* **30**, 185–203.

Tomita, T. & Watanabe, H. (1973). A comparison of the effects of adenosine triphosphate with noradrenaline and with the inhibitory potential of the guinea-pig taenia coli. *J. Physiol., Lond.* **231**, 167–77.

Walsh, J. V. & Singer, J. J. (1977). Electrical properties of freshly isolated single smooth muscle cells. *Proc. Int. Un. Physiol. Sci.* **8**, 799.

Woodbury, J. W. & Crill, W. E. (1961). On the problem of impulse conduction in the atrium. In *Nervous inhibition*, ed. E. Florey, pp. 124–35. Oxford: Pergamon Press.

The motor innervation of the mammalian muscle spindle

Y. LAPORTE

More than 30 years have passed since Leksell (1945) showed that the rate of firing of some muscle receptors – presumably muscle spindles – was considerably increased following the selective stimulation of slow-conducting motor axons. Since then the combined efforts of physiologists and histologists have succeeded in elaborating a rather detailed picture of the motor innervation of mammalian spindles, yet it must be acknowledged that it is still incomplete and partly controversial.

In this review the following points will be considered:

(*a*) the intrafusal muscle fibres and their motor endings;
(*b*) the fusimotor innervation;
(*c*) the skeleto-fusimotor innervation.

INTRAFUSAL MUSCLE FIBRES AND THEIR MOTOR ENDINGS

For a long time, only two types of intrafusal muscle fibres were distinguished, the nuclear-bag fibres and the nuclear-chain fibres. However, recent histochemical and ultrastructural studies (Ovalle & Smith, 1972; Banks, Barker, Harker & Stacey, 1975; Barker *et al.*, 1976*a*; Banks, Harker & Stacey, 1977) have shown that there are two types of nuclear-bag fibre, the bag_1 and the bag_2 fibres. Both types of bag fibre are present in every spindle, usually one of each type. In cat spindles, the bag_1 fibre as compared with the bag_2 fibre is shorter and thinner (see Fig. 1); the glycogen content of the bag_1 fibre is less and the alkaline ATPase reaction is weak. In the juxta-equatorial region the sarcomeres of the bag_1 are longer than those of the bag_2 and chain fibres; they lack an M line, or have a double M line, and their sarcoplasmic reticulum is poorly developed. The histochemical profile and the ultrastructure of the bag_2 fibre in general resemble those of the chain fibres which have the highest glycogen content and whose enzymatic activities are high. Bag_2 fibres are surrounded in their polar regions by many more elastic fibres than bag_1 fibres (Gladden, 1976).

Evidence for two types of bag fibre was also provided by cinemato-

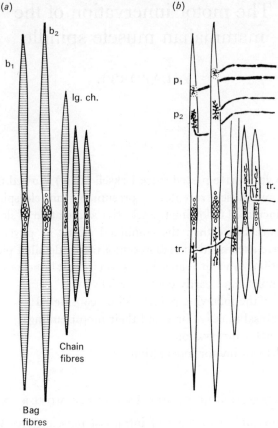

Fig. 1. Schematic representation of (a) the intrafusal muscle fibres of the cat and
(b) of their motor innervation b_1: bag_1 fibre; b_2: bag_2 fibre; lg.ch.: long chain; p_1:
p_1 plates; p_2: p_2 plates; tr.: trail ending.

graphical observations of living spindles. Bessou & Pagès (1975) were the
first to report that the stimulation of single static γ axons could activate
some bag fibres and that the contraction of these fibres was much stronger
and quicker than those of the bag fibres innervated by dynamic γ axons
(see Fig. 2). Boyd, Gladden, McWilliam & Ward (1977) also distinguish
two types of bag fibre which they call 'static nuclear-bag fibre' and 'dynamic nuclear-bag fibre', depending upon the type of fusimotor axon which
activates them. They found, as did Bessou & Pagès, that the time course of
tetanic contraction and relaxation of the bag fibre activated by static γ axons
was much shorter than those of bag fibres activated by dynamic γ axons and
that the amplitude of movement in a static-activated bag fibre was much
larger than that of a dynamic-activated fibre. Furthermore, they showed
that the dynamic bag fibre, previously called the 'slow' nuclear-bag fibre

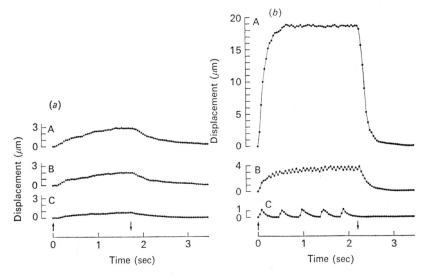

Fig. 2. Comparison of the contractions elicited in two nuclear-bag fibres by the repetitive stimulation of (*a*) a dynamic γ axon and (*b*) of a static γ axon. Cat tenuissimus spindle. Ordinates shows displacement of a striation in μm. In the dynamic-activated bag fibre the striation was located at 1.45 mm from a reference point situated in the equatorial region (junction of the spindle nerve with the bundle of intrafusal muscle fibres). Stimulation frequencies: A, 110 sec⁻¹; B, 75 sec⁻¹; C, 45 sec⁻¹. In the static-activated bag fibre the striation was located at 0.66 mm from the reference point. Stimulation frequencies: A, 33 sec⁻¹; B, 12.5 sec⁻¹; C, 2.2 sec⁻¹. (From Figs. 4 and 6, in Bessou & Pagès, 1975.) The arrows indicate the onset and the end of the periods of stimulation.

(Boyd, 1976) exhibits 'creep' following passive stretch which most probably accounts for the slow decay of the primary ending discharge after a ramp-and-hold stretch (see p. 51).

In cat spindles, in addition to the two (rarely three) nuclear-bag fibres, there are usually four to six nuclear-chain fibres which – in the same spindle – may differ considerably in length. One of them, the 'long chain', present in about one third of cat tenuissimus spindles, may sometimes be as long as the bag fibres (Barker *et al.*, 1976*a*).

There are three types of motor ending in cat spindles (Barker, Stacey & Adal, 1970): the p_1 plates, the p_2 plates and the trail endings (see Fig. 1). The p_1 plates, which resemble extrafusal motor endplates, are mainly confined to the bag fibres (of 100 p_1 plates in leg muscles, Barker *et al.* (1970) found that 75 were located on bag fibres and 25 on chain fibres); they are found in the polar region of the intrafusal fibres in the vicinity of the extremities of the capsule. The p_2 plates are larger than the p_1 plates; they lack a sole plate and have characteristically knob-like axon terminals.

Their location is very similar to that of p_1 plates except that fewer are found on chain fibres. It should be noted that the studies of Barker *et al.* on p_1 and p_2 plates were carried out before the two types of bag fibre were recognized. The trail endings are diffuse terminals which consist of elongated non-myelinated ramifications originating from the terminal or pre-terminal node of a motor axon; they occupy a very extensive area in the juxta-equatorial region. They are found in practically all spindles, mostly on chain fibres but also on bag fibres. Trail endings have been observed in electron micrographs on both types of bag fibre (Barker & Stacey, 1970).

FUSIMOTOR INNERVATION

It was in 1962 that Matthews demonstrated that fusimotor axons could be classified in two functional categories, static and dynamic, by means of their different actions on the response of primary endings to ramp-and-hold stretch. The dynamic axons increase the dynamic responsiveness of these endings whereas the static ones decrease it. He then put forward the hypothesis (1964, 1972) that the dynamic γ axons achieve their action through the bag fibres and the static ones through the chain fibres, encouraged by Boyd's (1962) claim that the two kinds of intrafusal muscle fibre received an independent motor innervation. This simple model was never accepted by Barker and his colleagues because they (Barker & Ip, 1965; Barker *et al.*, 1970) observed that individual γ axons often innervated both bag fibres and chain fibres.

Over the last five years three investigations combining experimental and histological approaches have vindicated Barker's position, since all of them have conclusively shown that static γ axons commonly supply bag fibres as well as chain fibres.

(1) Cat tenuissimus muscles were prepared in which the motor innervation was reduced to a single γ axon by cutting the other motor axons and allowing them to degenerate (Barker *et al.*, 1973). The distribution of the surviving axon – only static axons were prepared – and the type of its motor endings was ascertained in teased, silver preparation. No other endings but trails were found. The static axons distributed trail endings to both bag and chain muscle fibres in the spindle poles with about twice the frequency of supplying them to poles in which the distribution was restricted exclusively to either bag or chain muscle fibres.

(2) Tenuissimus intrafusal muscle fibres impaled by micro-electrodes and activated by single static γ axons were located after electrophoretic injections of Procion Yellow. Of thirteen muscle fibres, eight were shown to be nuclear-bag fibres and five nuclear-chain fibres (Barker, *et al.* 1972, 1978).

(3) Brown & Butler (1973), applying the glycogen-depletion technique of Edström & Kugelberg (1968) to the study of the distribution of fusimotor axons, found that prolonged stimulation of single γ static axons depleted glycogen in both bag and chain fibres.

Then Bessou & Pagès (1975), who made cinematographic recordings of intrafusal movements induced by the stimulation of single fusimotor axons, reported that in tenuissimus spindles the static γ axons evoked contractions either in bag fibres alone, or in chain fibres alone, or in both types of intrafusal muscle fibre. They found that at least one third of static axons supplied both bag and chain fibres.

It should be pointed out that most of these investigations were performed at a time when the distinction between bag_1 and bag_2 fibres had not been made or even when there was some confusion in the identification of the so-called 'typical' and 'intermediate' bag fibres, two terms which are now abandoned.

Very recently, Boyd et al. (1977), in a study comparable to that of Bessou & Pagès (1975), also reported that chain fibres were not the only fibres activated by static axons. Stimulation of single static γ axons could produce either contraction in the bundle of chain fibres only, or contraction in one bag fibre only or contraction in one bag fibre together with chain fibres. As already mentioned (p. 46) they named the bag fibre activated by static axons as the 'static nuclear-bag fibre' and the bag fibre activated by dynamic axons, the 'dynamic nuclear-bag fibre'.

It has thus become obvious that the selective distribution hypothesis of Matthews – bag fibres exclusively supplied by dynamic axons and chain fibres exclusively by static axons – should be abandoned. Unfortunately this does not mean that a general agreement on the intrafusal distribution of the two kinds of fusimotor axon has finally been reached. What seems firmly established is that the dynamic axons exert their action through bag_1 fibres. In their analysis based on the glycogen-depletion method, Barker et al. (1976b) found that dynamic axons depleted bag_1 fibre almost exclusively; additional depletion in a chain fibre or in a bag_2 fibre was only observed in 3 out of 16 spindles. Bessou & Pagès (1975) and Boyd et al. (1977) observed that the stimulation of a dynamic axon produced a very weak contraction in only one of the two bag fibres. However, no agreement has been reached on the distribution of the static axons. A new controversy has arisen because of the discrepancy between the observations made with the glycogen-depletion method and the cinematographical observations of living spindles.

Barker et al. (1976b) reported that stimulation of single static γ axons induced glycogen depletion – indicating neural activation – as much in bag_1 as in bag_2 fibres. The distribution of 8 static axons to cat tenuissimus spindles was studied. Zones of glycogen depletion were found in 27 spindles; chain fibres were depleted in 24 spindles, bag_2 fibres in 13 spindles and bag_1 fibres also in 13 spindles. The intracapsular location of the zones of depletion was very similar to that of the trail endings found in spindles deprived by degeneration of all motor innervation except that supplied by a single static axon.

These results are apparently in conflict with the observations of Bessou & Pagès (1975) and those of Boyd et al. (1977), who have never seen the same bag fibre to be supplied by both a static and a dynamic axon. One would expect this to happen sometimes since bag_1 fibres are present in every spindle and are supplied by dynamic axons.

Can this essentially negative finding be regarded as compelling evidence that static axons never supply bag_1 fibres? As Bessou & Pagès (1975) themselves carefully pointed out it is quite conceivable that in microscopic observations a weak contraction in an intrafusal fibre (e.g. a bag_1 fibre) may be masked by the occurrence of a strong contraction of nearby intrafusal fibres (e.g. several chain fibres and/or a bag_2 fibre). Emonet-Dénand, Laporte, Matthews & Petit (1977) also considered the possibility that the dynamic effect need not be associated with appreciable shortening of a muscle fibre but may depend upon localized changes in its mechanical properties.

Theoretically, if in a given spindle some static γ axons in addition to innervating chain and/or bag_2 fibres also supply the bag_1 fibre – i.e. the fibre responsible for dynamic action – one would expect that its activation should affect the response of the primary ending to a ramp-and-hold stretch carried out during the repetitive stimulation of these axons. There is now strong indirect evidence that it does so and that composite responses do occur. Emonet-Dénand et al. (1977) made a systematic survey – in cat peroneus brevis muscle – of the actions exerted by as many as possible of the fusimotor axons supplying a given spindle, in order to see whether intermediate types of fusimotor action could be recognized as falling between the well-known static and dynamic types. They classified the responses into six categories ranging from apparently pure dynamic action (category I) to apparently pure static action (category VI). Typical examples of each of the six categories are illustrated in Fig. 3. The features of the responses on which the classification is based are indicated in the caption. As shown by the records, the essential sign of dynamic action (i.e. of activation of bag_1 fibre) is the slow adaptative decay of firing with a time constant of about 0.5 sec which occurs on the plateau of the ramp stretch (see category I response), and the clearest sign of static action is firing on the releasing phase of the stretch. About one third of the responses are sug-

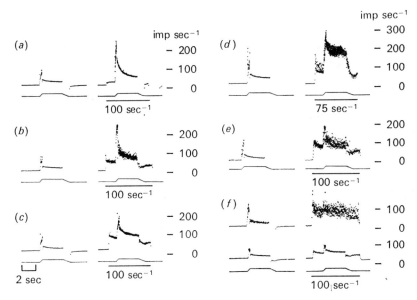

Fig. 3. Categories of fusimotor actions. Each pair of records shows: on the left, the passive response of a primary ending (cat peroneus brevis muscle) to a ramp stretch applied at 6 mm sec⁻¹ followed, a few seconds later, by a release at 2 mm sec⁻¹; on the right, the response during repetitive fusimotor stimulation at 100 sec⁻¹ or in one instance at 75 sec⁻¹. The bars indicate the periods of stimulation. Amplitude of movement, 2 mm in all cases. (a) Category I. Purely dynamic action. Characterized by marked increase in dynamic index with slow decay of firing after the dynamic phase of ramp-and-hold stretch, little or no firing during release and regular afferent discharge. (b) Category II. Dynamic action modified by static action. Strong dynamic features (increase in dynamic index with accompaniyng slow decay) but relatively great excitation at constant initial length, appreciable variability of discharge and firing on release. (c) Category III. Unclassifiable. Dynamic and static features equally balanced and/or changing with stimulation frequency or initial length. (d) Category IV. Static action modified by dynamic action. Strong static features (considerable excitation at constant initial length, considerable firing during release, irregularity of the afferent firing) but slow decay of firing on ramp plateau and dynamic index often slightly increased. (e) Category V. Static action with conceivable dynamic participation. Strong static features with faint signs of dynamic action. (f) Category VI. Purely static action. Excitation with usually a decrease of the dynamic index; no slow decay on completion of ramp; nearly always firing on release; sometimes gross variability of afferent discharge (upper record) or driving; sometimes regular afferent discharge (lower record). Percentage of the various categories of responses observed in a sampling of 153 responses: I, 23 %; II, 6 %; III, 5 %; IV, 7 %; V, 15 %; VI, 44 %. (From Emonet-Dénand, Laporte, Matthews & Petit, 1977.)

gestive of an admixture in various proportions of static and dynamic actions. Especially important for the present discussion are categories IV (static action modified by dynamic action) and category V (static action

3

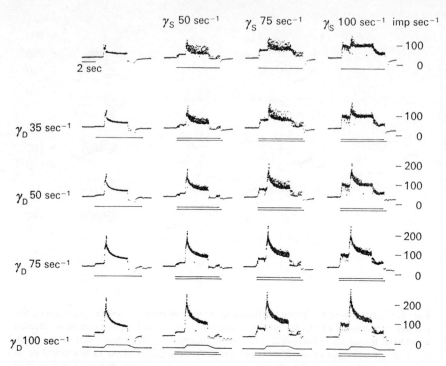

Fig. 4. Intermediate responses obtained by the simultaneous stimulation of a pair of single fusimotor axons, one with a pure dynamic action, the other with a pure static action. First vertical row: stimulation, at various frequencies, of a single dynamic (category I) γ axon (γ_D). First horizontal row: stimulation, at various frequencies, of a single static (category VI) γ axon (γ_S). All the other records were obtained by simultaneously stimulating both axons, as indicated by the double bar. By independently altering the rate of stimulation of each axon it was possible to obtain responses similar to the intermediate responses of categories II, III, IV, V that may be given by individual fusimotor axons. Amplitude of the movement: 2 mm. (From Emonet-Dénand, Laporte, Matthews & Petit, 1977.)

with conceivable dynamic participation); these include responses in which a slow decay was observed on the plateau of the ramp stretch in addition to typical static features.

A strong reason for thinking that the intermediate responses elicited in some spindles by certain single γ axons result from the common innervation of functionally different types of muscle fibre is given by experiments in which two γ axons, one belonging to category I (pure dynamic action) the other to category VI (pure static action) are simultaneously stimulated. As shown by Fig. 4, all intermediate categories of response can be obtained simply by altering the rate of stimulation of each of the two γ axons. Thus it appears that certain fusimotor axons do not restrict their innervation to the 'expected' kind of intrafusal muscle fibre (dynamic to bag_1 and static to bag_2 and chain).

The non-selective distribution of static axons seems to be too frequent – about one third of the static responses are of an intermediate type – to be dismissed as an ontogenic imperfection. It is likely that it has some functional consequence; for example, as pointed out by Emonet-Dénand et al. (1977), during strong central activation of the static fusimotor system it could help to preserve the responsiveness of the bag_1 terminals of individual primary endings by preventing slackness without the need for activation of the dynamic system.

It should be emphasized that the existence of intermediate types of *responses* does not in any way interfere with the classification of motor *axons* into two functionally distinct – dynamic and static – classes. This assertion rests on the observation that whenever a single motor axon was tested on more than one ending (up to seven) practically without exception its action proved to be either predominantly static or predominantly dynamic.

The fact that static control action is apparently exerted through two kinds of muscle fibre, the bag_2 and the chains, is very intriguing. These two kinds of fibre, although they have nearly similar ultrastructural and histochemical profiles, are morphologically too different to suppose that they exert the same influence on the primary endings. Presumably, the heterogeneity of its effectors must confer to the static system some functional advantage but for the time being one must admit that the complexity of this system is still a challenge for experimentalists.

SKELETO-FUSIMOTOR INNERVATION

Skeleto-fusimotor axons are axons that supply both extrafusal and intrafusal muscle fibres. In mammalian skeletal muscles, they are commonly referred to as β motor axons, regardless of their conduction velocity, in order to distinguish them from exclusively fusimotor axons, the γ axons, and exclusively skeletomotor axons, the α axons.

Physiological evidence of their existence was originally obtained by Bessou, Emonet-Dénand & Laporte (1963, 1965), who observed in the deep lumbrical muscles of the cat that the repetitive stimulation of some single motor axons elicited both the contraction of extrafusal muscle fibres and an increase in the rate of discharge of primary endings. The latter effect was attributed to the contraction of intrafusal muscle fibres mainly because it persisted after the contraction of extrafusal muscle fibres had been selectively eliminated by small amounts of a curarizing drug. Histological confirmation was provided by Adal & Barker (1965), who observed in teased lumbrical muscles that some motor axons supplied, both spindles and extrafusal muscle fibres. At one time it was suggested (Haase, Meuser & Tan, 1966; Haase & Schlegel, 1966) that there were some motor axons in the 50–110 m sec^{-1} conduction velocity range that exclusively supplied muscle spindles. This so-called α fusimotor innervation was ruled out after an extensive study carried out on motor axons supplying the muscles whose spindles were supposed to be α-innervated. It was found that all 1800 motor axons with conduction velocities above 50 m sec^{-1} (except a few damaged ones) activated extrafusal muscle fibres (Ellaway, Emonet-Dénand, Joffroy & Laporte, 1972).

Skeleto-fusimotor axons were first observed experimentally in several small muscles in which it is relatively easy to test the action of nearly all the motor axons on a large fraction of the spindle population, i.e. cat superficial lumbrical muscles (Ellaway, Emonet-Dénand & Joffroy, 1971); rabbit lumbrical muscles (Emonet-Dénand, Jankowska & Laporte, 1970); rat tail muscles (Kidd, 1964; Andrew & Part, 1974); cat tenuissimus and abductor digiti quinti medius muscles (McWilliam, 1975). It was only recently that following some technical improvements (see Emonet-Dénand & Laporte, 1974) β motor axons were demonstrated experimentally in several large hindlimb muscles of the cat: flexor hallucis longus, peroneus brevis, tibialis anterior and soleus (Emonet-Dénand, Jami & Laporte, 1975). In a quantitative analysis carried out on the peroneus brevis muscle (Emonet-Dénand & Laporte, 1975) it was found that 72 % of the spindles of that muscle are supplied by at least one β axon, and that 18 % of the motor axons activating extrafusal muscle fibres were β axons. Most of the cat β axons which were identified by the technique of selective blockade of extrafusal junctions were found to have a conduction velocity lower than 80 m sec^{-1} (in peroneus brevis one third of these relatively slow axons are β) and to exert a strong dynamic action on primary endings as illustrated by Fig. 5. Dynamic β axons and dynamic γ axons may supply the same spindle (Bessou et al., 1965; Emonet-Dénand & Laporte, 1975).

The types of intrafusal and extrafusal muscle fibre innervated by dynamic β axons have been recently determined with the glycogen-depletion technique (Barker et al., 1977). In each experiment a single dynamic β axon was prepared and repeatedly stimulated (for the regime and conditions of stimulation, see Barker et al., 1977). It was found that the intrafusal distribution of dynamic β axons was almost exclusively restricted to bag$_1$ fibres and that the extrafusal muscle fibres belong to the slow oxidative type as defined by Ariano, Armstrong & Edgerton (1973). In that study, carried out on tenuissimus and peroneus brevis muscles, 24 β-innervated spindle poles were analysed. In each of them the bag$_1$ fibre was depleted; in only

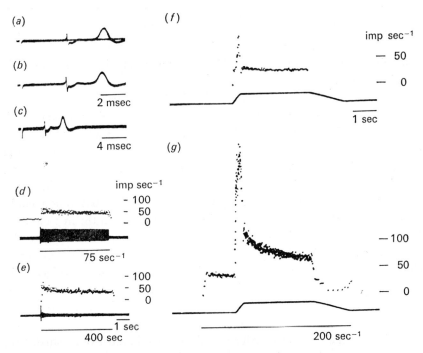

Fig. 5. Dynamic skeleto-fusimotor axon supplying a cat peroneus brevis muscle. Records (a), (b), (c) show that a single motor axon to the peroneus brevis muscle was activated by stimulation of a ventral root filament; the first recording electrode lifted a small nerve branch supplying the muscle while the second electrode was on the surface of the muscle. Superimposed records: (a) threshold stimulation of the axon showing the all-or-none behaviour of the nerve and muscle potentials. (b) and (c) Stimulation, at least 50 times the threshold for the axon, showing that the stimulated ventral root filament contained no γ axons. Note the slower sweep speed in (c). Conduction velocity of the motor axon, 56 m sec^{-1}. Records (d) and (e) show that this axon supplied both extrafusal and intrafusal fibres. Upper trace: instantaneous frequency of discharge of a primary ending recorded from an afferent fibre in a dorsal root filament; lower trace: muscle action potentials. When the single motor axon was stimulated at 75 sec^{-1} (record (d)) there was both an increase in the frequency of discharge of the ending and an activation of the extrafusal fibres as shown by their action potentials. Stimulation of the axon at a much higher rate, 400 sec^{-1}, rapidly resulted in the block of the extrafusal junction (see Emonet-Dénand & Laporte, 1974) but the spindle activation persisted (record (e)). Comparison of records (f) and (g) shows the strong dynamic effect of this β axon. The muscle was submitted to a ramp-and-hold stretch of 2 mm followed by a progressive return to the original length. (f) No stimulation. (g) Stimulation of the β axon at 200 sec^{-1}. Note the large increase of the dynamic index and the characteristic slow decay of firing after the completion of the dynamic stretch. (From Laporte & Emonet-Dénand, 1976.)

two poles was one fibre of another type (bag$_2$ and long chain) also depleted. The zones of depletion were located in the mid-polar region, i.e. in the region where most p$_1$ plates are found. The number of slow oxidative fibres belonging to the β motor units could not be accurately determined but it is certainly small.

Another type of β axon, whose distribution differs strikingly from that of the dynamic β axon, has been recently found in cat spindles (Harker, Jami, Laporte & Petit, 1977), as a result of studying the distribution of very fast conducting motor axons. This was undertaken because, when using the extrafusal selective blockade method for detecting skeleto-fusimotor axons, practically no β axons were found among the axons belonging to the upper conduction velocity range, i.e. 85–110 m sec^{-1}. This observation could mean either that there are no β axons in this range, or that the method used is not adequate for all kinds of β axon. An unexpected answer was given by a new way of applying the glycogen-depletion method. A certain number of single motor axons – 12 to 20 – supplying the same muscle and having in common a conduction velocity faster than 85 m sec^{-1} were stimulated together so as to elicit glycogen depletion in the muscle fibres they supplied. The spindles of that muscle were then systematically analysed for glycogen depletion. In the peroneus tertius muscle (chosen because of its small size) 25% of the examined spindles were found to have some glycogen depletion; the zones of depletion occurred almost exclusively in chain fibres and the chain fibre depleted was invariably the longest chain present. It was checked that the peroneus tertius muscle, as all other similarly tested muscles (see Ellaway et al., 1972), does not have exclusively fusimotor axons in the α conduction velocity range. Thus the presence of glycogen depletion in a significant number of spindles shows that among the very fast conducting axons there must be some skeleto-fusimotor axons supplying chain fibres. Similar observations were made on the tenuissimus by Jami, Lan-Couton, Malmgren & Petit (1978). The extrafusal muscle fibres supplied by the fast skeleto-fusimotor axons belong to the fast oxidative glycolytic type (L. Jami, personal communication). The function of these β axons is currently being investigated.

All these observations show that whenever sought for with adequate techniques β axons can be detected and that skeleto-fusimotor axons play a significant part in the motor supply of cat spindles. However, it is not yet possible to estimate the overall importance of this system. If, as suggested by Barker et al. (1970), p$_1$ plates are exclusively supplied by β axons, the skeleto–fusimotor innervation must be widespread since p$_1$ plates are present in as much as 20 to 80% of the spindles in various cat hindlimb muscles. It is hoped that further experiments will give unequivocal answers to two questions: are β axons supplying spindles through p_1 plates and are p_1 plates exclusively supplied by β axons?

Most of the recent investigations presented in this review were supported by grants from the Foundation for French Medical Research and from INSERM (ATP 76–61).

REFERENCES

Adal, M. N. & Barker, D. (1965). Intramuscular branching of fusimotor fibres. *J. Physiol., Lond.* **177**, 288–99.

Andrew, B. L. & Part, N. J. (1974). The division of control of muscle spindles between fusimotor and mixed skeleto-fusimotor fibres in a rat caudal muscle. *Q. Jl exp. Physiol.* **59**, 331–49.

Ariano, M., Armstrong, R. & Edgerton, V. (1973). Hindlimb muscle fiber populations of five mammals. *J. Histochem. Cytochem.* **21**, 51–5.

Banks, R. W., Barker, D., Harker, D. W. & Stacey, M. J. (1975). Correlation between ultrastructure and histochemistry of mammalian intrafusal muscle fibres. *J. Physiol., Lond.* **252**, 16–17P.

Banks, R. W., Harker, D. W. & Stacey, M. J. (1977). A study of mammalian intrafusal muscle using a combined histochemical and ultrastructural technique. *J. Anat.* **123**(3), 783–96.

Barker, D., Banks, R. W., Harker, D. W., Milburn, A. & Stacey, M. J. (1976a). Studies of the histochemistry, ultrastructure, motor innervation, and regeneration of mammalian intrafusal muscle fibres. In *Understanding the stretch reflex*, ed. S. Homma, *Prog. Brain Res.* **44**, 67–88.

Barker, D., Bessou, P., Jankowska, E., Pagès, B. & Stacey, M. (1972). Distribution des axones fusimoteurs statiques et dynamiques aux fibres musculaires intrafusales, chez le chat. *C.r. hebd. Séanc. Acad. Sci., Paris, D* **275**, 2527–30.

Barker, D., Bessou, P., Jankowska, E., Pagès, B. & Stacey, M. (1978). Identification of intrafusal muscle fibres activated by single fusimotor axons and injected with fluorescent dye in cat tenuissimus spindles. *J. Physiol., Lond.* **275**, 149–65.

Barker, D., Emonet-Dénand, F., Harker, D. W., Jami, L. & Laporte, Y. (1976b). Distribution of fusimotor axons to intrafusal muscle fibres in cat tenuissimus spindles as determined by the glycogen-depletion method. *J. Physiol., Lond.* **261**, 49–69.

Barker, D., Emonet-Dénand, F., Harker, D. W., Jami, L. & Laporte, Y. (1977). Types of intra- and extrafusal muscle fibre innervated by dynamic skeleto–fusimotor axons in cat peroneus brevis and tenuissimus muscles, as determined by the glycogen-depletion method. *J. Physiol., Lond.* **266**, 713–26.

Barker, D., Emonet-Dénand, F., Laporte, Y., Proske, U. & Stacey, M. (1973). Morphological identification and intrafusal distribution of the endings of static fusimotor axons in the cat. *J. Physiol., Lond.* **230**, 405–27.

Barker, D. & Ip, M. C. (1965). The motor innervation of cat and rabbit muscle spindles. *J. Physiol., Lond.* **177**, 27–8P.

Barker, D. & Stacey, M. J. (1970). Rabbit intrafusal muscle fibres. *J. Physiol., Lond.* **210**, 70–2P.

Barker, D., Stacey, M. J. & Adal, M. N. (1970). Fusimotor innervation in the cat. *Phil. Trans. R. Soc., B* **258**, 315–46.

Bessou, P., Emonet-Dénand, F. & Laporte, Y. (1963). Occurrence of intrafusal muscle fibres innervation by branches of slow motor fibres in the cat. *Nature, Lond.* **193**, 594–5.

Bessou, P., Emonet-Dénand, F. & Laporte, Y. (1965). Motor fibres innervating extrafusal and intrafusal muscle fibres in the cat. *J. Physiol., Lond.* **180**, 649–72.

Bessou, P. & Pagès, B. (1975). Cinematographic analysis of contractile events produced in intrafusal muscle fibres by stimulation of static and dynamic fusimotor axons. *J. Physiol., Lond.* **252**, 397–427.

Boyd, I. A. (1962). The structure and innervation of the nuclear bag muscle fibre system and the nuclear chain muscle fibre system in mammalian muscle spindles. *Phil Trans. R. Soc.*, B **245**, 81–136.

Boyd, I. A. (1976). The response of fast and slow nuclear bag fibres and nuclear chain fibres in isolated cat muscle spindles to fusimotor stimulation, and the effect of intrafusal contraction on the sensory endings. *Q.Jl exp. Physiol.* **61**, 203–54.

Boyd, I. A., Gladden, M., McWilliam, P. N. & Ward, J. (1977). Control of dynamic and static nuclear bag fibres and nuclear chain fibres by γ and β axons in isolated cat muscle spindles. *J. Physiol., Lond.* **265**, 133–62.

Brown, M. C. & Butler, R. C. (1973). Studies on the site of termination of static and dynamic fusimotor fibres within muscle spindles of the tenuissimus muscle of the cat. *J. Physiol., Lond.* **233**, 553–73.

Edström, L. & Kugelberg, E. (1968). Histochemical composition, distribution of fibres and fatiguability of single motor units. *J. Neurol. Neurosurg. Psychiat.* **31**, 424–33.

Ellaway, P., Emonet-Dénand, F. & Joffroy, M. (1971). Mise en évidence d'axones squeletto–fusimoteurs (axones β) dans le muscle premier lombrical superficiel du chat. *J. Physiol., Paris* **63**, 617–23.

Ellaway, P., Emonet-Dénand, F., Joffroy, M. & Laporte, Y. (1972). Lack of exclusively fusimotor α-axons in flexor and extensor leg muscles of the cat. *J. Neurophysiol.* **35**, 149–53.

Emonet-Dénand, F., Jami, L. & Laporte, Y. (1975). Skeleto-fusimotor axons in hindlimb muscles of the cat. *J. Physiol., Lond.* **249**, 153–66.

Emonet-Dénand, F., Jankowska, E. & Laporte, Y. (1970). Skeleto-fusimotor fibres in the rabbit. *J. Physiol., Lond.* **210**, 669–80.

Emonet-Dénand, F. & Laporte, Y. (1974). Blocage neuromusculaire sélectif des jonctions extrafusales des axones squeletto–fusimoteurs produit par leur stimulation répétitive à fréquence élevée. *C.r. hebd. Séanc. Acad. Sci., Paris*, D **279**, 2083–5.

Emonet-Dénand, F. & Laporte, Y. (1975). Proportion of muscle spindles supplied by skeleto–fusimotor axons (β axons) in the peroneus brevis muscle of the cat. *J. Neurophysiol.* **38**, 1390–4.

Emonet-Dénand, F., Laporte, Y., Matthews, P. B. C. & Petit, J. (1977). On the subdivision of static and dynamic fusimotor actions on the primary ending of the cat muscle spindle. *J. Physiol., Lond.* **268**, 827–61.

Gladden, M. (1976). Structural features relative to the function of intrafusal muscle fibres in the cat. In *Understanding the stretch reflex*, ed. S. Homma, *Prog. Brain Res.* **44**, 51–9.

Haase, J., Meuser, P. & Tan, U. (1966). Die Konvergenz fusimotorischer α-Impulse auf deeffertierte Flexor Spindeln der Katze. *Pflügers Arch. ges. Physiol.* **289**, 50–8.

Haase, J. & Schlegel, J. H. (1966). Einige funktionelle Merkmale von α-innervierten Extensor- und Flexor-Spindeln der Katze. *Pflügers Arch. ges. Physiol.* **287**, 163–75.

Harker, D. W., Jami, L., Laporte, Y. & Petit, J. (1977). Fast conducting skeleto–fusimotor axons supplying intrafusal chain fibres in the cat peroneus tertius muscle. *J. Neurophysiol.* **40**, 791–9.

Jami, L., Lan-Couton, D., Malmgren, K. & Petit, J. (1978). 'Fast' and 'slow' skeleto-fusimotor innervation in cat tenuissimus spindles; a study with the glycogen-depletion method. *Acta physiol. scand.* (In Press.)

Kidd, G. L. (1964). Excitation of primary muscle spindle endings by β-axon stimulation. *Nature, Lond.* **203**, 1248–51.

Laporte, Y. & Emonet-Dénand, F. (1976). The skeleto-fusimotor innervation of cat muscle spindle. In *Understanding the stretch reflex*, ed. S. Homma, *Prog. Brain Res.* **44**, 99–106.

Leksell, L. (1945). The action potentials and excitatory effects of the small ventral root fibres to skeletal muscle. *Acta physiol. scand.* **10** (Suppl. 31), 1–84.

Matthews, P. B. C. (1962). The differentiation of two types of fusimotor fibre by their effects on the dynamic response of muscle spindle primary endings. *Q.Jl exp. Physiol.* **47**, 324–33.

Matthews, P. B. C. (1964). Muscle spindles and their motor control. *Physiol. Rev.* **44**, 219–88.

Matthews, P. B. C. (1972). *Mammalian muscle receptors and their central actions.* London: Edward Arnold.

McWilliam, P. N. (1975). The incidence and properties of β axons to muscle spindles in the cat hind limb. *Q.Jl exp. Physiol.* **60**, 25–36.

Ovalle, W. K. & Smith, R. S. (1972). Histochemical identification of three types of intrafusal muscle fibres in the cat and monkey based on the myosin ATPase reaction. *Can. J. Physiol. Pharmac.* **50**, 195–202.

Mechanisms determining the response pattern of primary endings of mammalian muscle spindles

C. C. HUNT

The primary ending of the mammalian muscle spindle responds to an increase in muscle length by a discharge which depends in frequency on the rate of change as well as the amplitude of stretch. In response to ramp-and-hold stretch the pattern of impulse activity shows a number of characteristic and well-known features. These include the initial burst and the subsequent dynamic discharge at frequencies above the static level, an adaptive fall during the early portion of hold stretch, a sustained static discharge and a pause or reduction in the frequency of baseline discharge on release.

In this article the mechanisms underlying the response pattern of the primary ending to ramp-and-hold stretch will be considered. It is generally held that muscle stretch transmits force via the intrafusal muscle fibers to the sensory terminals of the spindle, producing conductance changes in the receptor membrane. This results in depolarization of the terminals, causing, in turn, the electrotonic spread of current to an impulse-initiating site in the sensory axon.

Since a series of steps are involved in the response to changes in muscle length, it is difficult to determine the several mechanisms involved by analysis of the impulse response alone. For this reason we have utilized an isolated preparation which permits recording of the receptor potential as well as microscopic measurement of length changes in various elements of the muscle spindle. Further, removal of the spindle capsule in the region of the sensory terminals by micro-dissection makes it possible to exchange rapidly the fluid around the nerve terminals and to study the effects of alterations in ionic composition of the bathing fluid on the receptor potential.

The techniques for the isolation of spindles from tail muscles of the cat have been described (Hunt & Ottoson, 1975). For studies of the effects of altering the composition of extracellular fluid, isolated decapsulated spindles were placed in a rapid-flow chamber and the primary axon drawn into heavy oil in an agar–Locke solution-filled pipette, and differential d.c.

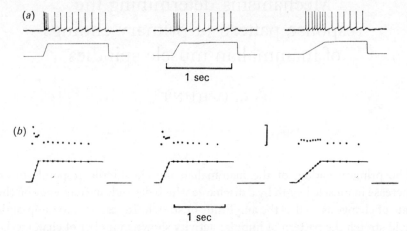

Fig. 1. (a) Response of a primary ending to ramp-and-hold stretch. Stretch amplitude 750 μm. (Modified from Hunt & Ottoson, 1975.) (b) Instantaneous frequency plot of response from another primary ending to ramp-and-hold stretch. (Modified from Hunt & Ottoson, 1976.) Vertical bar: frequency calibration, 0–250 sec^{-1}.

recording was employed between this pipette and another placed in a bathing solution (Hunt, Fukami & Wilkinson, 1977a,b).

The impulse response of one primary ending of an isolated spindle and an instantaneous frequency plot of the response of another to ramp-and-hold stretch are shown in Fig. 1. The responses are very similar to those recorded from primary endings of muscle spindles *in situ* (see Matthews, 1972). Impulse frequency rises early in the ramp, often evoking a series of impulses called the initial burst. This is followed by a later dynamic response during which frequency may increase or remain steady depending on stretch velocity. At the end of the ramp the frequency drops abruptly and there may be a pause in the discharge before impulse activity resumes during the hold phase. During the first few hundred milliseconds of hold stretch there is a small, slow, adaptive fall in frequency. In endings which show a baseline discharge the release from ramp-and-hold stretch produces a decrease in frequency or a cessation of discharge, as is well known from *in situ* studies.

After impulse activity is blocked by tetrodotoxin, the receptor potential may be recorded extracellularly from the primary axon near the ending. The response to ramp-and-hold stretch shows a sequence of potential changes which are, to a large extent, similar to the instantaneous impulse frequency response to the same stimulus. This may be seen in Fig. 2, which shows the response of the primary ending to ramp-and-hold stretch delivered at three different velocities. Early during ramp stretch there is a rapid depolariza-

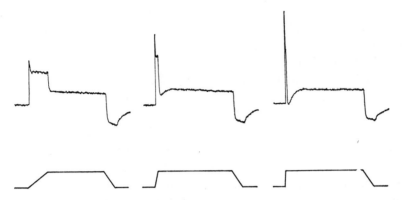

Fig. 2. Receptor potential of a primary ending in response to ramp-and-hold stretch at three different velocities of stretch. Upper traces – potential; lower traces – length change. Stretch duration 1 sec. Stretch amplitude 300 μm. Velocities 1, 3 and 9 mm sec^{-1}.

tion, the initial dynamic component of the receptor potential, which may rise to a distinct peak. This component occurs at the same time as, and is responsible for, the initial burst in the impulse response. The amplitude of this component is dependent upon stretch velocity. During the remainder of the ramp the late dynamic component of the receptor potential may remain constant or rise, depending on stretch velocity. Following the ramp, the potential falls towards and may undershoot the static level. This post-dynamic undershoot appears to be responsible, at least in part, for the pause in discharge which may occur at the end of high-velocity ramp stretch. There then follows a slow, adaptive fall in the receptor potential, during the first few hundred milliseconds of hold stretch, which is associated with the slow fall in impulse frequency that occurs during this period. On release from stretch there is, typically, a hyperpolarization relative to baseline followed by a slow return toward the baseline level which may take several seconds. This post-release undershoot must play a large role in causing the pause or reduction in discharge frequency on release. The mechanisms involved in the receptor potential and impulse response to ramp-and-hold stretch will be considered below.

MECHANICAL FACTORS

The role of mechanical factors in determining the response of primary endings to ramp-and-hold stretch is still not well understood. However, enough evidence is available to indicate that some features are attributable to such factors and others not. In addition to indirect evidence obtained by analysis of impulse response, two direct approaches have been employed: one is the recording of tension developed by the isolated spindle in response to stretch; the other is the measurement of length changes in elements of the spindle by microscopy.

The initial burst and the underlying initial component of the receptor potential appear to depend upon mechanical phenomena in the intrafusal muscle fibers. Since the initial burst is abolished by prior stretch and is enhanced by a previous period of fusimotor stimulation, Brown, Goodwin & Matthews (1969) suggested that it was caused by a short-range elasticity produced by cross bridges between thick and thin filaments in the intrafusal muscle fibers, bridges which are formed at rest and broken when stretch exceeds the limit of this elastic component. Direct evidence in support of this idea was provided by the recording of tension in the isolated spindle (Hunt & Ottoson, 1976). Fig. 3 shows the receptor potential response to a repeated ramp-and-hold stretch together with the tension developed in an isolated spindle. During the early part of ramp stretch there is a rapid rise in tension at the same time as the initial component of the receptor potential appears. Repeating the stretch at an interval of less than about 5 sec reduces both the initial component of the receptor potential and the early rise in tension. This indicates the presence of a short-range elastic component in the intrafusal fibers which takes several seconds to re-form.

The tension and the receptor potential responses to ramp-and-hold stretch of another primary ending are shown in Fig. 4. Tension rises to a peak at the end of ramp stretch and then declines smoothly toward the static level without any undershoot at the time of the postdynamic minimum. There is a slow fall in tension during the period of the adaptive fall in both static discharge and receptor potential, suggesting that back slippage in the intrafusal fibers may be responsible. Microscopic observations also show back slippage during this time, especially in bag fibers (Boyd, 1976). Thus, it seems reasonable to attribute the adaptive fall to mechanical factors on release of stretch. There is a small drop in tension below the baseline level on release. The time course of recovery of tension, and the relation between tension changes and the post-release undershoot have not been studied, but the evidence at hand suggests that stretch of the terminals may be reduced, at least briefly, on release.

Early mechanical models to explain the dynamic behaviour of primary endings contained viscous components in parallel with elastic elements in polar regions of the intrafusal muscle fibers (Matthews, 1933; Matthews, 1972).

Such models imply that the length changes of the sensory region depend on velocity as well as amplitude of stretch. Microscopic measurements of

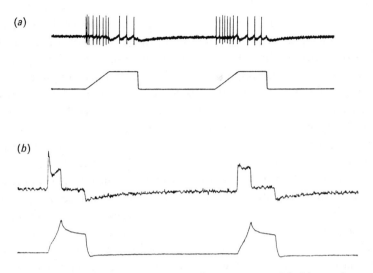

Fig. 3. (*a*) Impulse response of a primary ending to ramp-and-hold stretch repeated at an interval of 1 sec. (*b*) Receptor potential of a primary ending (above) and tension (below) responses of an isolated spindle to ramp-and-hold stretch repeated at an interval of 1 sec. (From Hunt & Ottoson, 1976.)

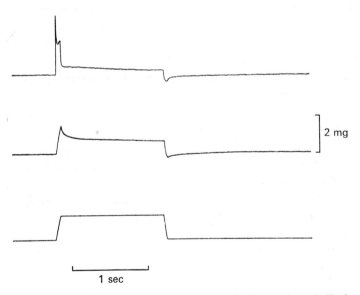

Fig. 4. Receptor potential (upper trace) tension (middle trace), and displacement (lower trace) in response to ramp-and-hold stretch. (Unpublished record of Hunt and Wilkinson.)

length changes in the region of sensory terminals in frog spindle (Ottoson & Shepherd, 1970) and in short capsule snake spindles (Fukami & Hunt, 1977) do not support this. Also, studies of the response of primary endings to sinusoidal stretch shows that sensitivity does not vary with frequency in a manner consistent with the response being determined simply by velocity and amplitude of stretch (Matthews & Stein, 1969; Poppele & Bowman, 1970; Hasan & Houk, 1975a). These studies have also shown clearly the non-linearity of the response to stretch and indicate that a model containing a non-linear frictional element in the polar regions of intrafusal fibers better predicts the behaviour of the primary ending (Hasan & Houk, 1975b).

It seems likely that short-range elastic components may permit the transmission of small amplitude forces to the region of the sensory terminals with relatively small losses whereas larger forces, exceeding the limits of the short-range elasticity, would suffer greater losses due to the extension of the intrafusal muscle fibers. Just how the non-linear response and the short-range elasticity underlying the initial burst may be related is not clear.

IONIC MECHANISMS

The conductance changes in mechanoreceptor terminals are difficult to study because the endings are usually encapsulated or embedded in tissue which provides a barrier to diffusion. However, previous studies have implicated sodium in the conductance changes of several mechano-receptors, such as the Pacinian corpuscle (Diamond, Gray & Inman, 1958), crustacean stretch receptor (Edwards, Terzuolo & Washizu, 1963; Obara, 1968) and frog spindle (Ottoson, 1964).

The mammalian spindle has a relatively large capsular space in the region of its sensory endings and the capsule in this region may be removed by micro-dissection. Hunt, Wilkinson, & Fukami (1978) have used this preparation to study the effects of changes in ionic composition of the bathing fluid. Such effects are rapid, in contrast with the much slower effects which occur when the capsule is intact. The receptor potential, recorded extracellularly from the primary axon, shows a rapid reduction when choline, TRIS (tris (hydroxymethyl) amino methane) or glucosamine is substituted for Na^+ in the bathing solution. The subsequent removal of Ca^{2+}, with the addition of ethyleneglycol bis (β-aminoethyl ether)-N,N'-tetra-acetic acid (EGTA), causes a further reduction in the response. These effects are readily reversible. Examples of the changes in response on removing Na^+ and Ca^{2+} are shown in Fig. 5.

When external Ca^{2+} concentration is varied between 0 and 10 mM in the presence of normal $[Na^+]_o$, no significant changes occur in the receptor

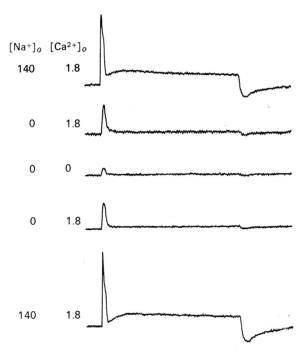

Fig. 5. Responses of a primary ending of an isolated decapsulated spindle to changes in ionic composition of bathing fluid (indicated at left). Ramp-and-hold stretches of 1-sec duration were identical. Tris(hydroxymethyl)amino methane Na^+ and Ca^{2+}. Sequence of records from above down. (Unpublished record of C. C. Hunt, R. S. Wilkinson and Y. Fukami.)

potential response to ramp-and-hold stretch. Yet, Ca^{2+} can, to a limited extent, substitute for Na^+ in the absence of the latter. Also, the Ca^{2+}-blocking agent D 600, a derivative of veratramil, abolishes the response in normal $[Ca^{2+}]_o$ and zero $[Na^+]_o$ but has no effect in the presence of normal $[Na^+]_o$. These studies indicate that the depolarization of the sensory terminals of the primary ending, produced by stretch deformation, results from an increase in sodium conductance and that, in the absence of Na^+, Ca^{2+} may partially substitute. It is of interest that Li^+ can substitute for Na^+ in the stretch-evoked response. In fact, the amplitude of the receptor potential increases by about 20 % when Li^+ is substituted for Na^+, suggesting that the conductance channel is more permeable to Li^+ than Na^+.

The postdynamic and post-release undershoots in the receptor potential response appear to be secondary to conductance changes in the primary ending. The ending becomes relatively more permeable to K^+ during these periods, causing its potential to be driven toward the potassium equilibrium potential (E_K). The effects of changing initial length, of altering the con-

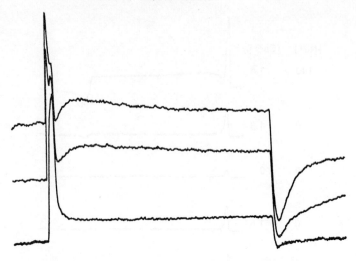

Fig. 6. Response of a primary ending of an isolated spindle to ramp-and-hold stretch delivered at different initial tensions. (Unpublished record of C. C. Hunt, R. S. Wilkinson and Y. Fukami.)

centration of K^+ in the bathing fluid, and of the addition of tetraethyl-ammonium indicate that K^+ conductance is increased, at least relatively during the postdynamic and post-release undershoots (Hunt *et al.* 1978). Chloride conductance changes do not seem to be involved.

When the initial length of the spindle is increased, the baseline potential shifts to a more depolarized level, which may be attributed to an increase in baseline sodium conductance (g_{Na}). While the dynamic and static responses show slight changes in configuration, the most striking changes occur during the postdynamic and post-release periods (Fig. 6). At low initial tension the potential falls smoothly from the end of the dynamic response toward the static level, but when initial tension is increased, at sufficiently high ramp velocities, the postdynamic undershoot becomes evident and further increases in initial length cause a greater undershoot relative to the static level. On release from stretch the ending becomes hyperpolarized relative to the baseline. With increases in initial length, the baseline progressively shifts to more depolarized levels but there is much less shift in the post-release level. The post-release minimum appears to approach an equilibrium level.

The extent to which the postdynamic minimum undershoots the static level depends upon the level of depolarization attained during the dynamic response. The latter may be varied by changing the velocity of stretch. As the level of depolarization late in the dynamic phase increases, the post-dynamic minimum reaches progressively greater levels of polarization. This

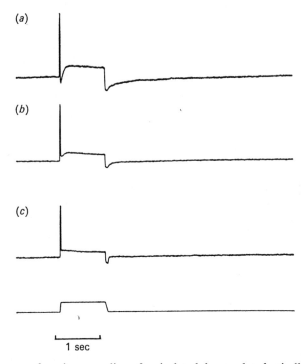

Fig. 7. Response of a primary ending of an isolated decapsulated spindle to changes in potassium concentration in the bathing fluid ($[K^+]_o$). (a) $[K^+]_o = 5$ mM; (b) $[K^+]_o = 10$ mM; (c) $[K^+]_o = 20$ mM. (Unpublished record of C. C. Hunt, R. S. Wilkinson and Y. Fukami.)

could result from a voltage dependent increase in potassium conductance (g_K) which is turned on during the dynamic depolarization and which turns off slowly when the potential falls toward the static level.

Increasing the concentration of K^+ in the bathing solution shifts the baseline potential, as might be expected if the baseline level were determined, in part, by a resting level of g_K. The changes in the postdynamic and post-release levels with increasing $[K^+]_o$ are greater than the changes in baseline level suggesting that during those phases the potential is determined to a greater extent by g_K. An example of the effect of increasing the K^+ concentration in the bathing fluid is shown in Fig. 7. There is a progressive reduction in the postdynamic and post-release undershoots relative to baseline as $[K^+]_o$ is raised from 5 to 10 to 20 mM.

Tetraethylammonium (TEA) is known to block certain potassium conductance, in particular the delayed rectifier channel. Hunt, Fukami & Wilkinson (1977b) have studied the effect of TEA on the receptor potential response in order to determine its effect on the components which appear to be related to potassium conductance changes. In concentrations of 7–10 mM or greater, TEA was found to reduce or abolish the postdynamic undershoot. It may also produce a small depolarization of the baseline level and a small increase in the amplitude of the dynamic and static responses. The post-release undershoot was not reduced in size but the time course of its return toward baseline was more rapid. These findings suggest that the postdynamic undershoot is due to a voltage-dependent g_K turned on during the dynamic depolarization. The effects on the amplitude of the dynamic and static response to ramp-and-hold stretch could also result from a reduction in g_K by TEA at the nodes of Ranvier in the myelinated branches of the primary ending. This could increase the length constant, causing a larger fraction of the potential changes in the nerve terminals to be recorded.

The persistence of the post-release undershoot in the presence of TEA suggests that it has another basis. It seems likely that during release g_{Na} is reduced compared to its baseline level, producing a relative increase in g_K and causing the potential to approach E_K. This may depend, at least to some extent, on a reduction of stretch of the terminals due to mechanical properties of the intrafusal muscle fibers (see above).

The site of the conductance change to Na^+ produced by stretch deformation is presumably in the terminal membrane, probably the membrane facing the intrafusal fiber. While voltage-dependent changes in g_K could also occur in this membrane, it is possible that they take place in the nodes of the myelinated branches of the primary axon.

IMPULSE INITIATION

To what extent the properties of the impulse-initiating site play a role in determining the response pattern of the primary ending is not clear. Earlier studies of the effect of externally applied rectangular current pulses on static discharge frequency indicated that there was no detectable accommodation at the impulse-initiating site, although a subsequent report by Emonet-Dénand & Houk (1969) showed some accommodation using this approach. This method can reveal accommodation with a fairly long time constant but may not detect a dependence of impulse frequency upon relatively rapid rates of change of generator current that occur during the dynamic response to high-velocity stretch.

Hunt & Ottoson (1975) found that the frequency of discharge during the dynamic response at rapid rates of extension was higher than predicted from the amplitude of the receptor potential and the steady-state relation between impulse frequency and static discharge. Similar discrepancies were noted in short capsule snake spindles (Fukami, 1975) and in isolated cat tendon organs (Fukami & Wilkinson, 1977). Such findings suggest that the impulse-initiating sites may be sensitive to rate of change as well as amplitude of receptor potential. A possible, but unlikely, alternative explanation is that the time course of the recorded receptor potential is slower, due to electrotonic decrement, than the course of generator current at the impulse-initiating site. This would require that the effective recording site of the receptor potential be a considerable distance from the impulse-initiating site. Preliminary studies on the effect of passing current of varying slope through the isolated primary axon indicates that impulse threshold is sensitive to rate of change of current (C. C. Hunt and D. Ottoson, unpublished).

Another interesting question is the extent to which impulses, once initiated, antidromically invade the branches of the primary axon. Impalement of the last myelinated segment of primary axon branches shows receptor potential responses to small-amplitude, sinusoidal stretch (about 5 to 10 μm at 25 Hz) of about 20 mV in amplitude whereas action potentials, coincident with those propagating up the parent axon, are only a few millivolts in amplitude (C. C. Hunt, R. S. Wilkinson and Y. Fukami, unpublished). This suggests that antidromic propagation into branches is limited, and that the action potential fails to invade the more distal nodes of the primary branches. At the present time there is no direct evidence as to the location of the impulse-initiating site or whether the impulse-initiating site may shift, although indirect evidence suggests it may do so.

In summary, the impulse response of the primary ending to ramp-and-hold stretch results mainly from the configuration of the receptor potential, although with high-velocity stretch, impulse frequency appears to depend to some extent on rate of change of the receptor current. The receptor potential results from depolarization of the terminals produced by an increased conductance to sodium. In the absence of Na^+, Ca^{2+} may substitute, in part, in this conductance channel. Whether this channel is also permeable to K^+ is not yet known. The receptor potential response to ramp-and-hold stretch is modified by changes in the relative contribution of g_K. In the postdynamic period a voltage-dependent g_K may produce an undershoot in receptor potential associated with a drop in impulse frequency. In the post-release period the receptor potential hyperpolarizes relative to the baseline. This results because g_K determines the potential to a greater extent in the post-release period than during baseline. It is possible that this may be produced by a diminution in g_{Na} on release, relative to the baseline, and to a lesser extent from an increase in g_K turned on by the static level of depolarization. The changes in g_K may be important in the responses of this ending to phasic stretch, particularly to repetitive stretch, as in sinusoidal length change at relatively high frequencies. Primary endings can follow sinusoidal length changes up to several hundred hertz. A rapid repolarization in the relaxation phase produced by an increased g_K may enable the ending to respond to repeated sinusoidal stretch without undergoing summation in receptor potential response.

Certain features of the response, such as the initial burst and the adaptive fall early in the static phase, may be attributed to mechanical factors, but we have little information on the contribution of other mechanical processes, such as the coupling between length changes of the intrafusal fibers and the overlying sensory terminals.

It is a great pleasure to contribute an article in this volume honoring Archie McIntyre. I have had the privilege of working with him in several studies and we have been friends for many years. I much admire him as a person and as a colleague who has unusual breadth of knowledge and experimental skill.

Support by grants from the Public Health Service (NS 07907) and from the Muscular Dystrophy Associations (Jerry Lewis Muscular Research Center, Washington University) is acknowledged.

REFERENCES

Boyd, I. A. (1976). The mechanical properties of dynamic nuclear bag fibres, static nuclear bag fibres and nuclear chain fibres in isolated cat muscle spindles. *Prog. Brain Res.* **44**, 33–49.

Brown, M. C., Goodwin, G. M. & Matthews, P. B. C. (1969). After effects of fusimotor stimulation on the response of muscle spindle primary afferent endings. *J. Physiol., Lond.* **205**, 677–94.

Diamond, J., Gray, J. A. B. & Inman, D. R. (1958). The relation between receptor potential and the concentration of sodium ions. *J. Physiol., Lond.* **142**, 382–94.

Edwards, C., Terzuolo, C. A. & Washizu, Y. (1963). The effect of changes of the ionic environment upon an isolated crustacean sensory neuron. *J. Neurophysiol.* **26**, 948–58.

Emonet-Dénand, F. & Houk, J. (1969). Some effects of polarizing current on discharges from muscle spindle receptors. *Am. J. Physiol.* **216**, 404–6.

Fukami, Y. (1975). Receptor potential and impulse initiation in two varieties of reptilian muscle spindle. *Nature, Lond.* **257**, 240–1.

Fukami, Y. & Hunt, C. C. (1977). Structures in sensory region of snake spindles and their displacement during stretch. *J. Neurophysiol.* **40**, 1121–31.

Fukami, Y. & Wilkinson, R. S. (1977). Responses of isolated Golgi tendon organs of the cat. *J. Physiol., Lond.* **265**, 673–89.

Hasan, Z. & Houk, J. C. (1975a). Analysis of response properties of deefferented mammalian spindle receptors based on frequency response. *J. Neurophysiol.* **38**, 663–72.

Hasan, Z. & Houk, J. C. (1975b). Transition in sensitivity of spindle receptors that occurs when a muscle is stretched more than a fraction of a millimeter. *J. Neurophysiol.* **38**, 673–89.

Hunt, C. C., Wilkinson, R. S. & Fukami, Y. (1978). Ionic basis of the receptor potential in primary endings of mammalian muscle spindles. *J. gen. Physiol.* (In Press.)

Hunt, C. C. & Ottoson, D. (1975). Impulse activity and receptor potential of primary and secondary endings of isolated mammalian muscle spindles. *J. Neurophysiol.* **39**, 324–30.

Hunt, C. C. & Ottoson, D. (1976). Initial burst of primary endings of isolated mammalian muscle spindles. *J. Neurophysiol.* **39**, 324–30.

Matthews, B. H. C. (1933). Nerve endings in mammalian muscle. *J. Physiol., Lond.* **78**, 1–53.

Matthews, P. B. C. (1972). *Mammalian muscle receptors and their central actions.* London: Edward Arnold.

Matthews, P. B. C. & Stein, R. B. (1969). The sensitivity of muscle spindle afferents to small sinusoidal changes in length. *J. Physiol., Lond.* **200**, 723–43.

Obara, S. (1968). Effects of some organic cations on generator potential of crayfish stretch receptor. *J. gen. Physiol.* **52**, 363–86.

Ottoson, D. (1964). The effect of sodium deficiency on the response of the isolated muscle spindle. *J. Physiol., Lond.* **171**, 109–18.

Ottoson, D. & Shepherd, G. M. (1970). Length changes within isolated frog muscle spindle during and after stretching. *J. Physiol., Lond.* **207**, 747–59.

Poppele, R. E. & Bowman, R. J. (1970). Quantitative description of linear behavior of mammalian muscle spindles. *J. Neurophysiol.* **33**, 59–72.

Joint receptors in the wrist of the cat

D. J. TRACEY

Of all the myelinated afferents on whose physiology Archie McIntyre's work has contributed so much, afferents from the capsules of joints are the least understood. There are several reasons for this. The receptors lack the complex and elegant structure of muscle spindles, and the accessibility of cutaneous receptors; and despite a number of studies, reflex effects of low-threshold joint afferents have not yet been unequivocally demonstrated (see Skoglund, 1973, for review). As a result they have been less attractive or amenable to study. Nevertheless, joint receptors have sporadically attracted physiologists since Goldscheider (1889) provided evidence that receptors in the joint capsule were largely responsible for position and movement sense of the limbs. This idea gained further support from psychophysical experiments (Browne, Lee & Ring, 1954; Gelfan & Carter, 1967) which showed that joint afferents, rather than muscle or skin afferents, were involved in kinaesthesia: and from physiological evidence that there are receptors in the cat's knee joint capsule capable of signalling joint movement and joint position (Boyd, 1954). Steady joint angle was shown to be signalled by a type of range fractionation, in which each slowly adapting receptor responds over a limited range, say 15–30°, and the ranges of the total population of receptors overlap to cover the full extent of joint movement, from full flexion to full extension.

Interest has been focussed on the slowly adapting joint receptors for the following reason: it is difficult to see how other receptors could signal joint position in a straightforward way. The most obvious alternative is the muscle spindle. While the spindle does signal muscle length, which indicates joint angle, it does so with a variable sensitivity which is set by the γ efferent system. So to get information about joint position from muscle spindles, the afferent signals from the spindle must be integrated with the γ bias for those spindles. This is by no means impossible, but it is complicated, and has tended to be eliminated by Occam's razor.

This is all very well, but an exclusive role for joint afferents in kinaesthesia can no longer be upheld. There is increasing evidence that muscle afferents contribute to conscious proprioception (Goodwin, McCloskey & Matthews, 1972a,b,c; Gandevia & McCloskey, 1976) and that in man, total replacement by prosthesis of a joint including its capsule impairs ability to perceive limb position rather little (Cross & McCloskey, 1973; Grigg, Finerman & Riley, 1973). Cutaneous receptors also seem capable of signalling steady joint position (Knibestöl, 1975). Furthermore, a re-examination of the receptor population in the cat's knee joint has shown that the vast majority of slowly adapting receptors only responds either close to full extension or to full flexion, and often at both (Clark & Burgess, 1975): those few receptors which do respond in the intermediate range do not respond to flexion–extension movement, and are almost certainly muscle spindles of the popliteus (Clark & Burgess, 1975; McIntyre, Proske & Tracey, 1977). This suggests that knee joint receptors cannot signal joint position over the whole range of joint movement, and these findings have been confirmed in the monkey (Grigg & Greenspan, 1977). In fact, only one case is now known where joint receptors signal over the whole range of naturally-occurring joint movement. This is the costo-vertebral joint (Godwin-Austen, 1969).

In spite of this challenge to the primacy of joint afferents in kinaesthesia, there is still reason to believe that they play a significant role in the control of movement. Low-threshold joint afferents (i.e. mechanoreceptors in the group II range) project to the cerebellum (Lindström & Takata, 1972; Belcari, Carli & Strata, 1974); and to the cerebral cortex (Clark, Landgren & Silfvenius, 1973). There is an interesting report that section of posterior and medial articular nerves to the knee joint of the cat produces disturbances in walking and in some reflexes such as the placing reaction (Freeman & Wyke, 1966), although this is difficult to reconcile with the evidence on joint replacement mentioned above, and has been disputed by Lindström & Norrsell (1971). More significant is recent evidence that mechanoreceptors from joints exert a strong influence on precentral neurones of the conscious monkey – in fact these inputs are much more frequent than those from muscle or skin. Lemon & Porter (1976) examined the natural stimuli which modulated the discharge rate of movement-related precentral neurones. Of 257 such neurones, 152 responded to joint movement, but not to palpation of muscles acting at the joint. Thirty-five cells responded to muscle palpation, and most of these also responded to movement of the joint at which the muscle was attached. Only 27 cells responded to tactile stimulation of the skin.

I think we can conclude that while information on limb position and movement is not signalled exclusively by joint afferents, they do play an important part. But several problems remain to be solved. One is the extent of integration of input from muscle, skin and joint receptors at different levels in the central nervous system. A second problem is whether or not the properties of joint receptors in the knee are typical. Both of these problems are currently being examined in Archie McIntyre's laboratory, and I will discuss the receptor study here.

We decided to look at afferents from joints in the forelimb. These are interesting because of the concentration on the forelimb in studies of the control of movement, and because the diversity of movement is greater at forelimb joints than in the hindlimb. There is a report which suggests that populations of mechanoreceptors from the elbow and the knee are similar (Millar, 1975). We examined receptors from the wrist joint, partly to see whether there is any difference in receptor characteristics for different joints in the same limb, and partly as a basis for a future study on ascending pathways. A preliminary report has appeared (Tracey, 1977).

PROPERTIES OF THE WRIST JOINT NERVE

The posterior interosseous nerve in man gives rise to an articular branch to the wrist (Gray & Gardner, 1965). The homologous nerve in the cat is the dorsal interosseous nerve, and dissection showed that this nerve gives off an articular branch which supplies joint capsules in the wrist from the dorsal aspect. The dorsal interosseous nerve is a branch of the deep radial, and supplies the supinator and extensor muscles of the forearm. Once it enters the abductor pollicis longus it branches into a superficial muscle nerve and a deep branch. This deep branch becomes the wrist joint nerve (WJN) although it may give off a further muscle branch first (Fig. 1). Stimulating the WJN at 10 times threshold produces no motor response. This was checked in every experiment. From five such nerves, segments were dissected out for histological study. The segments were fixed and stained with osmium tetroxide (Boyd & Davey, 1968), and embedded in paraffin. Transverse sections were cut at 4–7 μm, and fibre counts and diameter measurements were made using a Zeiss particle size analyser TGZ 3 according to the method of Westerman & Wilson (1968). Although the fibre count varied somewhat from nerve to nerve (range 137–251) the fibre diameter histograms consistently showed two peaks, one at 3 μm and one at 9 μm (Fig. 2). These peaks correspond to the group III and group II peaks already described for knee joint nerves (Boyd & Davey, 1968), although fibres larger than 13 μm in diameter were not found.

Bipolar recordings from the whole nerve show no tonic activity when the wrist is held at full extension. (Full extension of the cat's wrist is close to 180°, i.e. with the forepaw in line with the forearm. Full flexion is then about 40°.) If the wrist is flexed, a burst of activity is seen during the

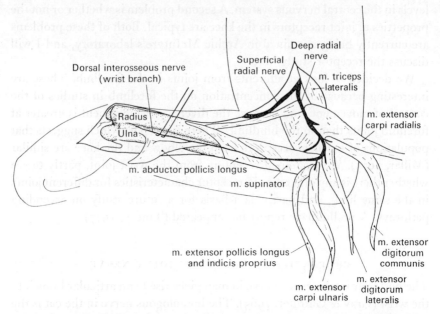

Fig. 1. Sketch of the dissection of the left forelimb of the cat (lateral view) to expose the wrist branch of the dorsal interosseous nerve. Electrodes are shown at the site of recording, stimulation, and excision for histology. The five extensor muscles shown on the right were removed before preparing a paraffin pool. The radius was clamped near the junction of m. abductor pollicis longus and m. supinator.

movement phase, and again if it is extended (Fig. 3a). Once the wrist is flexed to about 90°, i.e. within 45° of full flexion, tonic activity is seen whose intensity becomes greater as full flexion is approached (Fig. 3b). Tonic activity in the whole WJN is also seen if the wrist is pronated or supinated. Tapping gently on the back of the wrist produces brief, high-frequency bursts of activity probably arising from Pacinian corpuscles.

Careful stroking of hairs on the back of the wrist did not elicit any activity provided that the skin surface was not touched (which excites Pacinian corpuscles): and circumcising the skin at the level of the wrist had no effect on the response of the whole nerve to wrist movement – so it is unlikely that cutaneous afferents run in the WJN. During the dissection to expose WJN, all six of the carpal and digital extensor tendons were routinely cut at the level of the wrist, so responses to wrist flexion seen in WJN cannot be due to muscle spindles or tendon afferents.

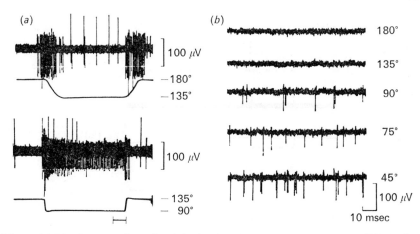

Fig. 3. (*a*) Bipolar recording of activity in the whole wrist joint nerve (WJN) during movements at the wrist joint. Time mark 0.2 sec for upper record, 2 sec for lower record. (*b*) Tonic activity in the whole WJN at successive angles of flexion.

SINGLE UNIT RESPONSES

Single functional afferents were isolated from the dorsal root filaments after an extensive denervation of the arm and shoulder. The radial nerve enters the spinal cord via roots C7, C8 and T1, but WJN units were found only in posterior C7 and anterior C8. The threshold of each unit to stimulation of the WJN was noted, and the conduction time at twice threshold measured* (Hunt & McIntyre, 1960). The response of the unit to natural stimulation of the wrist was studied, and classified according to the scheme of Clark (1975). A total of 101 WJN units were examined, and Table 1 shows the relative frequency of different types. Units were classified mainly according to their response to wrist movements (Fig. 4) but supplementary tests were also used, such as sinusoidal vibration applied to the back of the wrist, tapping on the back of the wrist, or tugging the distal ends of cut extensor tendons. Thus Pacinian corpuscle-like units responded to rapid movements of the wrist in any direction. They did not fire at any static wrist position, but responded if the back of the wrist was tapped very lightly, or if the distal ends of cut extensor tendons were tugged. They followed sinusoidal vibrations applied to localised spots on the back of the wrist at up to 400 Hz.

Phasic units had rather similar characteristics. Most of them responded to flexion, supination and pronation of the wrist: about half also responded to extension. Their response outlasted the movement phase, but never lasted as long as 10 sec: this was the basis of the distinction between phasic

* At the end of each experiment the conduction pathway was dissected out and its length measured for estimates of conduction velocity.

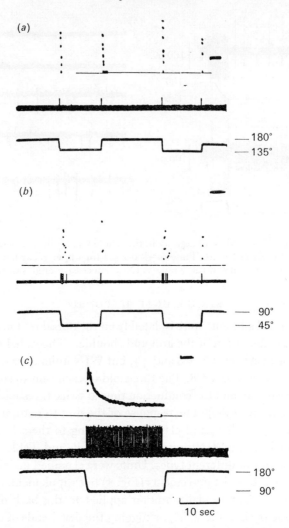

Fig. 4. Representative responses of singles units in the wrist joint nerve to wrist flexion and extension. In each record, instantaneous frequency is shown in the top trace. Calibration mark = 50 Hz. Middle trace shows the unit, while the bottom trace is the angle of flexion. (a) Pacinian corpuscle-like unit; (b) phasic unit; (c) slowly adapting unit.

and slowly adapting units. Like Pacinian corpuscle-like units, most of those tested responded to tugging on the distal ends of cut extensor tendons, and to taps on the back of the wrist, although with a higher threshold. They also responded to sinusoidal vibration, but could not be made to follow frequencies higher than 200 Hz.

Slowly adapting units were defined as those responding to a new wrist

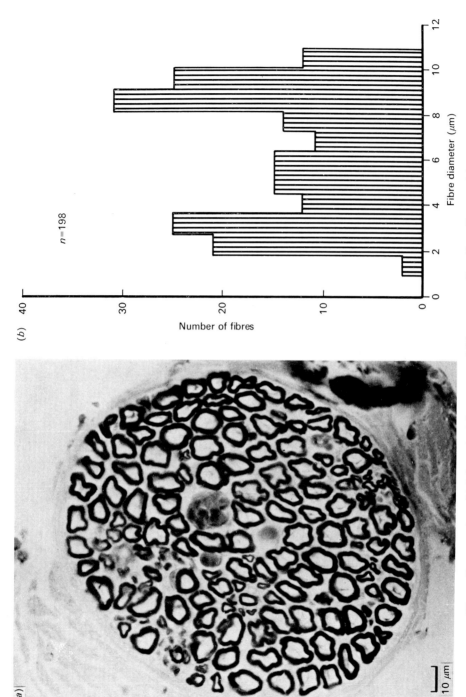

Fig. 2. Cross section of (a) the wrist joint nerve and (b) corresponding fibre diameter histogram.

Table 1. *Adequate stimuli*

Response type	No. of WJN* units	Carpal flexion	Carpal extension	Carpal rotation Radial side up	Carpal rotation Radial side down	Tap	Tendon tug
Slowing adapting	28	22	0	11	13	4/7	2/9
Phasic	37	26	15	21	27	10/11	8/11
Pacinian-like	8	8	8	8	8	8	2/2
Weakly activated	7	1	0	0	2	2	2
Not activated	21	—	—	—	—	—	—

* WJN = wrist joint nerve.

Table 2. *Conduction velocities*

Response type	No. of WJN* units	Mean	Standard error
Slowly adapting	28	52.9	2.2
Phasic	37	54.7	1.0
Pacinian-like	8	57.3	3.1
Weakly activated	7	54.0	2.3
Not activated	21	35.0	2.6

* WJN = wrist joint nerve.

position for longer than 10 sec. Such angles were either within 45° of full flexion, or at extreme pronation or supination – never at full extension.

Weakly activated units responded with only one or two spikes to extreme bending of the wrist or to firm taps on the back of the wrist. Even these stimuli were insufficient to activate some units – no natural stimulus was found to which they would respond, although noxious stimuli were not used, for fear of damaging other receptors in the population.

One might ask whether units which were not activated were simply mechanoreceptors which had been damaged, or whether they formed a different population. Conduction velocities throw some light on this (Table 2). While the first four classes of unit had mean conduction velocities which were all between 52 and 58 m sec^{-1}, the non-responsive units had a mean conduction velocity of 35 m sec^{-1}, suggesting that they belonged to a different population. This was confirmed by one-way analysis of variance: conduction velocities of the first four classes belonged to the same population ($P > 0.05$) while conduction velocities of the non-responsive group did not belong to this population ($P < 0.01$). Boyd & Davey (1968, p. 47)

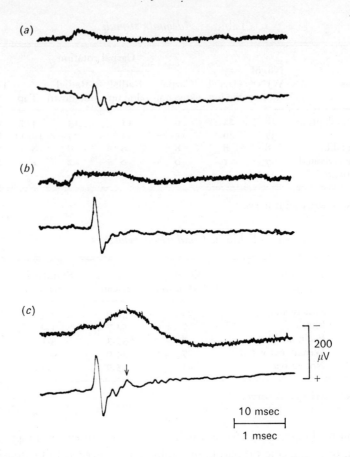

Fig. 5. Upper traces: cord dorsum potentials at the level of C8. Lower traces: wrist joint nerve (WJN) volley recorded proximally from the dorsal interosseous nerve. The arrow indicates the second component of the WJN compound action potential, which is related to the second peak in the cord dorsum potential. Threshold (T) refers to the nerve volley. (a) 1.3T; (b) 1.6T; (c) 4.2T.

state that 'About 40% of the myelinated fibres in all articular nerves terminate in free nerve endings and endings on blood vessels . . . they form the group III component in articular nerves.' It is probably these fibres which provide the units which do not respond to non-noxious mechanical stimulation. In fact this group is probably under-represented in our data, due to the sampling bias in favour of large spikes and therefore large fibres.

From the mean conduction velocities of the responsive units, one would expect a mean fibre diameter of about 9 μm, using a Hursh conversion factor of six. This corresponds well with the peak at close to 9 μm in Fig. 2b.

CORD POTENTIALS

In the course of these experiments, cord dorsum potentials were routinely recorded to check the integrity of the WJN along its course. A characteristic double peaked waveform was seen, somewhat similar to that shown by Bernhard (1953) for muscle nerves (Fig. 5). The first peak has a duration of about 5 msec and appears at stimulus strengths close to threshold (T) for WJN, and is maximal at 1.5–2T. At this stimulus strength the first component in the WJN volley, recorded proximally from the dorsal interosseous nerve, is also maximal. As the stimulus strength is increased above 1.5–2T, a second peak starts to appear both in the nerve volley and in the cord dorsum potential. These are both maximal by 3–5T. It seems reasonable to identify the two peaks in the cord potential as the N_1 potential and the late negative potential of Bernhard (1953), ascribed by him to activity of low- and high-threshold afferents, respectively.

One might expect to find a positive dorsal root potential (DRP) associated with the late negative wave, similar to that reported for high-threshold muscle afferents (Mendell, 1970). Such a positive DRP, which is taken to indicate presynaptic hyperpolarisation, has not been found in these experiments, presumably due to the blocking effect of barbiturate anaesthesia (Mendell & Wall, 1964).

DISCUSSION

Segregation of unit types

The compound action potentials of the elbow joint nerve and the posterior articular nerve of the knee joint show two low-threshold components, with a high-threshold component at around 2T (Clark, Landgren & Silfvenius, 1973). These three components are apparently correlated with three groups of fibre diameter in the elbow joint nerve (Andersen, Körner, Landgren & Silfvenius, 1967), and the authors suggest that the two low-threshold components correspond with different receptor types. No evidence was found in the present study to suggest that different mechanoreceptor types could be segregated on the basis of conduction velocity or threshold.

Mixed nature of the deep radial nerve

Several studies of the central pathways and effects of forelimb muscle afferents have been carried out with the assumption that the deep radial and dorsal interosseous are pure muscle nerves (e.g. Schmidt & Willis, 1963;

Rosén, 1969). It is now apparent that the deep radial is a mixed nerve, and some of the effects produced by its stimulation may be due to joint afferents.

Comparison of receptors from different joints

The slowly adapting receptors are perhaps of most interest, since they have appeared the most likely candidates for signalling joint position. Slowly adapting receptors from the wrist joint resemble slowly adapting receptors from the elbow (Millar, 1975) and from the knee (Clark & Burgess, 1975) in that they do not respond at intermediate angles. They differ, however, in responding only at the flexion end of the range. In both the other studies cited, a large proportion of receptors fired at both extreme flexion and extreme extension. Ambiguous information is thus signalled to the central nervous system, if this is the case under natural conditions.

Behavioural data show that joint position is more accurately perceived at proximal joints than at distal ones (Goldscheider, 1889). A comparison of the data from wrist and elbow does not suggest any basis for this in the properties of slowly adapting receptors. It may be that the relevant receptor properties have not been measured, e.g. the minimal change in joint angle for a change in firing rate: but it seems more likely that the explanation of behavioural data, like the basis of position and movement sense in general, will be found in the way in which joint, skin and muscle information is integrated rather than in the properties of joint mechanoreceptors alone.

I should like to acknowledge my indebtedness to Archie McIntyre, without whom I could not have done this work; to Rod Westerman for constructive criticism and advice; and to Greg Gordon and Jandri Hoggins for technical assistance. The work was supported by the ARGC grant No. D1-76/15103.

REFERENCES

Andersen, H. T., Körner, L., Landgren, S. & Silfvenius, H. (1967). Fibre components and cortical projections of the elbow joint nerve in the cat. *Acta physiol. scand.* **69**, 373–82.

Belcari, P., Carli, G. & Strata, P. (1974). The projection of the posterior knee joint nerve to the cerebellar cortex. *J. Physiol., Lond.* **237**, 371–84.

Bernhard, C. G. (1953). The spinal cord potentials in leads from the cord dorsum in relation to peripheral source of afferent stimulation. *Acta physiol. scand.* **29** (Suppl. 106), 1–29.

Boyd, I. A. (1954). The histological structure of the receptors in the knee joint of the cat correlated with their physiological response. *J. Physiol., Lond.* **124**, 476–88.

Boyd, I. A. & Davey, M. R. (1968). *Composition of peripheral nerves.* Edinburgh & London: E. & S. Livingstone.

Browne, K., Lee, J. & Ring, P. A. (1954). The sensation of passive movement at the metatarso-phalangeal joint of the great toe in man. *J. Physiol., Lond.* **126**, 448–58.

Clark, F. J. (1975). Information signalled by sensory fibers in medial articular nerve. J. Neurophysiol. 38, 1464–72.

Clark, F. J. & Burgess, P. R. (1975). Slowly adapting receptors in cat knee joint: can they signal joint angle? J. Neurophysiol. 38, 1448–63.

Clark, F. J., Landgren, S. & Silfvenius, H. (1973). Projections to the cat's cerebral cortex from low threshold joint afferents. Acta physiol. scand. 89, 504–21.

Cross, M. J. & McCloskey, D. I. (1973). Position sense following surgical removal of joints in man. Brain Res. 55, 443–5.

Freeman, M. A. R. & Wyke, B. (1966). Articular contributions to limb muscle reflexes. The effects of partial neurectomy of the knee joint on postural reflexes. Brit. J. Surg. 53, 61–9.

Gandevia, S. C. & McCloskey, D. I. (1976). Joint sense, muscle sense and their combination as position sense, measured at the distal interphalangeal joint of the middle finger. J. Physiol., Lond. 260, 387–407.

Gelfan, S. & Carter, S. (1967). Muscle sense in man. Exp. Neurol. 18, 469–73.

Godwin-Austen, R. B. (1969). The mechanoreceptors of the costo-vertebral joints. J. Physiol., Lond. 202, 737–53.

Goldscheider, A. (1889). Untersuchungen über den Muskelsinn. Arch. Anat. Physiol. 369–502. Cited by Howard, I. P. & Templeton, W. B. (1966). Human spatial orientation. London: Wiley.

Goodwin, G. M., McCloskey, D. I. & Matthews, P. B. C. (1972a). The contribution of muscle afferents to kinaesthesia shown by vibration induced illusions of movement and by the effects of paralysing joint afferents. Brain 95, 705–48.

Goodwin, G. M., McCloskey, D. I. & Matthews, P. B. C. (1972b). The persistence of appreciable kinesthesia after paralysing joint afferents but preserving muscle afferents. Brain Res. 37, 326–9.

Goodwin, G. M., McCloskey, D. I. & Matthews, P. B. C. (1972c). Proprioceptive illusions induced by muscle vibration: contribution by muscle spindles to perception? Science 175, 1382–4.

Gray, D. J. & Gardner, E. (1965). The innervation of the joints of the wrist and hand. Anat. Rec. 151, 261–6.

Grigg, P., Finerman, G. A. & Riley, L. H. (1973). Joint position sense after total hip replacement. J. Bone Jt Surg. 55A, 1016–25.

Grigg, P. & Greenspan, B. J. (1977). Response of primate joint afferent neurons to mechanical stimulation of knee joint. J. Neurophysiol. 40, 1–8.

Hunt, C. C. & McIntyre, A. K. (1960). Characteristics of responses from receptors from the flexor longus digitorum muscle and the adjoining interosseous region of the cat. J. Physiol., Lond. 153, 74–87.

Knibestöl, M. (1975). Stimulus response functions of slowly adapting mechanoreceptors in the human glabrous skin area. J. Physiol., Lond. 245, 63–80.

Lemon, R. N. & Porter, R. (1976). Afferent input to movement-related precentral neurones in conscious monkeys. Proc. R. Soc., B 194, 313–39.

Lindström, S. & Norrsell, U. (1971). A note on knee joint denervation and postural reflexes in the cat. Acta physiol. scand. 82, 406–8.

Lindström, S. & Takata, M. (1972). Monosynaptic excitation of dorsal spinocerebellar tract neurons from low threshold joint afferents. Acta physiol. scand. 84, 430–2.

McIntyre, A. K., Proske, U. & Tracey, D. J. (1977). Evidence for the presence of muscle spindle afferents in a knee joint nerve of the cat. Proc. Aust. physiol. pharmac. Soc. 8, 178P.

Mendell, L. M. (1970). Positive dorsal root potentials produced by stimulation of small diameter muscle afferents. Brain Res. 18, 375–9.

Mendell, L. M. & Wall, P. D. (1964). Presynaptic hyperpolarization: a role for fine afferent fibres. *J. Physiol., Lond.* **172**, 274–94.

Millar, J. (1975). Flexion–extension sensitivity of elbow joint afferents in cat. *Exp. Brain Res.* **24**, 209–14.

Rosén, I. (1969). Localization in caudal brain stem and cervical spinal cord of neurones activated from forelimb group I afferents in the cat. *Brain Res.* **16**, 55–71.

Schmidt, R. F. & Willis, W. D. (1963). Depolarization of central terminals of afferent fibers in the cervical spinal cord of the cat. *J. Neurophysiol.* **26**, 44–60.

Skoglund, S. (1973). Joint receptors and kinaesthesia. In *Handbook of sensory physiology*, ed. A. Iggo, vol. II, pp. 111–36. Berlin: Springer-Verlag.

Tracey, D. J. (1977). Characteristics of primary afferents from the wrist joint of the cat. *Proc. Aust. physiol. pharmac. Soc.* **8**, 177P.

Westerman, R. A. & Wilson, J. A. F. (1968). The fine structure of the olfactory tract in the teleost *Carassius carassius* L. *Z. Zellforsch.* **91**, 186–99.

Irregularity in the discharge of slowly adapting mechanoreceptors

U. PROSKE

Mechanoreceptors represent a very intensively studied group, yet despite continuing efforts by experimenters, using a wide range of preparations from both vertebrates and invertebrates, details of the all-important mechanism of impulse initiation remain obscure. We have recently become interested in this problem and have tried to obtain new insight by studying the pattern of discharge in an impulse train, in particular the rate of firing of the receptor and the variability in the intervals between successive impulses.

Considering a receptor in the most general terms it consists of a peripheral unmyelinated terminal. This may be simple and unbranched as for the Pacinian corpuscle or may show profuse branching as do tendon organs and some cutaneous receptors. Some of the more proximal branches may be myelinated but the ultimate terminations, which are thought to be the site of stimulus transduction, are unmyelinated. The unmyelinated terminal may or may not be associated with specialised accessory structures, such as the capsule of the Pacinian corpuscle or the receptor cells of cutaneous type I mechanoreceptors.

The stimulus energy triggers a conductance change in the distal terminals of the receptor which results in the production of the generator potential. The generator potential is then conducted passively up to the region of impulse initiation, commonly thought to be the first node of Ranvier, where the action potential is initiated. Whether the action potential is then able to invade antidromically the distal terminal structure remains controversial.

Many slowly adapting mechanoreceptors have large portions of their peripheral dendritic structure myelinated, which presumably means that such regions are able to conduct propagated impulses. Examples are tendon organs, muscle spindles, some joint receptors and cutaneous type I receptors. Fig. 1 illustrates the extensive branching shown by tendon organs in lizards. Such anatomical features lead to the suggestion of more than a single site of impulse generation in the dendritic tree and give rise to the possibility of interaction between several generators.

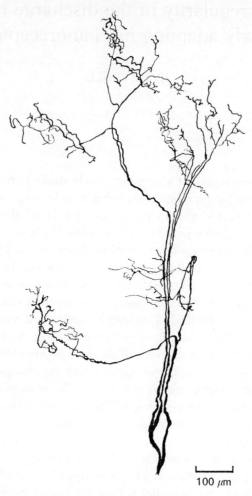

100 μm

Fig. 1. A camera lucida drawing of the extensive branching shown by the axon supplying a lizard tendon organ. With the silver-impregnated preparation chosen here it was not possible to determine which portions were myelinated and which were unmyelinated. Observations on other occasions showed that all of the larger branches were, in fact, myelinated. Silver stain according to Barker & Ip (1963).

The first account which tried to explain the steady, maintained discharge of slowly adapting mechanoreceptors was that of Adrian & Zotterman (1926). They suggested that following the initiation of an impulse, the receptor (the frog muscle spindle) fell into a period of decreased excitability and remained silent until recovery had proceeded to a point where the depolarisation was able to initiate the next impulse. This sort of proposition has been supported by many subsequent observations including recordings from frog muscle spindles (Katz, 1950b) and stretch receptors in crustaceans (Eyzaguirre & Kuffler, 1955).

If impulses are able to be initiated at more than one site, then an impulse from one terminal will conduct orthodromically up to a branch point and from there, as well as continuing orthodromically, will conduct anti-dromically to invade neighbouring terminals. Thus any one impulse-generating site may be thrown into a period of decreased excitability either by an impulse being generated at that site or by antidromic invasion from neighbouring sites.

In a discussion of this problem, Eagles & Purple (1974) have distin-guished between two situations. In the first, which they call 'simultaneous reset', an action potential generated at one site antidromically invades neighbouring sites to initiate there a 'recovery cycle' (Horch, Whitehorn & Burgess, 1974) with very little delay. This also includes the possibility of an adjacent site managing to initiate an impulse but which collides with the impulse from the first site. The important requirement for simultaneous re-set is that conduction paths are sufficiently short for all terminals to be at virtually the same level of excitability after the generation of an impulse. If, however, conduction paths are long, as for example with splanchnic mechanoreceptors (Floyd & Morrison, 1974), then impulses can be initiated at very different times yet still be able to collide at some point in the branched structure. This case called 'non-simultaneous reset' by Eagles & Purple is probably not applicable to the slowly adapting mechanoreceptors considered here.

Several situations can be envisaged by examining simultaneous re-set in more detail. First, if the rate of impulse generation at one site is very much faster than at neighbouring sites then the discharge recorded in the parent axon will be composed only of impulses from the faster generator. This is because the faster generator will be continually re-setting all other genera-tors. If, however, adaptation acts to produce a gradual fall in firing rate and the rate of adaptation of different generators is not the same then it could be envisaged that a previously suppressed generator, adapting less rapidly than the dominant generator, will at one point in time have a sufficiently high firing rate to take over domination of the discharge. Some evidence for generator switching was presented recently by Brokensha & Westbury (1974), recording the discharges of frog muscle spindles.

During the period of switching from one generator to another there will be occasions when both generators contribute to the discharge. This is possible in spite of re-setting because both generators show an inherent irregularity in their discharge, and sufficient time elapses, following the act of re-setting, for the variability to determine which generator will be the next to contribute to the discharge. This sort of interaction has been called 'probabilistic mixing'.

Eagles & Purple (1974) considered the case of up to four interacting generators each having a similar rate of firing and variability. From computer modelling they concluded that the discharge observed in the parent axon showed as a result of probabilistic mixing a progressive lowering of mean interval and standard deviation as more generators were included in the computation. Thus if several generators contribute to the discharge, the result will be a higher mean rate of firing and a more regular discharge. This sort of conclusion runs contrary to the interpretations of many receptor physiologists who have tended to believe that interaction between several generators acts to increase the irregularity of the discharge (see, for example, Chambers, Andres, Duering & Iggo, 1972).

A thought-provoking discussion of the irregularity of the discharge in cutaneous type I mechanoreceptors was recently made by Horch et al. (1974). They pointed out that interaction between generators can lead to an increase in the irregularity of the discharge only if re-setting does not take place. However, it is not a straightforward matter to decide whether re-setting is occurring. This is because if impulses are arriving at a branch point at a higher rate than they are leaving it, i.e. extensive 'frequency reduction' is occurring, then at that branch point an antidromic impulse will act to re-set the discharge. Thus although re-setting may appear to be occurring, in fact the distal impulse generating sites themselves are not being re-set.

Assuming, nevertheless, that for the majority of mechanoreceptors re-setting by invasion of distal terminals does take place, then if there are many possible generators and these are able to contribute to the parent discharge by probabilistic mixing, the possibility arises that at least some of the decline in firing rate and increase in variability of discharge observed during adaptation can be attributed to the progressive withdrawal of numbers of generators. We have attempted to test this possibility directly and have chosen as our example the tendon organ. Tendon organs lie in series with the muscle fibres of a muscle and respond during the rising phase of a twitch contraction with a burst of impulses (Fig. 2a). During passive muscle stretch they respond both to change and the rate of change in tension (Fig. 2b). Most of our experiments have used a group of tendon organs which lie in the tendon of the caudo-femoralis muscle of the lizard *Tiliqua* (Gregory & Proske, 1975; Proske & Gregory, 1976). We have been able to record the discharge in a single functional afferent fibre and at the same time plot the receptor's location in the tendon, using a fine probe. On several occasions two or more regions of maximum sensitivity, 'hot spots', could be detected lying adjacent to one another. These were interpreted by

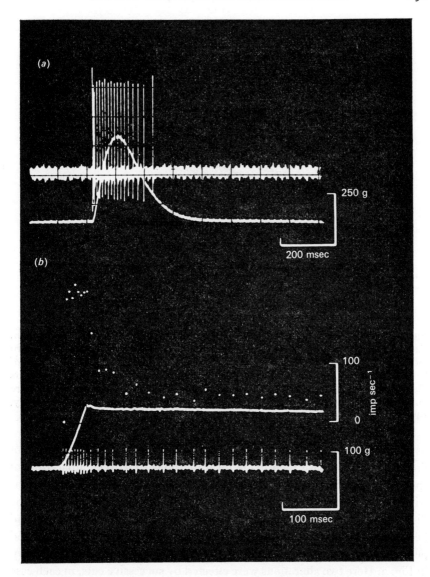

Fig. 2. The responses of tendon organs to muscle contraction and stretch. In (a) is shown the response of a tendon organ of the cat soleus muscle responding during contraction of the whole muscle. Impulses, upper trace, tension below. In (b) a tendon organ of the caudo-femoralis tendon of the lizard *Tiliqua* is shown to respond both to the rise in tension and the rate of rise during a 'ramp-and-hold' stretch. Here the dots are a display of the instantaneous frequency of the train of action potentials shown on the bottom trace. Each dot represents an action potential and the height of the dot above zero, indicated by the frequency calibration on the right of the figure, is proportional to the reciprocal of the interval between successive impulses. The middle trace shows the tension changes during the stretch.

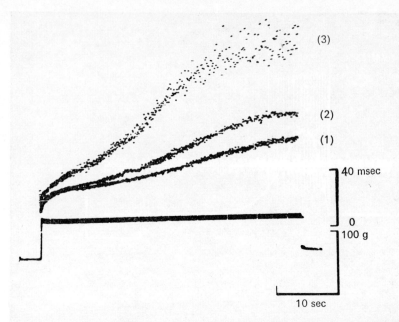

Fig. 3. The responses of a lizard tendon organ during a step-stretch of the tendon. Each action potential is represented by a dot and the height of the dot above zero indicated by the interval calibration on the right is proportional to the *interval* between successive impulses. Three traces have been superimposed: (1) represents the control response; (2) and (3) the response after each of two successive cuts were made within the region of receptor termination. (Redrawn from Proske & Gregory, 1976.)

us as representing different terminal branches of that receptor. By means of fine micro-scalpels we then made local incisions inbetween adjacent 'hot spots', hoping thereby to reduce the number of impulse generators contributing to the discharge. Frequently the cut had no effect at all, or the discharge suddenly fell silent, presumably because the parent axon had been cut. On several occasions, however, the cut resulted in a sudden and permanent alteration in the pattern of discharge and an example is shown in Fig. 3. Here two alterations were effected by successive cuts, in each case resulting in an increase in the rate of adaptation and increase in variability. We interpret these observations as providing direct support for an Eagles & Purple-type model and the possibility persists that the discharge recorded from a receptor such as a tendon organ is in fact a composite discharge including contributions from many generators. It must, however, be added that any manipulation resulting in damage to parts of a dendritic structure is likely to produce injury depolarisation at the cut terminals. While we do not believe that the frequently observed and reproducible result shown in

Fig. 3 can be simply dismissed in this way, we cannot consider our interpretation as being exclusive of other alternatives.

During the progress of these experiments which at one stage also included preparations of amphibian muscle spindles and mammalian muscle spindles and tendon organs, we noticed that different receptors showed substantial differences in the variability of their discharge. Furthermore, in response to a sustained stimulus the discharge of a slowly adapting mechanoreceptor showed a progressive increase in variability as adaptation proceeded and the mean firing rate fell. Thus the observed variability seemed to depend on the rate at which the receptor was firing. We decided to compare the variability of discharge for several receptors and in an attempt to achieve as irregular a discharge as possible we made our measurements at the lowest rate the receptor would maintain without stopping. Initially the preparation was stretched until the receptor began to respond and then held at the new length until adaptation had proceeded and the discharge reached a more or less steady level. If the stretch applied was too large then once the discharge had reached the adapted state the muscle was slackened off a little thereby producing a drop in discharge. This was repeated several times until the lowest possible rate of firing was achieved without the receptor actually falling silent. On several occasions when the discharge did stop, it was noticed that the amount of subsequent stretch required to restart it was surprisingly large. It seemed as though ongoing activity permitted the discharge to reach a lower rate of firing. Once the receptor had stopped, the rate at which it would resume discharging if the preparation was then gradually re-stretched was much higher than that just prior to falling silent. There seemed to be some sort of 'hysteresis-like' behaviour in the stretch responsiveness of the receptor.

These initial observations prompted what seemed at the time a rather bizarre manoeuvre. The stretch on the muscle was reduced gradually, thereby lowering the firing rate, until eventually the receptor fell silent. Then a small stretch was applied, insufficient to restart the discharge but reaching a level of tension at which the receptor had just previously been responding. Next, by means of a pair of stimulating electrodes, an electrically evoked action potential was initiated in the filament of nerve containing the afferent fibre whose discharge was being recorded. Thus an antidromic impulse was set up and this travelled back into the peripheral receptor terminals. By means of such an electrically evoked antidromic impulse it was possible to re-trigger the discharge. The receptor responded not with just one or two impulses but continued firing for the duration of the stretch. This experiment is shown in Fig. 4 for a tendon organ and a secondary ending of a muscle spindle of the cat soleus muscle.

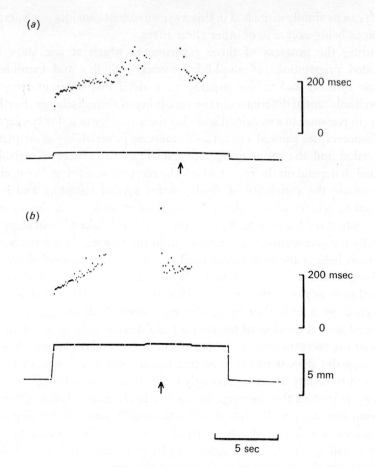

Fig. 4. Electrical re-triggering of the discharge of a secondary ending of a muscle spindle (*a*) and a tendon organ (*b*), both from the soleus muscle of the cat. In each set of records the dots represent action potentials, the height of a dot above zero being proportional to the interval between it and the previous action potential. The continuous traces below represent muscle length. The receptors were subjected to two successive, superimposed length changes. In (*a*) an initial step of 0.7 mm was succeeded by a second step (barely visible) of 0.02 mm. In (*b*) the initial step was 4.3 mm followed by 0.2 mm. At a point, shown by the arrow, the nerve fibre from which recordings were being made was stimulated antidromically. The antidromic impulse was able to restart the discharge which, following the first step, had fallen silent owing to adaptation and which the second stretch was unable to re-trigger by itself.

Antidromic re-triggering was tried with several different preparations of slowly adapting mechanoreceptors and it was concluded that, although common, it was not always possible to demonstrate. Muscle spindles of the frog sartorius muscle and primary endings of cat soleus spindles could not

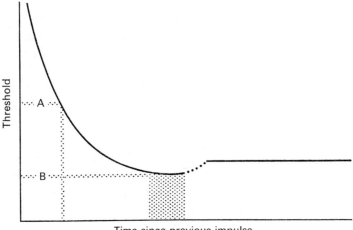

Fig. 5. A diagrammatic representation of the recovery of threshold at the region of impulse initiation following the passage of an action potential. Initially, threshold is very high, as represented by the refractory period, and then falls down to a minimum before rising again slightly to reach static threshold. The next action potential is initiated at a point on the curve at which threshold reaches the level of depolarisation produced by stretch. Two levels of depolarisation (A and B) are considered. The stippling shows the fluctuation in membrane potential. See text for details. (Redrawn from Gregory, Harvey & Proske, 1977.)

be made to re-trigger (as discussed by Gregory, Harvey & Proske, 1977), and they must be considered as a separate group. However, lizard tendon organs, cat tendon organs and cat spindle secondaries all showed the effect. It is now necessary to explain the phenomenon of antidromic re-triggering and to show how it can help to account for the changes in discharge variability that are observed during periods of adaptation.

As Adrian & Zotterman had originally described, following the initiation of an impulse, the region in which the impulses are generated, the 'trigger zone', falls into a period of reduced excitability. Furthermore, it is known that the frequency of discharge of a receptor is proportional to the degree of depolarisation (Katz, 1950b). Thus, as proposed by Buller (1965), recovery of threshold of the trigger zone would be expected to follow a hyperbolic time course, representing the absolute and relative refractory periods. Superimposed on the threshold recovery curve (Fig. 5) is the degree of depolarisation produced by the stretch. The point at which the level of depolarisation transects the threshold recovery curve determines where in time the next impulse will be initiated. When the amount of depolarisation produced by stretch is large the point at which the two curves transect is on the steep portion of the threshold recovery curve (A, Fig. 5).

However, the depolarisation is not steady but shows small random fluctuations due perhaps to molecular agitations or to ionic noise (Katz, 1950a). The fluctuation in membrane potential (as shown by the stippling in Fig. 5) has little effect on the variability between successive impulses when the point of initiation of the next impulse falls on the steep portion of the curve (A, Fig. 5). If, however, depolarisation due to stretch is much lower (B, Fig. 5) the point at which the next impulse is initiated falls on the flatter portion of the recovery curve and here the same amount of membrane noise is able to induce a much larger variability in the intervals between impulses.

We are able to provide an explanation of the electrical re-triggering experiment by proposing that following the previous impulse, threshold does not continue to fall asymptotically to the final static value, but after a certain period of time rises again, with an unknown time course, to a static level, that is, slightly higher than the previous minimum. This minimum in the threshold recovery curve we have called the 'late supernormal period' in contrast with the well-known supernormal period which occurs much earlier (see, for example, Lindblom & Tapper, 1966). If during adaptation of an ongoing discharge an interval occurs which is long enough to allow threshold to rise up from the minimum to the static value, then if the level of depolarisation due to stretch is somewhere below that represented by the static value, the discharge will stop. It is then possible to re-stretch the preparation to a point at which the level of depolarisation lies just below static threshold. If now an impulse is initiated, by electrical stimulation, the impulse trigger zone will go through the usual recovery period and as threshold approaches the minimum value it will meet the level of depolarisation and the next impulse is initiated. Thus it is the presence of the late supernormal period that permits electrical re-triggering.

Another interesting consequence of the late supernormal period is that although it restricts the lower end of the range of firing rates of the receptor it also reduces the effect of any noise on the variability of the discharge. If there was no minimum in the threshold recovery curve and if depolarisation due to stretch was such that membrane potential lay just below static threshold then it can be envisaged that occasional random fluctuations of potential of sufficient amplitude would be able to trigger an impulse. This would produce a completely random discharge (apart from the restrictions imposed by the refractory period) as can indeed be demonstrated for the frog muscle spindle (Gregory et al., 1977). Of course if the fluctuations in membrane potential are large enough, then even a late supernormal period will not stop the initiation of randomly triggered impulses. This is thought to be the case for the primary ending of cat spindles. Here the evidence for the presence of a late supernormal period is based on statistical analysis of the interval distribution in an impulse train (Gregory *et al.*, 1977). For all of the other receptor types studied, once intervals occur that are sufficiently long to exceed the supernormal period the receptor will fall silent.

In conclusion, we have shown that the irregularity observed in the discharge of some slowly adapting mechanoreceptors may be the result of a number of quite different causes and that an attempt at interpreting such variability inevitably leads to consideration of detailed aspects of the transduction process. We hope that propositions such as those presented here will lead to attempts at obtaining experimental support for the various hypotheses. Certainly it is our intention to continue to try to expose by direct testing the existence of a late supernormal period in the recovery of excitability. It should also be added that other more exotic possibilities for the transduction process, such as the suggestion by Horch *et al.* (1974) of the existence of synapses out in the peripheral receptor structure, have not been considered in this account. In spite of the restrictions, this article will have achieved its aim if it has done no more than point out some of the subtlety and complexity of the transduction process.

As one of Archie McIntyre's first graduate students here at Monash it gives me great pleasure to contribute to a Festschrift written in his honour, which includes contributions from so many others of his friends. During the 13 years that I have worked either under Archie's thoughtful guidance or more recently with him in direct collaboration, he has continued to infect all of us with his enthusiasm and scholarship. It is Archie's long-lasting interest in receptor mechanisms which has prompted me to choose my topic for this book. If our story makes even the most modest contribution to the field, it will largely be due to Archie McIntyre's continued support and encouragement.

I would also like to add that all of the experiments described here have been done in collaboration with Dr Ed Gregory. For the experiments on antidromic re-triggering we were joined by Dr Robin Harvey from Bristol.

REFERENCES

Adrian, E. D. & Zotterman, Y. (1926). The impulses produced by sensory nerve-endings. Part 2. The response of a single end-organ. *J. Physiol., Lond.* **61**, 151–71.

Barker, D. & Ip, M. C. (1963). A silver method for demonstrating the innervation of mammalian muscle in teased preparations. *J. Physiol., Lond.* **169**, 73–4P.

Brokensha, G. & Westbury, D. R. (1974). Adaptation of the discharge of frog muscle spindles following a stretch. *J. Physiol., Lond.* **242**, 383–403.

Buller, A. J. (1965). A model illustrating some aspects of muscle spindle physiology. *J. Physiol., Lond.* **179**, 402–16.

Chambers, M. R., Andres, K. H., Duering, M. & Iggo, A. (1972). The structure and function of the slowly adapting type II mechanoreceptor in hairy skin. *Q. Jl exp. Physiol.* **57**, 417–45.

Eagles, J. P. & Purple, R. L. (1974). Afferent fibres with multiple encoding sites. *Brain Res.* **77**, 187–93.

Eyzaguirre, C. & Kuffler, S. W. (1955). Processes of excitation in the dendrites and in the soma of single isolated sensory nerve cells of the lobster and crayfish. *J. gen. Physiol.* **39**, 87–119.

Floyd, K. & Morrison, J. F. B. (1974). Interactions between afferent impulses within a peripheral receptive field. *J. Physiol., Lond.* **238**, 62–3P.

Gregory, J. E., Harvey, R. J. & Proske, U. (1977). A late 'supernormal period' in the recovery of excitability following an action potential in muscle spindle and tendon organ receptors *J. Physiol., Lond.* **271**, 449–72.

Gregory, J. E. & Proske, U. (1975). Responses of tendon organs in a lizard. *J. Physiol., Lond.* **248**, 519–29.

Horch, K. W., Whitehorn, D. & Burgess, P. R. (1974). Impulse generation in type I cutaneous mechanoreceptors. *J. Neurophysiol.* **37**, 267–81.

Katz, B. (1950a). Action potentials from a sensory nerve ending. *J. Physiol., Lond.* **111**, 248–60.

Katz, B. (1950b). Depolarisation of sensory terminals and the initiation of impulses in the muscle spindle. *J. Physiol., Lond.* **111**, 261–82.

Lindblom, U. & Tapper, D. N. (1966). Integration of impulse activity in a peripheral sensory unit. *Exp. Neurol.* **15**, 63–9.

Proske, U. & Gregory, J. E. (1976). Multiple sites of impulse initiation in a tendon organ. *Exp. Neurol.* **50**, 515–20.

Muscle spindle projections to the cerebral cortex in monkeys

R. PORTER

That sensation and movement are inevitably interrelated is demonstrated by a large number of observations. In psychological studies the discriminative power for detection of shapes and textures is aided by movement of the receptive surface over a test object. Nervous information from receptors influenced by the object is assumed to be combined with that from efferent signals used to move the part containing those receptors. The combination of these two sorts of information is relevant to judgements about the nature of the object being felt. Other aspects of the spectrum of interactions between receptor function and movement response have been more traditionally studied by neurophysiologists. These concern the particular and specific reflex responses produced by activity in identified afferent fibres connected with identifiable receptors and the influences of induced muscle contraction on the activities of these receptors.

In spite of all the information that has been collected about the properties of mammalian muscle spindles (see Matthews, 1972, for review) the manner in which these receptors function during voluntary movement has only recently come under active investigation (Vallbo, 1970) and the contribution that this discharge makes to the control and regulation of movement performance is still largely a matter for speculation. In the past few years, it has been shown conclusively that the influences of muscle spindles are not confined to spinal levels and such coordinating centres as the cerebellum, but that relatively direct and secure projections conduct these influences to the cerebral cortex (Albe-Fessard & Liebeskind, 1966). Although a major component of the effect of activation of primary endings of muscle spindles is revealed in area 3a, transitional between classical motor and sensory areas of the cerebral cortex (Oscarsson & Rosén, 1963; Landgren & Silfvenius, 1969; Phillips, Powell & Wiesendanger, 1971), projections judged to arise from both primary and secondary endings have been demonstrated to reach neurones in the primary motor area 4 in the anaesthetized monkey (Lucier, Ruegg & Wiesendanger, 1975; Hore, Preston, Durkovic & Cheney, 1976). In this situation, neurones considered to be concerned in the production of movement performance because their discharges precede muscle contraction in conscious monkeys and change in relation to particular aspects of movement performance (Evarts, 1968) are also in receipt of feedback from sensitive detectors of length changes within the muscle (Phillips, 1969; Porter, 1976). Hence the possibility exists that some of the 'control loops' to which these receptors contribute in the overall regulation of muscle activity include the cerebral cortex and particularly its 'sensori-motor area' (Evarts, 1973; Marsden, Merton & Morton, 1973).

Attempts have been made to discover the locations of the muscle territories within which activation of muscle spindles will influence particular precentral neurones in conscious monkeys trained to perform movement tasks. This can be done only by using manipulations of the limb which the animal will tolerate without moving or resisting. So characterization of the influence is less certain than in experiments on anaesthetized animals. But, for a small number of precentral cells good evidence can be accumulated for an effect of muscle spindles on cortical cells in area 4 of animals trained with food rewards to sit quietly and accept passive manipulation of the limbs (Lemon & Porter, 1976).

Precentral neurones, including identified pyramidal tract neurones, have been shown to discharge in advance of muscle contraction during voluntary movements (Evarts, 1966), and only when particular movement performances, utilizing particular groups of muscles, were executed (Fetz & Finocchio, 1972). For each precentral neurone a motor field could be defined and the neurone regarded as having an influence on this motor field. In the awake, cooperating animal, natural, peripheral stimuli within this motor field frequently caused responses of the precentral neurone. For a small number of neurones the response was undoubtedly to cutaneous stimuli (Rosén & Asanuma, 1972) and Asanuma (1975) has speculated about the significance of this. But the majority of precentral neurones is affected by natural stimulation of structures other than the skin (Fetz & Baker, 1969).

The commonest peripheral stimulus capable of causing precentral neurones to discharge in the passive relaxed animal is joint movement (Lemon & Porter, 1976). Often, movement only of a single joint in a single direction caused the neurone under study to fire a burst of nerve impulses. Joint movement is, of course, an effective, natural stimulus for muscle spindles within muscles stretched by the movement. When subjected to passive increases in length, muscle spindle receptors demonstrate both dynamic and static responses; when passively shortened the receptors become silent. Parallels of these effects were seen in the responses of precentral neurones to joint movement in the conscious monkey and it was argued that these could not easily be explained using the known properties of receptors in joint capsules (Burgess & Clark, 1969).

Moreover, it was sometimes possible to elicit responses in the precentral neurone by palpation of the belly of the muscle which had been stretched, when natural stimulation of the skin overlying that muscle or prodding of the capsule of the joint which had been moved were ineffective. Prodding of the muscle belly was usually effective in only a limited region: but brief, gentle taps delivered within this zone caused regular, reproducible discharges of the precentral neurone after a very short latency (Lemon & Porter, 1976).

The final observation which relates to the demonstration of an input to particular precentral neurones of information from localized muscle spindle receptors relies on the possibility of driving nerve impulses in muscle spindle afferents by vibration of a muscle belly (Bianconi & van der Meulen, 1963). Conscious monkeys did not readily tolerate the application of a vibrating probe to those regions of muscle tissue from which responses of precentral neurones were obtained. Nevertheless a few precentral neurones could be examined for brief periods of time. Vibration at 100 Hz applied to the appropriate region of the muscle belly was effective in increasing the discharge of all the precentral cortical cells studied, but less than half of them were affected by vibration at 300 Hz.

Although the conclusion that the cortical influences in area 4 derive from muscle spindles in the conscious monkey must be less reliable than it is in those experiments where more rigid tests of receptor function can be employed and where conduction velocities of afferent fibres can be estimated, the conscious animal is able to make its own self-paced movements and the responses of the same neurones can be studied during natural movement performance. In this way the relationship between those discharges which occur during passive manipulation of joints can be compared with that which exists between natural discharges and self-initiated active movement (Lemon, Hanby & Porter, 1976). For just over one half of the precentral neurones the muscle which had to be stretched to produce the responses in the passive, relaxed animal was found to be contracting in association with the precentral neurone's natural activity during active movement. This could mean that fusimotor drive to the active muscle maintained an input to the cerebral cortex during active contraction of the muscle. But, of course, the discharge during the active state could have been driven from other sources altogether.

A passive response to muscle spindle influences of a second group of precentral neurones to joint movement under both passive and self-initiated active conditions was suggested by the finding that these cells responded when a particular muscle was stretched either actively in self-initiated movements or passively by the experimenter (Lemon *et al.*, 1976). It is also evident that in the work of Fetz, Finocchio, Baker & Soso (1974) some precentral neurones discharged during active and passive movements of a joint in the same direction, but in their experiments it was not at all clear whether the input was from muscle receptors, joint afferents or a combination of these.

It is very likely that the state of muscle spindles is dramatically modified during movement performance. Moreover, transmission through particular afferent pathways may be facilitated or depressed during voluntary movement (Tsumoto, Nakamura & Iwama, 1975). Hence the responses of precentral neurones which are demonstrated in the passive, relaxed animal should not be assumed to be occurring unchanged during movement performance. Evarts (1973) showed that a sudden disturbance imposed on a learned movement performance was capable of influencing precentral neurones whose discharges co-varied with that movement performance. A similar result has been reported for different movement tasks by Conrad *et al.* (1974) and by Porter & Rack (1976). The latencies of the responses in the cortical neurones were short and it was concluded that these responses were associated with long-loop automatic compensation for the imposed disturbances. Both the cortical responses and the compensatory muscle contractions could be modified by prior instruction about the response to be made (Evarts & Tanji, 1974).

It is not clear whether the muscle spindle component of the responses of precentral neurones to peripheral stimuli is of major or minor significance in the long-loop responses studied by Evarts (1973). The potential exists for a variety of other receptors to be involved and for complex interactions between afferent influences to occur in the central nervous system. Nor is it clear which pathways are utilized in the central nervous system by the afferents involved, although it can be suggested that the short-latency responses of precentral neurones in the passive, relaxed monkey are very dependent on intactness of dorsal column afferents (Brinkman, Bush & Porter, 1978). In parallel with this is the report that automatic compensation for sudden disturbances imposed on voluntary muscle contraction in man requires the intactness of dorsal column afferents (Marsden *et al.*, 1973).

These observations extend considerations of the functions of muscle spindles well beyond the need to understand the stretch reflex. What is the total sphere of influence within the central nervous system of a given muscle spindle afferent? How specific are the anatomical connections made by the branches of this afferent fibre? The small muscle regions from which individual precentral neurones may be influenced suggest highly specific connections within the pathway to the cerebral cortex, but very little information exists about the microscopic location of these connections.

REFERENCES

Albe-Fessard, D. & Liebeskind, J. (1966). Origine des messages somato-sensitifs activant les cellules du cortex moteur chez le singe. *Exp. Brain. Res.* **1**, 127–46.

Asanuma, H. (1975). Recent developments in the study of the columnar arrangement of neurones within the motor cortex. *Physiol. Rev.* **55**, 143–56.

Bianconi, R. & van der Meulen, J. P. (1963). The response to vibration of the end organs of mammalian muscle spindles. *J. Neurophysiol.* **26**, 177–90.

Brinkman, J., Bush, B. M. & Porter, R. (1978). Deficient influences of peripheral stimuli on precentral neurones in monkeys with dorsal column lesions. *J. Physiol., Lond.* **276**, 27–48.

Burgess, P. R. & Clark, F. J. (1969). Characteristics of knee joint receptors in the cat. *J. Physiol., Lond.* **293**, 317–35.

Conrad, B., Matsunami, K., Meyer-Lohmann, J., Wiesendanger, M. & Brooks, V. B. (1974). Cortical load compensation during voluntary elbow movements. *Brain Res.* **71**, 507–14.

Evarts, E. V. (1966). Pyramidal tract activity associated with a conditioned hand movement in the monkey. *J. Neurophysiol.* **29**, 1011–27.

Evarts, E. V. (1968). Relation of pyramidal tract activity to force exerted during voluntary movement. *J. Neurophysiol.* **31**, 14–27.

Evarts, E. V. (1973). Motor cortex reflexes associated with learned movement. *Science* **179**, 501–3.

Evarts, E. V. & Tanji, J. (1974). Gating of motor cortex reflexes by prior instruction. *Brain Res.* **71**, 479–94.

Fetz, E. E. & Baker, M. A. (1969). Response properties of precentral neurons in awake monkeys. *Physiologist, Wash.* **12**, 223P.

Fetz, E. E. & Finocchio, D. V. (1972). Operant conditioning of isolated activity in specific muscles and precentral cells. *Brain Res.* **40**, 19–24.

Fetz, E. E., Finocchio, D. V., Baker, M. A. & Soso, M. J. (1974). Responses of precentral 'motor' cortex cells during passive and active joint movements. *4th A. Meet. Soc. Neurosci.* p. 208, (abstract).

Hore, J., Preston, J. B., Durkovic, R. G. & Cheney, P. D. (1976). Responses of cortical neurons (Areas 3a and 4) to ramp stretch of hindlimb muscles in the baboon. *J. Neurophysiol.* **39**, 484–500.

Landgren, S. & Silfvenius, H. (1969). Projection to cerebral cortex of Group 1 muscle afferents from the cat's hindlimb. *J. Physiol., Lond.* **200**, 353–72.

Lemon, R. N., Hanby, J. A. & Porter, R. (1976). Relationship between the activity of precentral neurones during active and passive movements in conscious monkeys. *Proc. R. Soc., B* **194**, 341–73.

Lemon, R. N. & Porter, R. (1976). Afferent input to movement-related precentral neurones in conscious monkeys. *Proc. R. Soc., B* **194**, 313–39.

Lucier, G. E., Ruegg, D. C. & Wiesendanger, M. (1975). Responses of neurones in the motor cortex and in area 3a to controlled stretches of forelimb muscles in cebus monkeys. *J. Physiol., Lond.* **251**, 833–53.

Marsden, C. D., Merton, P. A. & Morton, H. B. (1973). Is the human stretch reflex cortical rather than spinal? *Lancet* i, 759–61.

Matthews, P. B. C. (1972). *Mammalian muscle receptors and their central actions.* London: Eward Arnold.

Oscarsson, O. & Rosén, I. (1963). Projection to cerebral cortex of large muscle-spindle afferents in forelimb nerves of the cat. *J. Physiol., Lond.* **169**, 924–45.

Phillips, C. G. (1969). Motor apparatus of the baboon's hand. *Proc. R. Soc., B* **173**, 141–74.

Phillips, C. G., Powell, T. P. S. & Wiesendanger, M. (1971). Projection from low-threshold muscle afferents of hand and forearm to area 3a of baboon's cortex. *J. Physiol., Lond.* **217**, 419–46.

Porter, R. (1976). Influences of movement detectors on pyramidal tract neurones in primates. *A. Rev. Physiol.* **38**, 121–37.

Porter, R. & Rack, P. M. H. (1976). Timing of the response in the motor cortex of monkeys to an unexpected disturbance of finger position. *Brain Res.* **103**, 201–13.

Rosén, I. & Asanuma, H. (1972). Peripheral afferent inputs to the forelimb area of the monkey motor cortex: input–output relations. *Exp. Brain. Res.* **14**, 257–73.

Tsumoto, T., Nakamura, S. & Iwama, K. (1975). Pyramidal tract control over cutaneous and kinaesthetic sensory transmission in the cat thalamus. *Exp. Brain Res.* **22**, 281–94.

Vallbo, Å. B. (1970). Slowly adapting muscle receptors in man. *Acta physiol. scand.* **78**, 315–33.

Climbing fibre inputs from the interosseous and knee joint nerves

J. C. ECCLES

In about 1950 in Dunedin, A. K. McIntyre became interested in a small nerve that branched off the nerve to the flexor longus digitorum in the cat hindlimb. We dissected it in several animals and found that it travelled downward along the interosseous fascia beyond the ankle joint. It was named the interosseous nerve. Subsequently, in preparing the nerve to the flexor longus digitorum for stimulation this nerve was carefully dissected away.

Some years later, Hunt & McIntyre (1960) showed that this interosseous nerve contained three main classes of fibres: tension receptors with slowly adapting discharges, in large fibres up to group I in size; vibration receptors particularly sensitive to frequencies of 40 to 700 Hz, being rapidly adapting and extremely sensitive to vibration; tap receptors, probably a sub-class of the vibration receptors. Hunt (1961) identified the highly sensitive vibration and tap receptors as Pacinian corpuscles. It was of interest that these vibration receptors projected to the cerebral cortex via the dorsal columns (McIntyre, 1962) mostly to the contralateral sensory area II.

When investigating the electrical responses induced in the cerebellar cortex by stimulation of limb nerves, it was found (Eccles, Provini, Strata & Táboříková, 1968b) that two very small nerves were extremely effective in producing responses that could be identified as due to climbing fibre action on Purkinjě cells (Eccles et al., 1968a). For example, in Fig. 1 there are assembled the responses evoked in six recording sites from the molecular layers by double stimulation of eight hindlimb nerves. The recording sites were in the lateral vermal region of the ipsilateral cerebellar cortex, that is, in zone b2 of Oscarsson (1976). The two very small nerves, the dorsal nerve of the knee joint and the interosseous nerve, are seen to be remarkably effective in evoking climbing fibre responses, being only exceeded by the very large plantar cutaneous nerve that carries all the sensory information from the foot pads. Since the preparation was anaesthetised by pentobarbital sodium, the mossy fibre responses (N_3 waves) were negligible. It was remarkable that both the knee joint and interosseous nerves were more effective than the nerves to quite large muscles.

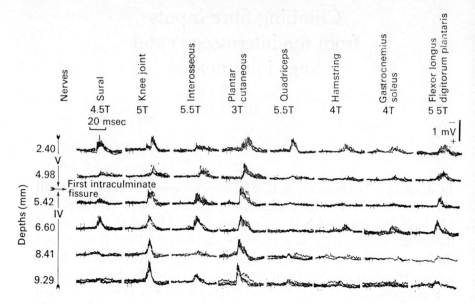

Fig. 1. Climbing fibre responses evoked in molecular layers at various depths by volleys in hindlimb nerves. See Fig. 1 of Eccles *et al.* (1968*a*) for the location of the recording micro-electrode track that passed from lobule V into lobule IV as indicated by the arrows to the left. The recording sites are indicated in that figure by filled circles in B. The depths of recording are shown to the left for the six recording sites and the eight nerves evoking the responses are indicated above with the strengths of stimulation relative to threshold (T), there being two stimuli at 4-msec intervals. There is superposition of about five responses with each trace.

In Fig. 2 there is a topographic plotting from another experiment. Four nerves are represented as indicated in the key to the right, which also gives the code for size of the climbing fibre-evoked responses. The recording loci are shown in four tracks through Larsell lobules V and IV in the lateral vermis. In this experiment also large responses were induced from knee joint and interosseous nerves. There was much variation in size between different loci, and on the whole the two small nerves were as effective as the large cutaneous nerves, sural and superficial peroneal. A remarkable feature was the differentiation between the relative effects of the four nerves, which appeared to have a patchy character.

The extraordinary effectiveness of these two small nerves in evoking climbing fibre responses can now be reconsidered in relation to the hypothesis that the climbing fibre input to a Purkinjě cell acts as an instruction for potentiation of the parallel fibre synapses that were activated at about the same time (Marr, 1969). Recently (Eccles, 1977) the empirical evidence in support of this hypothesis (Ito, 1975; Gonshor & Melvill Jones, 1976;

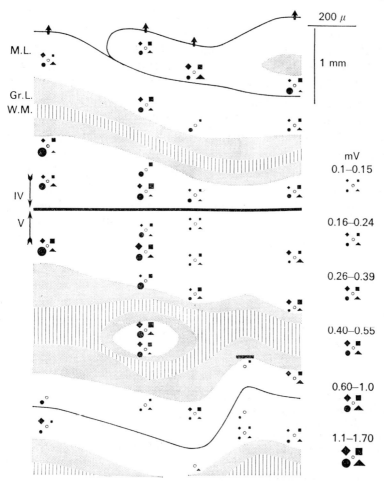

Fig. 2. Fine grain discrimination of climbing fibre inputs. This figure is constructed from the responses along five micro-electrode tracks in the lateral vermis passing through lobule V and into lobule IV, as indicated by the arrows to the left. See Fig. 5 of Eccles *et al.* (1968*b*) for the exact location. At each recording site the size of responses evoked from knee joint (◆), interosseous (■), superficial peroneus (●) and sural (▲) nerves are plotted according to the key symbols shown in the right margin. Arrows below show position of micro-electrode tracks and length scales show that the plotting is enlarged transversely. M.L. = molecular layer (white area); Gr.L. = granular layer (stippled area); W.M. = white matter (vertical lines).

Robinson, 1976) has been assembled and considered in relation to the highly specific topographic distribution of climbing fibre inputs to the cerebellar cortex, particularly to the anterior lobe.

In the light of this hypothesis the interosseous and knee joint nerves would have an importance in the learning of movement control that is much greater than would be expected from their size. There is as yet no information about the types of fibres concerned in the large climbing fibre responses evoked from these nerves. As indicated in Fig. 1 the stimulations were sufficiently strong to excite all group I and II fibres. It would be of interest to test the effectiveness of adequate stimulation, particularly of vibration of the foot, in evoking climbing fibre responses of the cerebellum. For example, it has been suggested by McIntyre that the sensitivity of the vibration receptors is so great that the cat could sense through its hindfeet the vibration caused by movements of other animals in its neighbourhood. This facility would probably be of great advantage in the dark. Further problems may eventually arise in attempting to discover how these climbing fibre inputs to the Purkinjě cells can be utilised in selectively increasing the mossy fibre-parallel fibre effectiveness on these cells. A further point of interest is that the projection of the interosseous nerve to the cerebral cortex (McIntyre, 1962) also may be of importance in the control of movement and in the cerebro-cerebellar circuitry concerned therein (Allen & Tsukahara, 1974).

The purpose of this note is to point out that this nerve discovered by McIntyre may have a special significance in the learning of control of movement.

REFERENCES

Allen, G. I. & Tsukahara, N. (1974). Cerebro-cerebellar communication systems. *Physiol. Rev.* **54**, 957–1006.

Eccles, J. C. (1977). An instruction-selection theory of learning in the cerebellar cortex. *Brain Res.* **127**, 327–52.

Eccles, J. C., Provini, L., Strata, P. & Táboříková, H. (1968a). Analysis of electrical potentials evoked in the cerebellar anterior lobe by stimulation of hindlimb and forelimb nerves. *Exp. Brain Res.* **6**, 171–94.

Eccles, J. C., Provini, L., Strata, P. & Táboříková, H. (1968b). Topographical investigations on the climbing fibre inputs from forelimb and hindlimb afferents to the cerebellar anterior lobe. *Exp. Brain Res.* **6**, 195–215.

Gonshor, A. & Melville Jones, G. (1976). Extreme vestibulo-ocular adaptation induced to prolonged optical reversal of vision. *J. Physiol., Lond.* **256**, 381–414.

Hunt, C. C. (1961). On the nature of vibration receptors in the hind limb of the cat. *J. Physiol., Lond.* **155**, 175–86.

Hunt, C. C. & McIntyre, A. K. (1960). Characteristics of responses from receptors from the flexor longus digitorum muscle and the adjoining interosseus region of the cat. *J. Physiol., Lond.* **153**, 74–87.

Ito, M. (1975). Cerebellar learning control of the vestibulo-ocular mechanisms. In *Mechanisms in transmission of signals for conscious behaviour*, 26th International Physiological Congress. New Delhi.

McIntyre, A. K. (1962). Cortical projection of impulses in the interosseous nerve of the cat's hind limb. *J. Physiol., Lond.* **163**, 46–60.

Marr, D. (1969). A theory of cerebellar cortex. *J. Physiol., Lond.* **202**, 437–70.

Oscarsson, O. (1976). Spatial distribution of climbing and mossy fibre inputs into the cerebellar cortex. *Exp. Brain Res.* Suppl. **1**, 36–42.

A study of some mechanisms operating in carotid body chemoreceptors

C. EYZAGUIRRE AND S. J. FIDONE

Sensory receptors can be classified into two categories: (1) simple receptors, where the 'first order' sensory neurone is excited directly by the natural stimulus; or (2) composite receptors, where a non-neural element is the primary target of the stimulus. These two receptor types have been called 'primary' and 'secondary', respectively. The carotid body, a model of arterial chemoreceptors, has traditionally been relegated to the second category. Applying this schema, the carotid nerve endings on the glomus cells correspond to the peripheral part of the sensory neurone, and the glomus cells themselves represent the accessory cells. This organization implies that a 'sensory synapse' should be operating at the glomus–nerve junction.

The morphology of the carotid body receptors is complex. The glomus (type I) cells (about 10 μ) are surrounded by capsular or sustentacular (type II) cells which in most cases separate glomus cells from capillaries. Nerve endings terminate on glomus cells, forming a 'synapse' (de Castro, 1951; Hess, 1968). Thus, the ending is separated from the cells by a double membrane layer (200–300 Å apart) with typical membrane thickenings which are found on either side of the junction (McDonald & Mitchell, 1975). Serial reconstruction of cells and endings from electron micrographs has revealed both bouton-type and calyx-type endings (Nishi & Stensaas, 1974). They can be identified also using high-power, Nomarski-type optics (Eyzaguirre & Gallego, 1975). The bouton-type endings contain abundant clear-core vesicles. The calyx-type envelop a large part of the glomus cell, and at times send processes to neighboring cells so that one calyx-type ending may terminate on more than one glomus cell (Eyzaguirre, Nishi & Fidone, 1972).

One of the difficulties in studying chemoreceptor mechanisms is that the carotid nerve fibers respond with an increased discharge when the preparation is exposed to a variety of stimuli. Thus, it is activated by a fall in pO_2 and pH and by an increase in pCO_2. Chemicals such as acetylcholine (ACh), nicotine, and sodium cyanide also increase the sensory discharge.

Other substances, such as dopamine (DA), reduce or block the sensory discharges of the cat carotid body *in vivo* (Sampson, 1972). In addition, it has become clear in recent years that other forms of stimulation, such as osmotic and temperature changes, also affect chemosensory discharges (McQueen & Eyzaguirre, 1974; Gallego & Eyzaguirre, 1976). Consequently, it is difficult to determine in this 'omnireceptor' what mechanisms of action operate at the receptor level to generate the sensory impulses.

We have postulated over the years, for the following reasons, that the glomus or type I cells are the primary transducer sites. (*a*) Type I cells are intimately apposed to nerve endings and the junction has the appearance of a synapse (de Castro, 1951); (*b*) the tissue contains putative neurotransmitter substances, namely ACh (Eyzaguirre, Koyano & Taylor, 1965; Fidone, Weintraub & Stavinoha, 1976) and DA (Chiocchio, Biscardi & Tramezzani, 1966); (*c*) these substances are able to modify chemoreceptor activity. For example, small amounts of ACh depolarize the nerve endings and increase the sensory discharges. This effect is potentiated after the tissue cholinesterase is inactivated (Eyzaguirre & Zapata, 1968*a*), and ACh is released from the carotid body during electrical stimulation and during interruption of flow of the bathing medium (Eyzaguirre & Zapata, 1968*a*). This substance is also released during natural (nitrogen, carbon dioxide) and chemical (sodium cyanide) stimulation (S. J. Fidone, unpublished). Against this excitatory neurotransmitter hypothesis for ACh is the fact that cholinergic blockers, while clearly blocking the effects of exogenously applied ACh, only depress the sensory discharge elicited by natural stimulation, but they do not block it (Nishi & Eyzaguirre, 1971). With regard to DA, fluorescence microscopy has shown that the glomus cells contain appreciable amounts of this agent (Chiocchio, King & Angelakos, 1971). DA is now thought to be an inhibitory transmitter in the carotid body since its exogenous application in the cat blocks or depresses the discharge *in vivo* (Sampson, 1972). However, its effects *in vitro* can be either inhibitory or excitatory (Zapata, 1975). Finally, there is now direct evidence that DA is released during carotid body stimulation (Gonzalez & Fidone, 1977).

In summary, our present view of the processes of chemoreceptor excitation would suggest that stimuli affect the resting state of the glomus cells, which in turn release a 'transmitter' or 'generator' substance (possibly ACh) into the enclosed synaptic cleft to depolarize the terminals (see Eyzaguirre & Zapata, 1968*b*; Eyzaguirre *et al.*, 1972). Ending depolarization would in turn initiate the sensory discharge by acting, presumably, on the first node of the myelinated sensory fibers. The chemosensory C fibers would be activated by depolarization at or near the sensory terminal. These ideas have been challenged by some who have advocated that glomus cells are not the primary elements in transduction (Biscoe, Lall & Sampson, 1970; Biscoe, 1971). They have suggested that the nerve endings impinging upon glomus cells are *efferent* and that the cells are the final element in an inhibitory pathway (see later). The sensory terminals would then be 'free' nerve endings in the pericellular spaces.

The purpose of this article is to present new information on the functional organization within the carotid body and the mechanisms which may be operating at the receptor level; specifically, some recent physiological, morphological and biochemical studies of glomus cells and nerve endings. It is hoped that these studies will contribute to a better understanding of the complicated mechanisms operating in these receptors. Most of these experiments have been performed in cats but in some we have used rabbits.

NATURE OF THE NERVE ENDINGS

The sensory nature of fibers in the carotid body, first suggested by the studies of de Castro (1928), has been the object of renewed interest and speculation in recent years. De Castro showed that fibers in the cat carotid nerve, terminating upon glomus cells are sensory because such nerve endings were unaltered following intracranial section of the glossopharyngeal nerve root central to its sensory ganglion (de-efferentation). Section of the nerve peripheral to the ganglion induced total degeneration of the endings. These findings have been challenged recently by Biscoe *et al.* (1970). They repeated de Castro's original experiments and claimed that synapses between carotid nerve terminals and glomus cells are *efferent*, since it appeared to them that the nerve endings degenerated with de-efferentation of the carotid nerve. Their interpretation was reinforced by fibers passing along the carotid nerve (Neil & O'Regan, 1971). Hess & Zapata (1972) and Nishi & Stensaas (1974) repeated the de-efferentation experiments and confirmed de Castro's original findings, in direct contradiction with the results from Biscoe's group. The long degeneration times reported by Biscoe *et al.*, led Hess & Zapata to suggest an alternative explanation. They drew attention to the fact that two sensory ganglia are commonly located along the IXth nerve, the petrosal ganglion and the smaller superior, or Ehrenritter's, ganglion. The latter is located intracranially and might easily have been damaged when Biscoe *et al.* removed a portion of the nerve roots. In seeking to resolve this issue, axoplasmic flow of labelled material was utilized to demonstrate the distribution of nerve terminals from neurones in the sensory (petrosal) ganglion (Fidone, Stensaas & Zapata, 1975, 1977b,c; Stensaas & Fidone, 1977). Axonal transport of ^{3}H-amino acid to axon terminals following its perikaryal uptake and incorporation provides a precise method for identifying neuronal projections. In this study, ³H-proline was administered to the petrosal ganglion and the peripheral distribution of the labelled material was determined by means of liquid scintillation spectrometry and electron microscope autoradiography.

Following administration of ³H-proline to the petrosal ganglion the time course of net accumulation of radioactivity in the carotid body showed an upward trend in total carotid body counts min⁻¹. This suggested that migration of radioactivity out of the ganglion cells with time resulted in increased accumulation of radioactivity in the carotid body. Accurate measurements of the rates of axoplasmic transport in the carotid or glosso-pharyngeal nerves were precluded by the short lengths of nerve (20–25 mm) involved in these experiments, but the data suggested the presence of fast, intermediate and slow components. The fast component had a velocity estimated to be greater than 200 mm per day, while the slow and inter-mediate components had velocities ranging from 1–70 mm per day. Between one and two days after ³H-proline injection there appeared a trough in the curve of cumulative radioactivity in the carotid body which probably represented the interim between the arrival of the fast and the slower components of axonal transport.

The autoradiographic distribution of label in the carotid body was examined at one-day intervals from one to seven days after ganglionic injection of ³H-proline. The grains were localized almost exclusively over nerve fibers and nerve terminals (Fig. 1). Few grains were seen over glomus cells, sustentacular cells, or other constituents of the carotid body. Nerve terminals on glomus cells were identified by the presence of clear-core vesicles and mitochondria. The labelled axon profiles included bouton endings, calyciform endings and terminal enlargements of intermediate shape and size. We observed that 60–90 % of the nerve endings in a given ultra-thin section autoradiograph were labelled (mean number of tracks per ending, 3.6 ± 2.9 s.D. with a range from 1 to 21). Only an occasional cell was labelled, and then only with a single overlying track. The selective labelling of sensory nerve terminals in cat carotid body is dependent on several conditions. These are: (1) that fibers of passage through the ganglion do not incorporate and transport labelled amino acid to the carotid body; (2) that passive movement of label from the injection site is negligible; and (3), that no appreciable number of autonomic (efferent) ganglion cells are present in the petrosal ganglion whose axons enter the carotid body.

In consideration of the first condition, ³H-proline was administered to a desheathed region of the IXth nerve immediately distal to the petrosal ganglion. No significant migration of labelled material occurred along the nerve after 3 and 24 h. This observation argues against the possibility that fibers of passage through the petrosal ganglion can incorporate and transport labelled amino acid. This also agrees with observations by others that axons lack or have limited capacity for incorporation and transport of amino acids (Cowan et al., 1972).

With regard to passive movement of label from the injection site, we found that application of ^{3}H-proline to the nerve resulted in movement of label only 4–6 mm distally along the nerve after 3 or after 24 h. Following ganglionic injection, passive transfer of the labelled material should then be no more than 4–6 mm after 24 h. There was always at least 13–15 mm between the carotid body and the ganglion and at 24 h labelled endings were already identifiable in carotid body autoradiographs. When the ganglion was crushed prior to injecting ^{3}H-proline, radioactivity decreased to 1–2 % of that normally present along the nerve after three days in an uninjured preparation. It should also be noted that the prominent localization of label to nerve endings in the carotid body is not consistent with passive spread of significant amounts of radioactivity from the ganglion to the carotid body. If this were to occur, uniform labelling of the cellular constituents of the carotid body would be expected, including glomus cells. Certainly one would not predict nerve endings to be the exclusive site of protein synthesis in the carotid body, since it is well established that amino acid incorporation into proteins takes place primarily within the cell body.

With respect to the presence of autonomic ganglion cells in the carotid body, the early anatomical studies of the glossopharyngeal and carotid nerves revealed the presence of a small number of autonomic ganglion cells scattered along these nerves in the cat (de Castro, 1926). Torrance (1969) indicated that these ganglion cells might be strays from a larger population of autonomic neurones present in the petrosal ganglion. This arrangement has sometimes been seen in nerve trunks near autonomic ganglia. Although there are no reports of efferent neurones in the petrosal ganglion, studies with the nodose ganglion have indicated that a synapse might exist between preganglionic fibers arising from the brainstem and autonomic efferent neurones in the ganglion (see Daly & Hebb, 1966). Certainly the petrosal ganglion is predominantly a sensory ganglion, but if a small population of autonomic ganglion cells was present in the petrosal ganglion it might constitute an efferent pathway from brainstem to carotid body. Then the conclusion that nerve endings on glomus cells arise exclusively from sensory neurones would be nullified, since transport of labelled material into the terminals of visceromotor neurones would occur following ganglionic injection of ^{3}H-proline. With this problem in mind, the petrosal ganglion was examined for the presence of autonomic ganglion cells. Ultrastructural analysis of 902 neurones at 17 levels of two cat petrosal ganglia provided no evidence for the presence of synapses on cell somata or on their processes. The results were thus consistent with the concept that the petrosal ganglion is strictly a sensory ganglion, containing only unipolar nerve cells.

We conclude, therefore, that labelled nerve endings observed in this study are the distal processes of sensory neurones whose cell bodies are located in the ganglion. Furthermore, the high percentage of labelled endings in some autoradiographs suggests that most nerve endings on glomus cells in the carotid body arise from sensory neurones in the petrosal ganglion. Of course, the possibility cannot be ruled out that efferent terminals arising from neuronal perikarya located elsewhere might account for the remaining small percentage of unlabelled endings in an autoradiograph.

ARE THE GLOMUS CELLS NECESSARY FOR CHEMORECEPTION?

The idea that preneural elements are essential for chemosensory transduction is supported by the observations of de Castro (1951) and Zapata, Hess & Eyzaguirre (1969), in which carotid bodies were reinnervated with vagal or superior laryngeal fibers. The fact that the newly formed pathway responded to carbon dioxide, asphyxia and nitrogen inhalation was interpreted to mean that the glomus cells are important for the chemoreceptor transduction process. This interpretation has been challenged by Mitchell, Sinha & McDonald (1972), who sectioned the carotid nerve and buried it in muscle or connective tissue at some distance from the carotid body. Axons in the neuroma at the cut end of the carotid nerves were reported to have chemoreceptive properties similar to those of intact carotid nerves, supporting the hypothesis that carotid nerve endings themselves are chemosensitive transducers. These discrepancies prompted us (Zapata, Stensaas & Eyzaguirre, 1976) to use another approach to see if isolated nerve endings could respond to the action of chemical or 'natural' stimuli.

The carotid nerve in the cat was crushed and allowed to regenerate in order to study whether recovery of chemosensory function depends on the re-establishment of apposition between regenerating carotid nerve fibers and glomus cells in the carotid body. When the nerve was crushed close (1–2 mm) to the carotid body, chemosensory activity in the nerve was re-initiated after a delay of six days. Recovery was further delayed when the crush was made at successively greater distances (5–6 and 10–12 mm) from the carotid body. Ultrastructural analysis showed that the reappearance of nerve endings on the glomus–sustentacular cell complex coincided in time

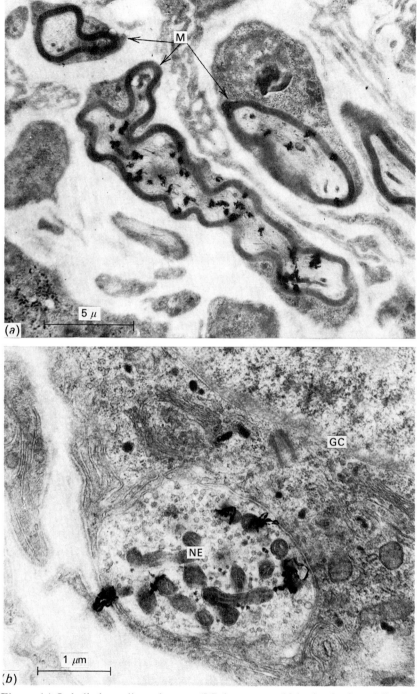

Fig. 1. (a) Labelled myelinated axons (M) in cat carotid body six days following administration of ^{3}H-proline to the petrosal ganglion. (b) A labelled bouton-like nerve ending (NE) on a glomus cell (GC) in cat carotid body contains a characteristic accumulation of synaptic vesicles and mitochondria. Six days after ^{3}H-proline treatment of the petrosal ganglion. (From Fidone, Stensaas & Zapata, 1975.)

with the re-establishment of chemosensory activity. No responses were elicited before apposition of nerve endings with glomus cells, which suggests that the growing tips of the regenerating axons *within* the carotid body are incapable of responding directly to chemical stimuli before contacting the cells. This means that the cells are essential in the transduction process. This poses an interesting question; namely how is restoration of function provided by apposition between regenerating axons and glomus cells? One possibility is that sensory nerve fibers physically apposed to glomus cells may be excited by neurotransmitter substance(s) released from these cells (see Zapata *et al.*, 1969; Zapata, 1975). Another possibility is that apposition between nerve endings and these cells leads to 'maturation' of the endings which are then capable of responding *directly* to chemoreceptor stimuli. Thus, these experiments alone do not establish that the glomus cells are the immediate transducer element in the carotid body. Nevertheless, it is clear that apposition between nerve endings and glomus cells is essential for chemoreceptor function in the carotid body. Our results are thus complementary to those of Verna, Roumy & Leitner (1975), who also found that carotid body glomus cells are essential for the chemoreceptor response.

These observations clearly disagree with those of Mitchell *et al.* (1972). Their conclusion that carotid nerve terminals alone are the chemoreceptive transducers assumes that in their experiments the regenerated carotid nerve terminals were not apposed to isolated islets of glomus tissue. This is an important consideration since miniature glomera are present along the carotid artery and the vago-sympathetic trunk (Matsuura, 1973). Since 12–18 months were allowed for regeneration of nerve fibers in their experiments, there was ample time for the axons to have reached accessory glomera.

ACETYLCHOLINE IN THIS CAROTID BODY

In earlier work, Eyzaguirre and his colleagues (see Eyzaguirre & Zapata, 1968*b*) suggested that ACh might be a sensory transmitter released from the glomus cells to excite neighboring afferent nerve terminals. But ACh has never been chemically identified in carotid body tissue, although bioassays have suggested the presence of an ACh-like substance in carotid body extracts (Eyzaguirre *et al.*, 1965; Jones, 1975). Furthermore, Osborne & Butler (1975), proceeding on Biscoe's (1971) suggestion that the synapses on the glomus cells are efferent, not afferent, have recently proposed that ACh is the transmitter at this junction and that consequently this substance is contained in the nerve terminals, not in the glomus cells.

To resolve this issue, we undertook to determine whether chemically identifiable ACh is present in the carotid body, and in what cellular constituents of the organ it is contained (Fidone *et al.*, 1976; Fidone *et al.*, 1977*a*). We measured the endogenous ACh levels in the tissue, using pyrolysis gas chromatography and mass spectrometry (GC/MS). We also determined whether the ACh in the tissue was lost following chronic denervation by section of the carotid nerve, as might be expected if the ACh were contained in the carotid nerve terminals. Finally, we have studied the kinetics of choline uptake by the carotid body, and the autoradiographic localization of a high-affinity component of this uptake process.

For GC/MS measurements of carotid body ACh, the organs were removed from the animals and quickly cleaned of surrounding connective tissue in a chamber filled with ice-cold Locke's solution equilibrated with 100 % O_2 and containing 30-μM eserine. Denervation of the carotid body on one side had been performed in some animals 14 days prior to removal of the organ for ACh analysis. The tissue was weighed (mean carotid body weight = 719 μg) and frozen in liquid nitrogen until analysis. It was concluded that significant degradation of tissue ACh did not occur during sample preparation, because carotid bodies, cooled *in situ* with ice-cold Locke's solution prior to removal from the animal, or cooled and then quickly frozen in liquid nitrogen, gave ACh values which were not significantly different from those obtained without these procedures ($P > 0.2$, non-paired, double-tailed t-test). The ACh determinations in normal and denervated carotid bodies yielded values in each group which were not significantly different from one another (normal, 11.7 ± 1.1 pmol per organ; denervated, 12.0 ± 2.0 pmol per organ with $P > 0.2$).

The ACh content of cat carotid bodies obtained previously using bioassays have claimed values as much as 10 times higher than these obtained using pyrolysis GC/MS (Eyzaguirre *et al.*, 1965; Jones, 1975). We are unable to explain this large discrepancy; however, pyrolysis GC/MS is a sensitive and highly specific method for measuring tissue ACh levels and is free of many of the difficulties and uncertainities which accompany bioassays. Also, in this regard, it is interesting to note that Christie (1933) reported many years ago that extracts from human carotid body tumours induced hypotension in decapitated cats. He showed that this effect was not blocked by atropine in doses which completely abolished the effects of injected ACh. These extracts also produced contraction of the virgin guinea pig uterus. By other tests the effects of histamine-like reactions were ruled out. This unindentified substance was labelled carotidin, and it was also found to be present in the carotid body of the elasmobranch. Whether carotidin or some other substance may have interfered with the bioassay determinations of ACh is unknown.

The pre-terminal fibers of the carotid nerve are either completely degenerated or in a very advanced state of degeneration 14 days after denervation (Hess & Zapata, 1972). Since carotid body ACh is unchanged at this time our data suggest that ACh is not contained in the carotid nerve innervation to the this organ. The remaining possible sources of carotid body ACh include the glomus cells, sustentacular cells, ganglion cells and the sympathetic innervation of the blood vessels. In cat carotid bodies the ganglion cells are too few in number and can be discounted. Also, ganglion cells are usually located around the perimeter of the organ and it is our experience that after carotid bodies have been thoroughly cleaned of their connective tissue capsules, they rarely show histological evidence of ganglion cells. Finally, preliminary data from our laboratories show that the ACh levels in the carotid body are also unchanged after chronic superior cervical ganglionectomy.

Numerous reports have appeared in recent years describing two kinetically different processes for choline (Ch) uptake: a low-affinity, saturable uptake common to all neurones, and a high-affinity, saturable uptake which is reported to be specific for cholinergic neurones (see Yamamura & Snyder, 1973). To examine whether a high- and low-affinity uptake of Ch is present in the carotid body, tissue samples were incubated at 37 °C in ^{3}H-Ch (1–70 μM) and the time course of accumulation of total radioactivity was determined (Fidone et al., 1977a). The uptake was linear for approximately 10 min, then gradually plateaued over the next 15–20 min. Thereafter, carotid bodies were incubated for 10 min in 1–70 μM ^{3}H-Ch; the uptake showed saturation with increasing ^{3}H-Ch concentrations. Data were corrected for a small, passive, non-saturable component of Ch entry which was determined in separate experiments by incubating carotid bodies in 1–70 μM ^{3}H-Ch together with a much higher concentration (10 mM) of unlabelled Ch. In this situation, the uptake of radioactivity by the specific, high- and low-affinity uptake processes is nearly completely abolished by the presence of the large excess of unlabelled Ch, whereas non-specific leakage of the labelled Ch into the cells is unaffected (Yamamura & Snyder, 1973). Kinetic analysis of the corrected data, using a Lineweaver–Burk double reciprocal plot, resulted in a curvilinear distribution of the data points which could be resolved into two distinct components. The Michaelis constant K_m values obtained for Ch uptake in the carotid body were $K_{m_H} = 3.6\,\mu$M for the high-affinity component and $K_{m_L} = 49.5\,\mu$M for the low-affinity component, which fall within the ranges of 1–8 μM for K_{m_H} and 25–100 μM for K_{m_L} reported for other putative cholinergic systems. Since at low Ch concentrations negligible accumulation of Ch occurs via the low-affinity component (Dowdall & Simon, 1973; Yamamura

& Snyder, 1973; Simon, Atweh & Kuhar, 1976), it has been suggested that the high-affinity uptake of Ch might furnish an empirical approach to the labelling of cholinergic cells. We have found that chronic total denervation of the carotid body (by section of the carotid nerve and the ganglio-glomerular sympathetic nerve) for periods of 1 week to three months did not result in the loss of the high-affinity component of Ch uptake. This suggests, in agreement with our GC/MS data described above, that these nerve fibers may not be cholinergic, but that one or more of the remaining cellular elements in the carotid body might have this property.

To determine which cells of the carotid body have the capacity for the high-affinity uptake of Ch, carotid bodies were incubated in 1-2 μM ^{3}H-Ch and were prepared for autoradiography, using rapid freezing and freeze-drying techniques in order to maintain histological localization of the labelled Ch. At low ^{3}H-Ch concentrations, when only the high-affinity component should contribute significantly to the uptake (Dowdall & Simon, 1973; Yamamura & Snyder, 1973; Simon et al., 1976), the auto-radiographic label was concentrated principally over the glomus cells. The degree of labelling of sustentacular cells is uncertain since these are difficult to distinguish with the light microscope. The high-affinity component was reduced or abolished in low, external Na$^+$ and by low concentrations of hemicholinium-3 whereas the low-affinity component was unchanged. Low Na$^+$ and low concentrations of hemicholinium-3 also reduce or abolish the synthesis of ^{3}H-ACh from ^{3}H-Ch in the carotid body. This was shown in separate experiments in which ^{3}H-ACh was separated and measured using high-voltage paper electrophoresis, radiochromatogram scanning, sample oxidation and liquid scintillation spectrometry. It was also found that as the ^{3}H-Ch concentration is reduced, the percentage of total radioactivity found as ^{3}H-ACh is increased. Similar findings have been reported in other cholinergic systems (Dowdall & Simon, 1973; Yamamura & Snyder,1973; Simon et al., 1976). These observations suggest that in the carotid body, as elsewhere, the high-affinity component of Ch uptake is coupled to the synthesis of ACh. Furthermore, it is reasonable to conclude that the glomus cells are capable of ACh synthesis, since autoradiography of carotid bodies incubated in low concentrations of ^{3}H-Ch indicates that the glomus cells may be the principal site of the high-affinity component of Ch uptake. However, our data cannot rule out the possibility that other cellular con-stituents of the carotid body may also have this property.

THE RECEPTOR (GENERATOR) POTENTIAL OF NERVE ENDINGS

In vivo studies

In order to elucidate the complex mechanisms of impulse initiation in the carotid body, it was felt that it was essential to record the electrical events of the nerve endings. A first step in this direction was accomplished when a slow depolarization was recorded concomitantly with an increased discharge following either natural (asphyxia, acid) or chemical (sodium cyanide, ACh, potassium chloride) stimulation (Eyzaguirre, Leitner, Nishi & Fidone, 1970; Eyzaguirre & Nishi, 1974). This was accomplished by placing a non-polarizable electrode on a nerve filament at its entry into the glomus and another, similar electrode at the cut end of the nerve 0.8 cm away. Depolarization was evoked by asphyxia and by intravenous or intra-arterial injections of different agents, such as Tyrode's equilibrated with different gas mixtures, saline at pH 2, sodium cyanide and ACh. The mass receptor potential was graded and relatively insensitive to the action of local anesthetics such as procaine and dibucaine. It appeared to be local (non-propagated) since it decreased in amplitude if the proximal recording electrode (close to the glomus) was moved along the nerve away from the carotid body. It disappeared completely when the proximal electrode was 1.5 mm away from the glomus.

Since exogenous ACh depolarized the receptors with a latency consistently shorter than that of other stimuli (see also, Nishi & Eyzaguirre, 1971), the depolarization in this case probably resulted from *direct* action of ACh on the endings. The other agents may have acted *indirectly* through release of a transmitter substance (possibly ACh) from the glomus cells. Small doses of hemicholinium were more effective in antagonizing receptor depolarization and sensory discharges evoked by acid, flushing and sodium cyanide than they were in antagonizing the depolarization induced by ACh. Atropine and mecamylamine depressed or blocked effects induced by flushing, acid, sodium cyanide and ACh.

In vitro studies

As a further step in trying to elucidate the mechanisms responsible for the generation of chemosensory discharges, we studied the effects of Na^+, K^+, Ca^{2+} and Mg^{2+} on the 'resting' polarization of chemoreceptors and on the mass receptor potential evoked by either ACh or sodium cyanide (Eyzaguirre & Nishi, 1976). In other receptors, the generator (receptor) potential is dependent on the external concentration of these ions. Furthermore, some of our previous studies (where impulse discharges were recorded

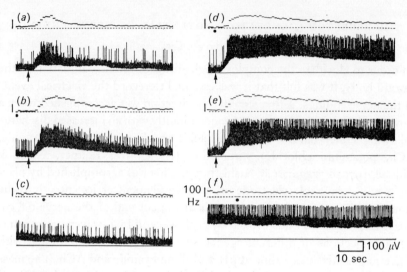

Fig. 2. Effect of eserine salicylate (10^{-6} w/v) on mass receptor potential and sensory discharge evoked by acetycholine (ACh) delivered at arrows. Upper traces, sensory discharge frequency; lower traces, receptor depolarization and impulse discharge. Receptor potentials and discharge frequency increase evoked by 50 μg ACh in (a) and 100 μg in (b). (c) follows immediately after (b). (d) Effect of 50 μg ACh during eserinization (10^{-6} w/v). (e) Effect of 100 μg ACh during eserinization. (f) follows immediately after (e). Dots indicate resetting of counter to 0 Hz. (From Eyzaguirre & Nishi, 1976.)

from the carotid nerve during different forms of stimulation) showed that Ca^{2+} and Mg^{2+} were important in the generation of chemoreceptor impulses (Eyzaguirre & Zapata, 1968c).

The carotid body and its nerve were removed from cats anesthetized with sodium pentobarbital and placed in an air gap system (resistance, 2–4 MΩ). Non-polarizable (Ag–AgCl) electrodes were placed across the gap between two chambers filled with Locke's solution. The carotid body was bathed in modified Locke's solution equilibrated with 50 % O_2 in N_2, pH 7.43 at 35 °C. The sensory discharges, changes in 'resting' receptor polarization and the mass receptor potential evoked by ACh or sodium cyanide were recorded simultaneously. Receptor potentials and sensory discharges evoked by ACh showed an appreciable increase in amplitude and frequency when the preparation was bathed in eserinized Locke's solution (Fig. 2). Eserine did not change appreciably the responses evoked by sodium cyanide. Excessive depolarization elicited by either ACh or sodium cyanide was accompanied by block of sensory discharges.

Removal of K^+ from the bathing solution induced receptor hyperpolarization and an increase in the amplitude of the evoked receptor potentials. An increase of K^+ concentration had the opposite effect. Reduction of Na^+ or sodium chloride to one half, or total removal of this salt, induced an initial reduction and later disappearance of the sensory discharges, some receptor hyperpolarization and a reduction in the amplitude of the evoked receptor potential.

Low or zero external Ca^{2+} produced receptor depolarization, a marked depression of the evoked receptor potentials, an increase in the frequency of the sensory discharges and a reduction in the amplitude of the nerve action potentials. High Ca^{2+} or Mg^{2+} had little or no effect on action potential amplitude or resting polarization, but decreased the sensory discharge frequency and the evoked receptor potentials. Total or partial replacement of Ca^{2+} with Mg^{2+} induced complex effects: (1) receptor depolarization which occurred in low Ca^{2+} was prevented by addition of Mg^{2+} ions; (2) the amplitude of the evoked receptor potentials was depressed; (3) the nerve discharge frequency was reduced as in high Mg^{2+} solutions; (4) the amplitude of the nerve action potentials was reduced as in low Ca^{2+} solutions.

INTRACELLULAR RECORDINGS FROM CAROTID BODY CELLS
in vitro

In these studies (Baron & Eyzaguirre, 1975, 1977; Eyzaguirre, Baron & Gallego, 1977), the carotid body of the cat was removed from the animal and placed in flowing mammalian saline at 36–37 °C. The nerve discharges were recorded from the carotid nerve with a suction electrode while glass micropipettes (10–40 MΩ) filled with 3M-KCl were used for intracellular recording. Thus, simultaneous records of intracellular activity, nerve discharges and bath temperature (monitored with a thermistor probe placed very close to the organ) were routinely obtained in each experiment. The impaled cells were identified by ejecting 6% Procion Navy Blue from the recording micropipette with direct cathodal current (2–4 nA). In the majority of impalements (90%) the cells were shown to be glomus or type I cells. In some instances, however, sustentacular or type II cells, and an occasional glomerulus (cluster of small cells) were stained. But the majority of the observations reported here refers to the behavior of single glomus cells.

The membrane potential (MP) of glomus cells (measured at 36–37 °C) varied from 10–55 mV, with a mean of about 20 mV. These low values may be due to injury of the cell membrane by the micro-electrode, since these cells are small, usually 10 μ in diameter. However, the low MPs may also be due to the ionic species involved in the maintenance of the potential. The input resistance (R_0) of these cells varied between 10 and 170 MΩ, with a mean value of 42 MΩ. These values are large, probably due to the small size of these cells. In fact, calculating the membrane resistance (R_m) from the mean R_0 value gives 132.5 Ωcm^2 which is, in fact, low compared with electrically excitable cells. The membrane capacity (C_m) gave a mean of 5.5 μF cm^{-2} $\pm$ 0.3, which is in line with values obtained by others using electrically excitable tissues.

Effects of temperature

Glomus cells proved to be extremely sensitive to temperature changes. Thus, a drop in temperature from 37 °C to 30 °C induced marked cell depolarization and a loss of R_0. These changes were accompanied by a marked reduction in sensory discharge frequency (Fig. 3). Also, the faster the cooling the larger the effects.

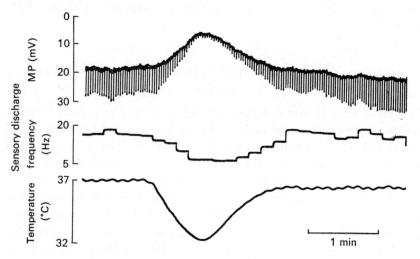

Fig. 3. Effect of cooling on the membrane potential (MP) and input resistance (R_0) of the glomus cell. Upper trace, R_0; middle trace, sensory discharge frequency; lower trace, temperature change. Initial MP was 18 mV and the initial R_0 was 40 MΩ. (From Baron & Eyzaguirre, 1977.)

In order to analyze the influence of different ions on the temperature effects, it was necessary to see whether or not depolarization by cooling had a reversal or equilibrium potential. For this purpose, the MP of the cells was displaced by currents injected through the micro-electrode. A systematic analysis made in 38 cells showed that there was a reversal potential for the cooling effect at about -5 mV. These results indicated that there was perhaps a simultaneous increase in conductance to more than one ionic species during a drop in temperature. Ionic substitution experiments were therefore conducted.

Removal of Na^+ from the bathing medium at normal (i.e. 37 °C) temperature induced cell hyperpolarization of about 4 mV and an increase in R_0 of about 25 MΩ. These changes still occurred in the cold (i.e. 30 °C). However, in sodium-free solutions, cooling was less effective in inducing cell depolarization. These results indicate that Na^+ ions are probably important in the maintenance of the cell MP at normal and low temperatures.

Ca^{2+} and Mg^{2+} ions proved important in the maintenance of MP and R_0 in these cells. Thus, removal of calcium from the bathing medium induced cell hyperpolarization of 3–4 mV and an unstable R_0 at normal temperatures. In the cold, lack of Ca^{2+} still induced cell hyperpolarization and an increase in R_0 of 4 to 5 MΩ. This is interesting since at low temperatures the instability in R_0, observed at normal temperatures, seems to disappear. One interpretation of this phenomenon is that lack of Ca^{2+} creates some membrane instability which allows other ions to move across the cell membrane. This then disappears when the temperature is lowered. An excess (10.8 mM) of Ca^{2+} or Mg^{2+} ions induced cell depolarization of about 4 mV in both cases and a decrease in R_0. The effects of Ca^{2+} were more pronounced than those of Mg^{2+}. At low temperatures only an excess of Ca^{2+} changed the MP and R_0 (in the same direction). Mg^{2+} ions were ineffective in the cold. The effects of Ca^{2+} on glomus cells, i.e. cell hyperpolarization in zero Ca^{2+} and depolarization in high Ca^{2+}, are quite different from what happens in electrically excitable tissues. Nerve and muscle cells are usually depolarized and their resistance becomes low in Ca^{2+}-free media.

Cl^- proved to be important in the maintenance of the MP of glomus cells. Thus, reducing the external Cl^- to 11.2 mM induced cell depolarization of about 8 mV and a decrease of R_0 of about 23 MΩ. These effects persisted at low temperatures, but in terms of MP changes were more marked in Cl^- than in normal solutions. Changes in R_0 due to cold were more marked in low Cl^- only in 66% of the cases.

Glomus cells showed remarkable insensitivity to changes in external K^+. Thus, removal of this ion from the bathing solution did not significantly change either the MP or R_0 of these cells. Also, adding an excess of K^+ (46.9 mM total concentration) only induced changes in MP and R_0 which could be accounted for by the reduction of Na^+ in the bathing solution. Reducing Na^+ was necessary, as shown later, to keep the osmolarity of the bathing fluid constant. Likewise, total replacement of Na^+ (154 mM) with K^+ induced MP and R_0 changes that were exclusively due to the removal of Na^+.

Ouabain (5×10^{-5} M) was used to investigate whether or not there is a Na^+ pump operating in these cells. This might be the case since ouabain induced cell depolarization at normal temperatures and this effect was potentiated during exposure of the cells to cold. However, the data are still too incomplete to postulate the presence of such a pump.

That the activity of glomus cells is highly dependent on temperature is not surprising since the carotid body tissue has a high metabolism, which is reflected in its high oxygen consumption (Daly, Lambertsen & Schweitzer, 1954; Leitner & Liaubet, 1971). The fact that changes in the MP and R_0 of glomus cells depend greatly on changes in temperature makes these cells good candidates to be the site of this high metabolic activity.

OSMOLARITY EFFECTS

When performing ionic substitution experiments it became clear that the carotid body is quite sensitive to osmotic changes. This prompted us to begin a systematic analysis of this phenomenon which later proved to have physiological implications (Gallego & Eyzaguirre, 1976).

Hypo-osmotic solutions were first prepared by removing part of the Na^+ salts from the control solution. This bathing medium had the disadvantage that part of the normal Na^+ content was reduced. To avoid this problem, control solutions were made with lower than normal Na^+ and the osmolarity was brought to normal levels (305 mosmol litre^{-1}) by addition of sucrose. Test solutions were then made hypo-osmotic by removing or reducing sucrose without changing the ionic composition of the medium. Hypo-osmotic solutions decreased the sensory discharge frequency and induced a clear increase in both MP and R_0. These effects were observed with a decrease in osmolarity of as little as 5 % of the control. More hypo-osmotic solutions (10 to 33 % of the control) induced greater increases in both MP and R_0. After exposing the preparations to hypo-osmotic solutions, a return to the control medium induced a slow recovery toward baseline levels. The fact that similar changes were observed by reducing either the Na^+ salts or sucrose in the bathing medium indicates that these changes were due to differences in osmotic strength and not to a reduction of Na^+ ions. Furthermore, the increase in MP and R_0 induced by hypo-osmotic solutions far exceeds similar changes induced by total removal of Na^+ ions. This phenomenon is still under investigation.

When the osmolarity of the test solution was increased by as little as 5 % by adding sucrose, sodium glutamate or glycerol, the sensory discharge increased while the MP and R_0 of the cells decreased. Using more hyper-osmotic solutions (15 % or more), the cells depolarized to values close to o mV in a few minutes. Concomitant with this change was a decrease in R_0 to very low values. The cells remained depolarized for the duration of flow of the hyperosmotic solution (up to 15 min). When the control solution was again allowed to flow, the MP and R_0 slowly recovered to control levels.

Osmotic changes may modify cell permeability to some ions. At present we do not know what ionic species are involved in this phenomenon. It may be that osmolarity variations change the membrane configuration, with consequent permeability changes; also, this effect may be partly due to movements of water across the cell membrane that may modify the intracellular concentration of some crucial ionic species.

EFFECTS OF 'NATURAL' STIMULI

In preparations bathed with modified Tyrode's solution equilibrated with 50% O_2 in N_2, cells were impaled with a micro-electrode and the nerve discharges were recorded with a suction electrode. After baselines were obtained, Tyrode's equilibrated with 6% CO_2 in 50% O_2 and 44% N_2 was allowed to flow through the preparation for 15 to 20 min.

Carbon dioxide had interesting effects. When the pH of the Tyrode's was buffered to 7.43 (identical to that of the control solution containing 50% O_2 in N_2), there was an increase in the frequency of the sensory discharges but no changes in either MP or R_0 (27 trials). However, when the pH of the bathing medium was allowed to fall to 6.0–6.6, there was a more marked increase in sensory discharge frequency, with a significant cell depolarization (3.6 mV) but with no significant changes in R_0 (50 trials). Thus, the effects of carbon dioxide on the MP appeared to be due exclusively to the fall in external pH. This was further tested by bathing the preparations in an acid medium (pH 6.5–6.8), which also induced a significant cell depolarization of 3.4 mV, although R_0 once again did not change significantly (27 trials). Conversely, when the pH of the bathing solution was allowed to increase to 8.2–8.5, there was an increase in MP by about 3.0 mV and no significant change in R_0.

These experiments indicate that H^+ ions do have a role in maintaining the cell MP. It is puzzling, however, that MP changes were not accompanied by R_0 changes. The reason for this apparent paradox is under investigation.

When flow of the bathing solution was interrupted, it invariably produced a marked increase in sensory discharge, as had been shown repeatedly with the *in vitro* studies. Usually, interruption of flow induced marked cell depolarization of about 15 mV and a loss of R_0 of about 49 MΩ. It is not known what mechanisms are involved in the response to lack of flow. However, when present, the phenomenon had a reversal or equilibrium potential of o to -8 mV, which may indicate simultaneous changes in permeability to more than one ion.

Results presented here are the beginning of an extensive study concerned with the properties of carotid body cells. They indicate that certain ions, such as Na^+, Ca^{2+}, Mg^{2+} and Cl^-, contribute to the maintenance of both MP and R_0. H^+ ions are also important since the cell membrane seems to be somewhat permeable to this ion. The latter, however, may only have an indirect role in modulating the permeability of the membrane to other ionic species. A number of questions still remains unresolved. For instance, acidity, flow interruption and hyperosmolarity induce cell depolarization and an increase in sensory discharge frequency. An increase in temperature has the same effect on the sensory discharge but it induces cell hyperpolarization and an increase in R_0. Carbon dioxide (at normal pH), which is a good stimulus of these receptors, does not change the MP and R_0 of the glomus cells. Other stimulating agents, such as nitrogen, sodium cyanide and ACh, while clearly increasing the sensory discharge frequency, do not change the resting MP of glomus cells in a consistent manner. Thus, at times, no changes are seen while in other instances a depolarization or hyperpolarization of the cells is observed (Eyzaguirre, Fidone & Nishi, 1975; Eyzaguirre et al., 1977; also, M. Baron, Y. Hayashida and C. Eyzaguirre, unpublished results). Some stimuli may have a different primary locus for their action than on the glomus cell. It is also possible that if the glomus cells are the primary transducer elements in the carotid body, their activity may not necessarily involve changes in MP and R_0. Instead, a situation may exist similar to that found at the neuromuscular junction where osmotic changes induce ACh release from the nerve endings without changes in either MP or R_0 (Hubbard, Jones & Landau, 1968).

In summary, our electrophysiological results have shown that heightened chemosensory nerve activity appears concomitantly with a receptor (generator) potential which has all the characteristics of a local, non-propagated response. We hope to analyze in depth the ionic components of this response and the effects of putative transmitters. Furthermore, our studies have shown that the membrane of the glomus cell has some interesting characteristics, but the role of this cell as the primary transducer element in the carotid body is still uncertain. Nonetheless, it does seem to have an essential role in the generation of chemosensory discharges. The crucial question is, how is this done? Is it release of a 'transmitter' substance, or is it conditioning of the nerve endings to become chemosensitive?

This work was supported by grants NS 05666, NS 12636 and NS 07938 from the US Public Health Service, and by a grant from the Utah Heart Association.

REFERENCES

Baron, M. & Eyzaguirre, C. (1975). Thermal responses of carotid body cells. *J. Neurobiol.* **6**, 521–7.

Baron, M. & Eyzaguirre, C. (1977). Effects of temperature on some membrane characteristics of carotid body cells. *Am. J. Physiol.* **233**(1), C35–46.

Biscoe, T. J. (1971). Carotid body: structure and function. *Physiol. Rev.* **51**, 437–95.

Biscoe, T. J., Lall, A. & Sampson, S. R. (1970). Electron microscopic and electro-physiological studies on the carotid body following intracranial section of the glossopharyngeal nerve. *J. Physiol., Lond.* **208**, 133–52.

Chiocchio, S. R., Biscardi, A. M. & Tramezzani, J. H. (1966). Catecholamines in the carotid body of the cat. *Nature, Lond.* **212**, 834–5.

Chiocchio, S. R., King, M. P. & Angelakos, E. T. (1971). Carotid body catechol-amines: histochemical studies on the effects of drug treatments. *Histochemie* **25**, 52–9.

Christie, R. V. (1933). The function of the carotid gland. I. The action of extracts of a carotid gland tumour in man. *Endocrinology* **17**, 421–32.

Cowan, W. M., Gottlieb, D. I., Hendrickson, A. E., Price, J. L. & Woolsey, T. A. (1972). The autoradiographic demonstration of axonal connections in the central nervous system. *Brain Res.* **37**, 21–51.

Daly, I. & Hebb, C. (1966). Innervation of the lungs. In *Pulmonary and bronchial vascular systems: their reactions under controlled conditions of ventilation and circulation*, Mongr. Physiol. Soc. (Lond.), ed. H. Barcroft, H. Davson & W. D. M. Paton, pp. 89–117. Baltimore, Maryland: Williams & Wilkins.

Daly, M. de B., Lambertsen, C. J. & Schweitzer, A. (1954). Observations on the volume of blood flow and oxygen utilization of the carotid body in the cat. *J. Physiol., Lond.* **125**, 67–89.

de Castro, F. (1926). Sur la structure et l'innervation de la glande intercarotidienne (glomus caroticum) de l'homme et des mammifères, et sur un nouveau système d'innervation autonome de nerf glossopharyngien. Etudes anatomiques et expérimentales. *Trab. Lab. Invest. biol. Univ. Madr.* **24**, 365–432.

de Castro, F. (1928). Sur la structure et l'innervation du sinus carotidien de l'homme et des mammifères. Nouveaux faits sur l'innervation et la fonction du glomus caroticum. Etudes anatomiques et physiologiques. *Trab. Lab. Invest. biol. Univ. Madr.* **25**, 331–80.

de Castro, F. (1951). Sur la structure de la synapse dans les chemocepteurs: leur mécanisme d'excitation et rôle dans la circulation sanguine locale. *Acta physiol. scand.* **22**, 14–43.

Dowdall, M. J. & Simon, D. J. (1973). Comparative studies on synaptosomes from squid optic lobes. *J. Neurochem.* **21**, 969–82.

Eyzaguirre, C., Baron, M. & Gallego, R. (1977). Intracellular studies of carotid body cells: effects of temperature, 'natural' stimuli and chemical substances. In *Tissue hypoxia and ischemia*, ed. M. Reivich, R. Coburn, S. Lahiri & B. Chance, pp. 209–23. New York: Plenum.

Eyzaguirre, C., Fidone, S. & Nishi, K. (1975). Recent studies on the generation of chemoreceptor impulses. In *The peripheral arterial chemoreceptors*, ed. M. J. Purves, pp. 175–94. Cambridge, New York & London: Cambridge Univ. Press.

Eyzaguirre, C. & Gallego, A. (1975). An examination of de Castro's original slides. in *The peripheral arterial chemoreceptors*, ed. M. J. Purves, pp. 1–23. Cambridge, New York & London: Cambridge Univ. Press.

Eyzaguirre, C., Koyano, H. & Taylor, J. R. (1965). Presence of acetylcholine and transmitter release from carotid body chemoreceptors. *J. Physiol., Lond.* **178**, 463–76.

Eyzaguirre, C., Leitner, L. M., Nishi, K. & Fidone, S. (1970). Depolarization of chemosensory nerve endings in carotid body of the cat. *J. Neurophysiol.* **33**, 685–96.

Eyzaguirre, C. & Nishi, K. (1974). Further study on mass receptor potential of carotid body chemosensors. *J. Neurophysiol.* **37**, 156–69.

Eyzaguirre, C. & Nishi, K. (1976). Effects of different ions on resting polarization and on the mass receptor potential of carotid body chemosensors. *J. Neurobiol.* **7**, 417–34.

Eyzaguirre, C., Nishi, K. & Fidone, S. (1972). Chemoreceptor synapses in the carotid body. *Fedn Proc. Fedn Am. Socs exp. Biol.* **31**, 1385–93.

Eyzaguirre, C. & Zapata, P. (1968a). The release of acetylcholine from carotid body tissues. Further study on the effects of acetylcholine and cholinergic blocking agents on the chemosensory discharge. *J. Physiol., Lond.* **195**, 589–607.

Eyzaguirre, C. & Zapata, P. (1968b). A discussion of possible transmitter or generator substances in carotid body chemoreceptors. In *Arterial chemoreceptors*, ed. R. W. Torrance, pp. 213–47. Oxford: Blackwell.

Eyzaguirre, C. & Zapata, P. (1968c). Pharmacology of pH effects on carotid body chemoreceptors *in vitro*. *J. Physiol., Lond.* **195**, 557–88.

Fidone, S. J., Stensaas, L. J. & Zapata, P. (1975). Sensory nerve endings containing 'synaptic' vesicles: an electron microscope autoradiographic study. *J. Neurobiol.* **6**, 423–7.

Fidone, S., Weintraub, S. & Stavinoha, W. (1976). Acetylcholine content of normal and denervated cat carotid bodies measured by pyrolysis gas chromatography/ mass fragmentometry. *J. Neurochem.* **26**, 1047–9.

Fidone, S., Weintraub, S., Stavinoha, W., Stirling, C. & Jones, L. (1977a). Endogenous acetylcholine levels in cat carotid body and the autoradiographic localization of a high affinity component of choline uptake. In *Function and functional significance of the carotid body*, ed. H. Acker, S. Fidone, D. Pallot, C. Eyzaguirre, D. Lübbers & R. Torrance, pp. 106–13. Springer-Verlag.

Fidone, S. J., Zapata, P. & Stensaas, L. J. (1977b). The origin of nerve terminals on glomus cells in cat carotid body: a study of axoplasmic movement of labeled material along sensory neurons of the petrosal ganglion. In *Function and functional significance of the carotid body*, ed. H. Acker, S. Fidone, D. Pallot, C. Eyzaguirre, D. Lübbers & R. Torrance, pp. 9–16. Springer-Verlag.

Fidone, S. J., Zapata, P. & Stensaas, L. J. (1977c). Axonal transport of labeled material into sensory nerve endings of cat carotid body. *Brain Res.* **124**, 9–28.

Gallego, R. & Eyzaguirre, C. (1976). Effects of osmotic pressure changes on the carotid body of the cat *in vitro*. *Fedn Proc. Fedn Am. Socs exp. Biol.* **35**, 404.

Gonzalez, C. & Fidone, S. (1977). Increased release of ^{3}H-dopamine during low O_2 stimulation of rabbit carotid body *in vitro*. *Neurosci. Lett.* **6**, 95–9.

Hess, A. (1968). Electron microscopic observations of normal and experimental cat carotid bodies. In *Arterial chemoreceptors*, ed. R. W. Torrance, pp. 51–6. Oxford: Blackwell.

Hess, A. & Zapata, P. (1972). Innervation of the carotid body: normal and experimental studies. *Fedn Proc. Fedn Am. Socs exp. Biol.* **31**, 1365–82.

Hubbard, J. I., Jones, S. F. & Landau, E. M. (1968). An examination of the effects of osmotic pressure changes upon transmitter release from mammalian motor nerve terminals. *J. Physiol., Lond.* **197**, 639–57.

Jones, J. V. (1975). Localization and quantitation of carotid body enzymes: their relevance to the cholinergic transmitter hypothesis. In *The peripheral arterial chemoreceptors*, ed. M. J. Purves, pp. 143–62. Cambridge, New York & London: Cambridge Univ. Press.

Leitner, L.-M. & Liaubet, M.-J. (1971). Carotid body oxygen consumption of the cat *in vitro*. *Pflügers Arch. ges. Physiol.* **323**, 315–22.

Matsuura, S. (1973). Chemoreceptor properties of glomus tissue found in the carotid region of the cat. *J. Physiol., Lond.* **235**, 57–73.

McDonald, D. M. & Mitchell, R. A. (1975). The innervation of glomus cells, ganglion cells and blood vessels in rat carotid body: a quantitative ultrastructural analysis. *J. Neurocytol.* **4**, 177–230.

McQueen, D. S. & Eyzaguirre, C. (1974). Effects of temperature on carotid chemoreceptor and baroreceptor activity. *J. Neurophysiol.* **37**, 1287–96.

Mitchell, R. A., Sinha, A. K. & McDonald, D. M. (1972). Chemoreceptive properties of regenerated endings of the carotid sinus nerve. *Brain Res.* **43**, 681–5.

Neil, E. & O'Regan, R. (1971). The effects of electrical stimulation of the distal end of the cut sinus and aortic nerves on peripheral arterial chemoreceptor activity in the cat. *J. Physiol., Lond.* **215**, 15–32.

Nishi, K. & Eyzaguirre, C. (1971). The action of some cholinergic blockers on carotid body chemoreceptors *in vivo*. *Brain Res.* **33**, 37–56.

Nishi, K. & Stensaas, L. J. (1974). The ultrastructure and source of nerve endings in the carotid body. *Cell Tiss. Res.* **154**, 303–19.

Osborne, M. P. & Butler, P. J. (1975). New theory for receptor mechanism of carotid body chemoreceptors. *Nature, Lond.* **254**, 701–3.

Sampson, S. R. (1972). Mechanism of efferent inhibition of carotid body chemoreceptors in the cat. *Brain Res.* **45**, 266–70.

Simon, J. R., Atweh, S. & Kuhar, M. J. (1976). Sodium-dependent high affinity choline uptake: a regulatory step in the synthesis of acetylcholine. *J. Neurochem.* **26**, 909–22.

Stensaas, L. J. & Fidone, S. J. (1977). An ultrastructural study of cat petrosal ganglia: a search for autonomic ganglion cells. *Brain Res.* **124**, 29–39.

Torrance, R. W. (1969). The idea of a chemoreceptor. In *The pulmonary circulation and interstitial space*, ed. A. P. Fishman & H. H. Hecht, pp. 223–37. Chicago: Univ. Chicago Press.

Verna, A., Roumy, M. & Leitner, L.-M. (1975). Loss of chemoreceptive properties of the rabbit carotid body after destruction of the glomus cells. *Brain Res.* **100**, 13–23.

Yamamura, H. & Snyder, S. (1973). High affinity transport of choline into synaptosomes of rat brain. *J. Neurochem.* **21**, 1355–74.

Zapata, P. (1975). Effects of dopamine on carotid chemo- and baroreceptors *in vitro*. *J. Physiol., Lond.* **244**, 235–51.

Zapata, P., Hess, A. & Eyzaguirre, C. (1969). Reinnervation of carotid body and sinus with superior laryngeal nerve fibers. *J. Neurophysiol.* **32**, 215–28.

Zapata, P., Stensaas, L. J. & Eyzaguirre, C. (1976). Axon regeneration following a lesion of the carotid nerve: electrophysiological and ultrastructural observations. *Brain Res.* **113**, 235–53.

Type J receptors in the gills of fish

G. H. SATCHELL

The term J receptor is an abbreviation of juxta pulmonary capillary receptor; it was introduced by Paintal in 1969. His earlier studies of vagal afferents in the cat (1955, 1957) had shown the existence of a group of fine fibres which were normally inactive, played no part in eupnoeic breathing, could sometimes be discharged by deflation and conducted at velocities mainly below 3 m sec^{-1}. Further study of these afferents was facilitated by the discovery that they could be discharged by the injection of the synthetic amidine drug, phenyl diguanide (PDG) into the right atrium and hence into the pulmonary circulation. The insufflation of volatile anaesthetics, such as halothane, into the lung also caused them to discharge. Measurements of the discharge latencies of individual fibres stimulated via these two routes led Paintal (1969) to conclude that the receptors were located in the alveolar interstitial tissue between the pulmonary capillary and the alveolar wall. Histological studies by Fillenz & Widdicombe (1972), Meyrick & Reid (1971) and Hung, Hertweck, Hardy & Loosli (1972) demonstrated the existence of fine nerve fibres and of structures suggestive of afferent endings in the alveolar wall.

Earlier work on the pharmacology of synthetic amidines and isothioureas had shown that in the cat they evoke bradycardia, hypotension and apnoea, followed by rapid shallow breathing (Dawes & Fastier, 1950; Dawes & Mott, 1950; Dawes, Mott & Widdicombe, 1951). The circulatory and respiratory responses were first thought to be two separate, vagally mediated reflexes; but Paintal (1955) was able to study them simultaneously with the display of vagal impulse activity and concluded that they were visceral components of a single reflex of which the afferent impulses travelled in these fine vagal fibres.

Further studies showed that the reflex had a somatic component. Both Deshpande & Devanandan (1970) and Schiemann & Schomburg (1972) reported that intra-right atrial PDG inhibited the monosynaptic reflex, an effect abolished by vagotomy. The latter authors also noted a reduction in lumbar fusimotor discharge, as also had Ginzel, Eldred, Watanabe &

Grover (1971). Schmidt & Wellhöner (1970) studied the γ efferent outflow to respiratory muscles and found that this was inhibited in nerves to both expiratory and inspiratory muscles. To this somatic component Paintal (1969) gave the term J reflex, and he developed the thesis (1969, 1970) that it served to protect the alveoli against oedema. The overactivity of venous muscle pumps in vigorous exercise could, he suggested, increase pulmonary capillary pressure to levels at which oedema threatened; the discharge of J receptors would then reflexly inhibit spinal motor output and diminish the action of muscle pumps. J receptors are known to be stimulated by agents and manoeuvres which all cause pulmonary oedema; examples are alloxan, microemboli and occlusion of the left atrioventricular junction. Later studies have added laryngeal constriction to the list of the somatic responses reflexly evoked by J receptor discharge (Stransky, Szereda-Przestaszewska & Widdicombe, 1973).

Paintal's hypothesis has not escaped criticism and Widdicombe (1974) has pointed out some of the gaps in our knowledge of the postulated sequence. There is no direct evidence that J receptors are activated in exercise or that exercise inhibits spinal reflexes. Widdicombe suggests that J receptors should be grouped, along with lung irritant receptors, as components of a more generalised pulmonary nociceptive system. Nevertheless, the contention that lung receptors can reflexly influence muscle tone is an interesting one, and extends the earlier observation of Schweitzer & Wright (1937) that electrical stimulation of the cat vagus inhibits the knee jerk and diminishes general muscle activity. Paintal's hypothesis might well be expected to apply to other classes of vertebrates and he suggested (1970) that a reflex arising from the gills of fish might inhibit swimming. The idea is an intriguing one, for the circulatory system of a fish differs from that of a mammal in ways which, it might be argued, make a J reflex of particular value. The 'in series' arrangement of the branchial and systemic circulations subjects the gills to the undiminished pressure of the single ventricle (Satchell, 1971). Venous pressures are very low because one third to one quarter of the pressure generated by the single ventricle is lost in the branchial circulation. No class of vertebrates is equipped with more varied and elaborate venous pumps (Birch, Carre & Satchell, 1969) and the movements both of swimming and of ventilation have been harnessed to operate muscle pumps which augment venous return. The arrays of secondary lamellae in fish provide closely spaced parallel pathways for blood and water flow. In such a system an increase in the thickness of the lamella wall by the accumulation of oedema fluid would not only increase the water–blood pathway; it would encroach on the channel between the lamellae through which the water flows. It was with ideas of this sort in mind that we, at Otago, decided to explore Dr Paintal's interesting suggestion.

THE RESPONSE OF THE DOGFISH *SQUALUS ACANTHIAS* TO PHENYL DIGUANIDE

Squalus acanthias has proved to be a convenient experimental animal, for it is easily caught and maintained in captivity and can be briefly anaesthetised by immersion in benzocaine 1:1000 w/v. Its circulatory anatomy is well described in elementary zoology texts. In this research, PDG was injected into the ductus cuvieri. From here the heart pumps it directly to the gills. When respiration and blood pressure were to be recorded the fish were restrained by two clamps on the dorsal fins. Fish held by a single clamp attached by rubber bands to a bar above the tank were able to swim; they did so for many hours at a time, remaining in the same position in the tank.

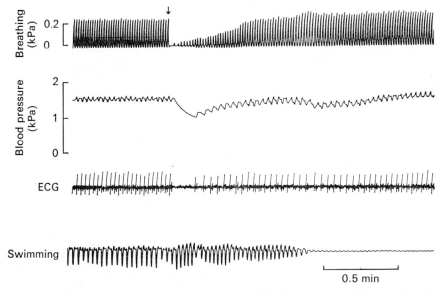

Fig. 1. The response of an unanaesthetised dogfish (*Squalus acanthias*) to phenyl diguanide. Traces from above down: breathing, measured as pressure in the first parabranchial cavity; dorsal aortic blood pressure; electrocardiogram (ECG); swimming. The upper three traces are from one stationary fish; the bottom trace has been aligned with the upper three and is from a separate freely mounted fish. At the arrow 60 μg kg^{-1} PDG was injected into ductus cuvieri.

When 50–200 μg kg^{-1} body weight of PDG, dissolved in dogfish saline, was injected into a dogfish there was an almost immediate apnoea, brady-cardia and hypotension (Fig. 1, Satchell, 1978). The apnoea, which

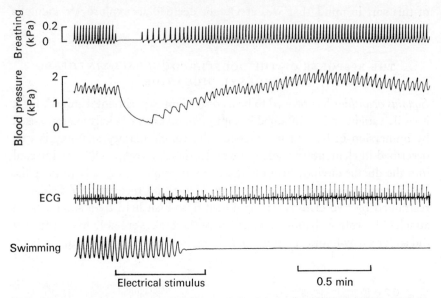

Fig. 2. The response of an unanaesthetised dogfish to electrical stimulation of the central end of the cut right vagus nerve. Traces from above down: breathing, measured as pressure in the first parabranchial cavity; dorsal aortic blood pressure; electrocardiogram (ECG); swimming. Horizontal bar on bottom left shows period of tetanic stimulation, 40 Hz, pulse width 0.1 msec.

resulted in the omission of two or three breaths, was followed by a period of shallow breathing. Heart rate fell from the resting rate of 38.4 ± 2.7 beats min^{-1} to 6.4 ± 0.7 beats min^{-1} and slowly recovered during a period of 15 min. Blood pressure, recorded in the dorsal aorta, fell from a resting level of 2.48 ± 0.31 kPa to 1.77 ± 0.3 kPa after 23 sec: it then recovered slowly and overshot the resting level. Prior injection of atropine, 0.14 mg kg^{-1} body weight, accelerated the heart and abolished the bradycardia; the fall of blood pressure caused by PDG was much less in atropinised fish but was still evident. This hypotension in the absence of bradycardia suggests that gut vessels are reflexly dilated, for elasmobranch fish lack sympathetic outflows to the skeletal muscles and skin.

In flexibly mounted specimens PDG in similar dosage inhibited swimming: the response was variable. Sometimes there was a reduction in the rate and intensity of trunk movements. Often there was a total inhibition of 2–3 min duration which occurred 0.75–1 min after injection. Sometimes the inhibition was rapid and dramatic. As apnoea commenced, the fish made two or three stronger strokes of the trunk and then slumped immobile in its harness only to recommence swimming a few minutes later.

These experiments thus established that in the dogfish, as in mammals,

PDG perfused through the respiratory circulation causes apnoea, brady-cardia, hypotension and an inhibition of motor activity. On the assumption that the receptors were likely to be in the pharynx the central ends of cut branchial nerves were electrically stimulated. A careful choice of stimulus parameters showed that the pattern of visceral and somatic inhibition elicited by PDG could be duplicated rather precisely with electrical stimulation (Fig. 2). Again the inhibition of swimming was variable in onset and duration. The fact that visceral and somatic inhibition could be elicited by stimulating the central ends of cut branchial nerves suggests that the afferent fibres ran in them and that the receptors were likely to be in the gills.

BRANCHIAL RECEPTORS

At Otago, C. A. Poole has developed an isolated, perfused dogfish gill preparation. A gill, with its afferent and efferent arteries and its nerve intact, is dissected from a stunned dogfish, supported in a moist, oxygenated chamber at 14 °C and perfused with elasmobranch saline. This is delivered through the afferent artery by a roller pump at a pressure of 3–5 kPa; drugs can be injected into the perfusion line. From the branchial nerve fine strands yielding unitary discharges can be dissected; such preparations last for at least 6 h and often longer.

Poole has focussed his attention on those fibres which discharge in response to an intra-arterial injection of PDG; some 200 units have been studied. The work will be reported in detail elsewhere (C. A. Poole and G. H. Satchell, unpublished). When first isolated they are found to be either inactive or discharging slowly (0–1 sec^{-1}). Most discharged in response to mechanical stimulation of the pharyngeal face of the gill filaments (Fig. 5). Some had a limited receptive field confined to part of one filament; others extended across several filaments. Fine mechanical stimulation precisely restricted to the secondary lamellae did not fire the receptors; the bristle had to make contact with the surface of the filament to elicit discharge.

An intra-arterial injection of PDG caused a burst of discharge at rates which could achieve 27 sec^{-1} but seldom exceeded 6 sec^{-1}. Discharge in all receptors responding to PDG was evoked by 5-hydroxytryptamine. When small islands of filter paper, 1 mm × 1 mm, which had been soaked in PDG and dried, were placed on the surface of the gill filaments, there was often a sustained discharge with a latency which varied, depending on the site of stimulation, from a few milliseconds to several seconds (Fig. 3). Control islands of filter paper elicited at most, only a few impulses. It was

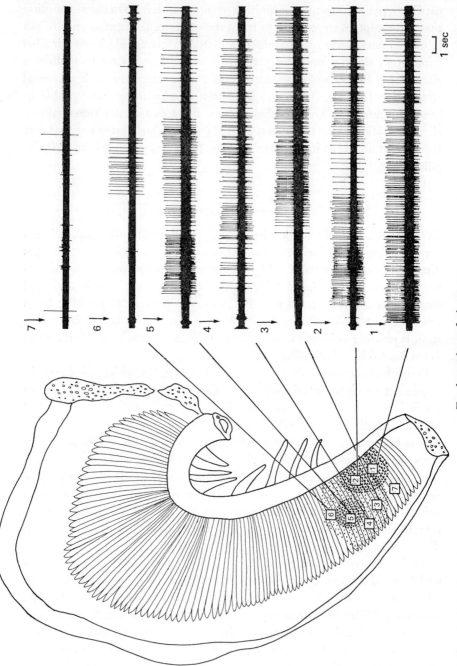

For legend see facing page.

thus possible to map the receptive fields of particular units. Often these proved to be very extensive and included as many as 15 filaments. Mechanical stimulation was only effective within the receptive field delineated by PDG. The great difference in the latencies of discharge to externally applied PDG suggests that some of the dendritic terminals are nearer to the surface of the epithelium than others.

When 10 mg of alloxan was injected intra-arterially (Fig. 4) there was a brief short-latency discharge like that caused by PDG and presumably due to the direct chemical stimulus of the drug. Within 1–2 min the perfusion pressure began to rise: as the pump delivered a constant volume the gill resistance must have increased. The secondary lamellae became noticeably oedematous as the pressure rose to 4–5 kPa. Some 1–5 min following the pressure rise (Fig. 4) the discharge rate began to increase; this increase was sustained for 15–30 min before returning to the resting level. Poole has recorded the response to alloxan of 24 units; a few showed less than a 1-min delay between the rise in pressure and the increased discharge. This may reflect only that there was an uneven perfusion and some filaments received alloxan ahead of the majority. The response of units to a direct increase in perfusion pressure caused by elevating the pump output was routinely assessed. Ten units were found to increase their discharge to this rise in pressure with delays of 10–20 sec.

Much remains to be learnt about these branchial afferent fibres. It is clear, however, that a single receptor may fire impulses in response to external mechanical stimulation, to PDG injected through the artery and to PDG applied on the surface of the gill filament. In addition it discharges to the chemical stimulus of alloxan and to the oedema it causes. How are we to picture the location and form of such a receptor? The fact that receptive fields include as many as 15 gill filaments suggests that the dendrites must branch extensively. Kempton (1969) reports small nerve bundles in both the septum and the gill filaments. Fig. 5 shows one possible arrangement amongst many, and serves only to focus attention on the problem. Silver-stained sections of gills prepared by Poole show fine nerve fibres close to the surface of the filaments and others in the core tissue between the upper and lower lamellae. Any one gill bears receptors on both its rostral and caudal

---

Fig. 3. The right fourth gill of a dogfish showing the sites of application of 1 mm-square pieces of filter paper previously soaked in phenyl diguanide, and the discharges of a branchial nociceptor they evoked. In the figure the islands have been shown enlarged. The receptive field has been stippled; the two darker areas yielded discharges of greater intensity. Site 7 is outside the receptive field and only the background discharge is observed. The arrows indicate the time of application of the filter paper squares.

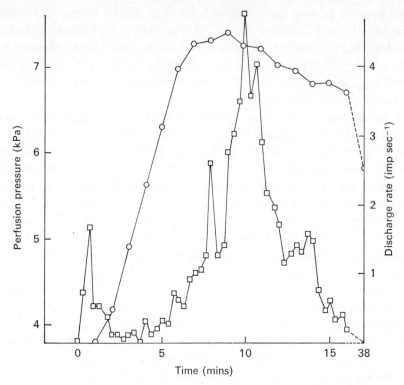

Fig. 4. Changes in the perfusion pressure of a dogfish gill, and of the discharge rate of a gill nociceptor in response to an injection of 10 mg alloxan. Pressure, O—O; discharge rate, □—□. Each point shows the mean rate over the previous minute.

faces and the possibility cannot be dismissed that a single receptive field might include areas on the two sides of the gill pouch, with dendrites in the pre- and post-trematic branches of a branchial nerve. The method of studying one gill at a time precluded examining this.

The ability of these receptors to discharge in response to PDG injected intra-arterially as well as applied to the gill surface was at first thought to parallel Paintal's (1969, 1970) findings that mammalian type J receptors can be fired both by PDG injected into the right atrium and by halothane insufflated into the lung. Perhaps in some gill receptors a dendrite is so placed as to be accessible to PDG from the blood stream and from the exterior. If this is so, a site somewhere in the central core at the base of the lamella is a possibility. However, the conclusion of the previous paragraph that the dendritic fields of these receptors are richly branched weakens this conclusion. There may well be a receptor with one dendrite close to a blood space and another beneath the surface of the filamentary epithelium.

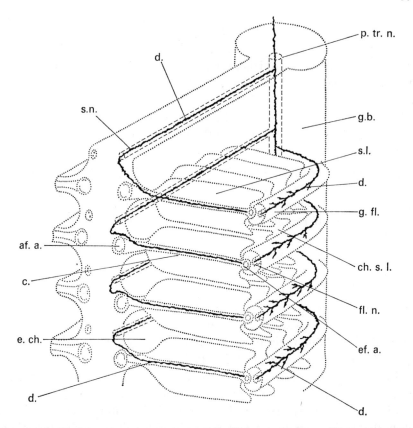

Fig. 5. Simplified drawing of part of a dogfish gill showing one possible arrangement of the dendrites of a branchial nociceptor. af.a., afferent filamentar artery; c., central core; ch.s.l., channels between secondary lamellae; d., dendrites of nociceptor; e.ch., efferent water channel; ef.a., efferent filamentar artery; fl.n., filamentar nerve; g.b., gill bar; g.fl., gill filament; p.tr.n., pre-trematic branch of branchial nerve; s.l., secondary lamella; s.n., septal nerve.

The variety of stimuli which will discharge these receptors suggests that they function as unspecialised nociceptors. The dendrites close to the surface of the gill filaments are well situated to monitor mechanical and chemical stimuli in the incoming water. Fish ventilation propels a current of water in one direction; it is not tidal. The gill filaments (Fig. 5) extend as a series of closely spaced parallel bars between which the respiratory stream must pass on its way to the channels between the secondary lamellae. The incoming water is thus bound to impinge against the filaments. In this aspect of their function these branchial nociceptors resemble the lung irritant receptors of mammals (Mills, Sellick & Widdicombe, 1969). Both lie in the pathway of the incoming respiratory medium.

In their response to intravascular PDG, 5-hydroxytryptamine and alloxan these receptors recall the type J receptors of mammals. Alloxan is known to cause pulmonary oedema by increasing the permeability of the alveolocapillary barrier (Aviado, 1965, Goetzman & Visscher, 1969). In cats it causes a delayed and sustained discharge of type J receptors (Paintal, 1969). Alloxan produced a visible oedema in the dogfish gill and a similar delayed discharge of these branchial receptors. Perhaps here also they serve to signal an increased hydration of the interstitial tissue of the lamella as oedema threatens.

We cannot of course be certain that the branchial receptors described here are those that mediate the reflex response to PDG in the unanaesthetised dogfish. We know only that both are discharged by PDG and that both send their afferent fibres up the branchial nerves. Apnoea and shallow breathing, bradycardia, hypotension and cessation of swimming together constitute an appropriate protective response to the arrival in the pharynx of irritant or toxic material. The inflow of water into the pharynx of a fish is commonly the result of its ventilatory pumping and of its forward movement through the water. There are many fish in which ventilation and locomotion are coupled (Satchell, 1968). Some ventilate when resting but successively inhibit their breathing movements as they gather forward speed. Others, such as tuna, have lost their capacity to ventilate and depend entirely on forward movement. Clearly, a reflex which serves to still the flow of water within the interlamellar spaces must inhibit both ventilation and locomotion. Furthermore, the 'in series' circulatory system makes it likely that toxic substances which reach the gills will be carried directly to the tissues, and, of particular importance, to the brain. Both the carotid and vertebral arteries arise directly from epibranchial vessels or their continuations. An inhibition of heart beat and a fall in blood pressure will temporarily slow the onward movement of the blood through the branchial circulation and diminish this hazard.

A case can thus be made that in the gills of dogfish there exists a primitive system of nociceptors which share some of the functions of the J receptors and of the lung irritant receptors of mammals. The nociceptors can, it seems, respond to mechanical and chemical stimuli in the inspiratory stream and to changes in the interstitial volume of the gills caused by oedema. The discharge of these receptors reflexly evokes visceral and somatic responses which inhibit circulation, ventilation and locomotion. We may speculate that it is from similiar, unspecialised nociceptors in primitive vertebrates that the specific nociceptors of the alveoli and airways of mammals have differentiated.

REFERENCES

Aviado, D. M. (1965). *The lung circulation.* New York: Pergamon Press.

Birch, M. P., Carre, C. G. & Satchell, G. H. (1969). Venous return in the trunk of the Port Jackson Shark, *Heterodontus portusjacksoni. J. Zool., Lond.* **159**, 31–49.

Dawes, G. S. & Fastier, F. N. (1950). Reflex actions of some isothiourea derivatives on circulation and respiration. *Br. J. Pharmac. Chemother.* **5**, 323–34.

Dawes, G. S. & Mott, J. C. (1950). Circulatory and respiratory reflexes caused by aromatic guanidines. *Br. J. Pharmac. Chemother.* **5**, 65–76.

Dawes, G. S., Mott, J. C. & Widdicombe, J. G. (1951). Respiratory and cardio-vascular reflexes from the heart and lungs. *J. Physiol., Lond.* **115**, 258–91.

Deshpande, S. S. & Devanandan, M. (1970). Reflex inhibition of monosynaptic reflexes by stimulation of Type J pulmonary endings. *J. Physiol., Lond.* **206**, 345–57.

Fillenz, M. & Widdicombe, J. G. (1972). Receptors of the lungs and airways. In *Handbook of sensory physiology*, ed. E. Neil, vol. 3, pp. 81–112. Berlin: Springer-Verlag.

Ginzel, K. H., Eldred, E., Watanabe, S. & Grover, F. (1971). Drug induced depression of gamma efferent activity – III. Viscero-somatic reflex action of phenyl-di-guanide, veratridine and 5-hydroxytryptamine. *Neuropharmacology* **10**, 77–91.

Goetzman, B. W. & Visscher, M. B. (1969). The effects of alloxan and histamine on the permeability of the pulmonary alveolocapillary barrier to albumin. *J. Physiol., Lond.* **204**, 51–61.

Hung, K. S., Hertweck, M. Z., Hardy, J. D. & Loosli, C. G. (1972). Innervation of pulmonary alveoli of the mouse lung: an electron microscopic study. *Am. J. Anat.* **135**, 477–96.

Kempton, R. T. (1969). Morphological features of functional significance in the gills of the spiny dogfish, *Squalus acanthias. Biol. Bull. mar. biol. Lab. Woods Hole* **136**, 226–40.

Meyrick, B. & Reid, L. (1971). Nerves in rat intra-acinar alveoli: an electron microscopic study. *Resp. Physiol.* **11**, 367–77.

Mills, J. E., Sellick, H. & Widdicombe, J. G. (1969). Activity of lung irritant receptors in pulmonary microembolism anaphylaxis and drug induced broncho-constrictions. *J. Physiol., Lond.* **203**, 337–57.

Paintal, A. S. (1955). Impulses in vagal afferent fibres from specific pulmonary deflation receptors. The response of these receptors to phenyl-di-guanide, potato starch, 5-hydroxytryptamine and nicotine and their role in respiratory and cardiovascular reflexes. *Q. Jl exp. Physiol.* **40**, 89–111.

Paintal, A. S. (1957). The location and excitation of pulmonary deflation receptors by chemical substances. *Q. Jl exp. Physiol.* **42**, 56–71.

Paintal, A. S. (1969). Mechanisms of stimulation of Type J pulmonary receptors. *J. Physiol., Lon.* **203**, 511–32.

Paintal, A. S. (1970). The mechanism of excitation of Type J receptors and the J reflex. In *Breathing* ed. R. Porter, pp. 59–76. Ciba Foundation Hering–Breuer Centenary Symposium. London: Churchill.

Satchell, G. H. (1968). A neurological basis for the co-ordination of swimming with respiration in fish. *Comp. Biochem. Physiol.* **27**, 835–41.

Satchell, G. H. (1971). *Circulation in fishes.* Cambridge: Cambridge Univ. Press.

Satchell, G. H. (1978). The J reflex in fish. In *Respiratory adaptations, capillary exchange and reflex mechanisms*, ed. A. S. Paintal & P. Gill-Kumar, pp. 432–41. Delhi: Vallabhbhai Patel Chest Research Institute.

Schiemann, B. & Schomburg, E. D. (1972). The inhibitory action of Type J pulmonary receptor afferents upon the central motor and fusimotor activity and responsiveness in cats. *Exp. Brain. Res.* **15**, 234–44.

Schmidt, T. & Wellhöner, H. H. (1970). The reflex influence of a group of slowly conducting vagal afferents on a and y discharges in cat intercostal nerves. *Pflügers Arch. ges. Physiol.* **318**, 333–45.

Schweitzer, A. & Wright, S. (1937). Effects on the knee jerk of stimulation of the central end of the vagus and of various changes in the circulation and respiration. *J. Physiol., Lond.* **88**, 459–75.

Stransky, A., Szereda-Przestaszewska, M. & Widdicombe, J. G. (1973). The effect of lung reflexes on laryngeal resistance and motor neurone discharge. *J. Physiol., Lond.* **231**, 417–38.

Widdicombe, J. G. (1974). Reflexes from the lungs in the control of breathing. *Recent Adv. Physiol.* **9**, 239–78.

Central actions of impulses in Pacinian corpuscles

P. T. YEO

The abundance of Pacinian corpuscles in the vicinity of the interosseous membrane of the cat's hindlimb and forelimb is well known (Hunt, 1961; Silfvenius, 1970). Impulses from these receptors, which are ultra-sensitive to high-frequency vibration (Hunt, 1961), ascend the rapidly conducting dorsal column and lemniscal pathway (Perl, Whitlock & Gentry, 1962; Petit & Burgess, 1968) to reach the somatosensory cortex (McIntyre, 1962).

At the spinal segmental level, little is known about the action of Pacinian corpuscles. They do not evoke a flexor reflex response or affect the size of test monosynaptic reflexes used to gauge reflex potency (McIntyre & Proske, 1968). However, antidromic testing using the method of Wall (1958) has revealed that a small input from Pacinian corpuscles decreases the electrical thresholds of the central terminals of afferent fibres from rapidly adapting cutaneous mechanoreceptors (Jänig, Schmidt & Zimmermann, 1968).

On the basis of their findings, these authors postulated that impulses from phasic mechanoreceptors preferentially depolarised the afferent terminals of other like receptors and that these effects were generated within a so-called phasic primary afferent depolarisation ('phasic PAD') system. The extent of such effects in relation to the receptive field origins of such fibres, however, is not clear.

The present study was undertaken to examine further the 'presynaptic' and postsynaptic effects, at the spinal level, of an input from Pacinian corpuscles. Presynaptic effects were studied by recording dorsal root potentials, testing for changes in excitability of the central terminals of certain afferent fibres and charting of target fibres selected for dorsal root reflex discharges. Effects at the postsynaptic level were deduced from observations of their actions on test monosynaptic reflexes and the excitability of group I afferent terminals.

METHOD

Adult cats were made acutely spinal by transection of the spinal cord at the first cervical segment and put on artificial respiration. The tidal volume was adjusted to give an expiratory carbon dioxide of 4–5 %, as monitored by a Beckman LB1 meter. The temperature of the animal's body was maintained at about 37 °C by an electric blanket (Electrophysiological Instrument Homeothermic blanket) wrapped closely around it. After the hips were fixed with pins, the lumbosacral spinal cord was exposed by a laminectomy extending from L5 to the S2 segments and covered with paraffin oil maintained at a temperature of 36–38 °C by the rays of an infrared lamp. The dorsal and ventral roots from L6 to S1 were freed and a few dorsal filaments, usually from L7 or S1 were dissected free of their pia arachnoid sheath and cut peripherally. The tail, left hip and left hindlimb were completely denervated, except for the interosseous nerve in the leg. The ankle was then fixed with bone pins and the exposed tissues immersed in a paraffin bath fashioned from the hindlimb skin flaps. The nerves that were prepared included the sural, that to the posterior knee joint and the nerves supplying the flexor digitorum longus, triceps surae, plantaris and tibialis posterior muscles.

Stimulation and recording techniques

Pacinian corpuscles innervated by the interosseous nerve were selectively stimulated by weakly vibrating the medial malleolus with a blunt probe which was attached to a Goodmans V47 vibrator; the amplitude of vibration was adjusted at between 20–50 μm and its frequency from 300–500 Hz.

For stimulation of peripheral nerves, platinum or silver–silver chloride electrodes were used. A gated pulse generator (Devices Digitimer D4030) provided the master trigger pulse to drive a pulse position unit (Neurolog Delay Width D402) and a pulse buffer unit (Neurolog Pulse Buffer NL510). Square wave pulses were obtained from isolation units (Neurolog NL800) capable of delivering constant current outputs with a choice of maxima of 10 and 100 μA and 1 and 100 mA, respectively. The cathode used for the antidromic stimulation of primary afferent terminals was a fine tungsten needle, insulated with varnish except for its tip. With the aid of a micromanipulator, it was advanced slowly into the spinal cord at the dorsal root entry zone of the mid L7 segment to a depth of 2–3 mm. Current was applied across the anode which was a platinum electrode placed on the back muscles. Nerve action potentials were amplified by a Field Effect Transistor probe unit with a gain of 10 before being fed differentially into the 3A9 amplifier of a Tektronix RM 565 oscilloscope. Signals from the RM 565 oscilloscope were simultaneously stored on magnetic tape for the purpose of later replay by a FM Tandberg Instrumentation recorder and additionally displayed on a Tektronix D13 Dual Beam oscilloscope from which responses could be photographed by a Polaroid camera.

Dorsal root potentials (DRPs) were led off from the cut dorsal root filaments of L7 or S1 segments, care being taken to make sure the proximal lead did not touch the cord dorsum. The signal to noise ratio of the small DRP signals was improved by averaging 25–100 of such responses with a computer of average transients (Mnemotron, C.A.T. 400B). Where possible the input signals were d.c. coupled; otherwise a.c. coupling with a time constant of 0.35 sec was used. The analysis time of the C.A.T. was set at either 125 or 250 msec. The signal of a satisfactorily averaged DRP was written out on graph paper by a Moseley 7030A X–Y plotter.

Oscilloscope displays of both monosynaptic reflex responses and antidromic afferent discharges, either unconditioned or conditioned by a burst of vibration, were filmed directly by a Grass Kymograph camera. Effects due to the vibratory stimulus were measured by calculating the ratio of the averaged amplitude of 20–30 conditioned responses to that of an identical number of test responses.

RESULTS

Dorsal root potentials

Following a burst of vibration applied to the medial malleolus, DRPs distributed over L6 to S1 segments (but most prominently found in L7) were consistently seen. The maximum amplitude of depolarisation varied from 100 to 200 μV, with an occasional one measuring 600 μV.

Despite biological variations in animals, the time course of DRPs recorded from different experiments was remarkably constant (Fig. 1). Depolarisation began 10–15 msec after onset of vibration, reached maximum amplitude at 35–40 msec and then declined into a hyperpolarisation at 100–200 msec. The maximum amplitude of the hyperpolarisation was never more than one quarter that of the depolarisation. In a few experiments the hyperpolarisation was hardly detectable.

Within any one experiment, as long as the amplitude of vibration was greater than about 15 μm, the time course of the DRPs remained relatively constant even when frequency and duration of vibration were altered. For the experiment, the records of which are illustrated in Fig. 2a, the duration of vibration was 100 msec, its amplitude was about 20 μm and 10 frequencies between 100 to 1000 Hz in steps of 100 Hz were used. The 10

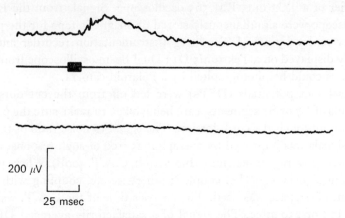

200 μV

25 msec

Fig. 1. Dorsal root potential (DRP) in response to a burst of vibration applied to the medial malleolus. Top Trace: record of 25 DRP responses (in a L7 dorsal root filament) as averaged by the Mnemotron 400B computer of average transients. Middle Trace: bar represents a burst of vibration with amplitude of 20 μm, frequency of 300 Hz and duration of 10 msec. Lower Trace: absence of response to an identical vibratory stimulus, 10 min after sectioning of the interosseous nerve.

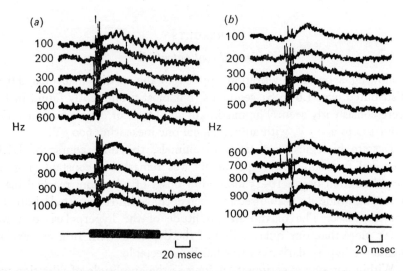

Fig. 2. Lack of effect of duration and frequency of vibration on the time course of the dorsal root potential (DRP) response. (a) shows 10 DRP records, each being a single response to a 100-msec burst of vibration (represented by bar) at the frequencies indicated on the left-hand column. (b) shows 10 similar DRP responses following the application of one cycle of vibration (represented by bar) at the frequencies indicated. The duration of vibration ranged from 10 msec (for 100 Hz) to 1 msec (for 1000 Hz).

corresponding DRP responses thus obtained were very similar in their time course and the maximum depolarisation. About 15 msec after the onset of vibration, depolarisation, followed closely by strong dorsal root reflex discharges, was clearly evident. Maximum depolarisation occurred at the interval of 40 msec. Hyperpolarisation was not always seen.

A set of 10 similar DRPs is illustrated in Fig. 2b. For each DRP response only one cycle of vibration for each of the frequencies indicated was applied; thus the duration of vibration ranged from 10 msec (for 100 Hz) to 1 msec (for 1000 Hz). The lack of effect of duration of vibration on the time course of the DRP response is demonstrated by the close similarity of this set of DRP responses to that shown in Fig. 2a.

That an input from the interosseous nerve was crucial in generating the DRP was shown by comparing the effects of vibration before and after sectioning the interosseous nerve in one experiment; the results are shown in Fig. 1. The upper trace shows the DRP response averaged from 25 trials, each produced by a 10-msec burst of vibration at 300 Hz (middle trace). The interosseous nerve was then sectioned and after all injury discharges had died down, the ankle was similarly vibrated 25 times. As illustrated in the lower trace of Fig. 1 no more DRP responses were obtained. Further systematic search failed to demonstrate any DRP responses in the ipsilateral dorsal root filaments from the L5 to S1 segments.

Dorsal root reflexes

Often in the course of recording DRPs, dorsal root reflex discharges occurring in the initial part of the slow DRP were seen. That such antidromic discharges were preferentially distributed to afferent fibres from certain receptive fields was confirmed when their destination was systematically searched for by recording from various nerves peripherally, when a burst of vibration was applied.

Typical results are shown in Fig. 3. In response to a 50-msec burst of vibration at a frequency of 200 Hz and amplitude of 20 μm, a strong discharge was recorded in the sural nerve and a weaker one, consisting of a few action potentials, in the posterior knee joint nerve. The interval between the onset of vibration and the appearance of the dorsal root reflex in both instances was about 15 msec; the discharge lasted about 20 msec. Such discharges were not seen in hindlimb muscle nerves, including those to the popliteus, plantaris, tibialis posterior, triceps surae and the flexor digitorum longus muscles. Absence of discharge in the latter two nerves is illustrated in the lower two traces of Fig. 3.

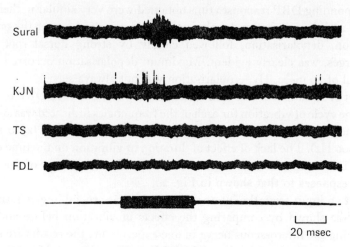

Fig. 3. Dorsal root reflexes. The vibration represented by the bar in the lowest trace lasted 50 msec; its frequency was 200 Hz and the amplitude 20 μm. The strongest discharges were recorded in the sural and posterior knee joint (KJN) nerves. No discharges were found in the hindlimb muscle nerves, of which that to the triceps surae (TS) and flexor digitorum longus (FDL) muscles are shown.

Changes in the excitability of primary afferent terminals

In concordance with the recording of DRPs and with the findings that dorsal root reflexes were preferentially distributed, antidromic testing revealed an increase in excitability of the intraspinal projections of the fastest conducting afferent fibres in the sural and posterior knee joint nerves.

Fig. 4 illustrates the time course of changes in excitabilities of fibre groups represented by the first and second elevations in the antidromic compound action potentials of the sural and posterior knee joint nerves, respectively. The most profound changes were detected in the fastest afferent group of the sural nerve. For this group, an increase in excitability was detected some 15 msec after the onset of a 30-msec burst of vibration at 500 Hz. It reached a maximum of 220 % of the control value at 37 msec and then gradually returned to the control value at 100 msec. An identical burst of vibration produced in the fastest afferent group of the posterior knee joint nerve, similar but much smaller excitability changes. Increased excitability was apparent after an interval of 10 msec, reached a peak of 125 % of the control value at about 20 msec and then declined to the control value at about 60 msec.

No changes were seen when the excitabilities of the second fastest conducting afferent groups of both sural and posterior knee joint nerves were

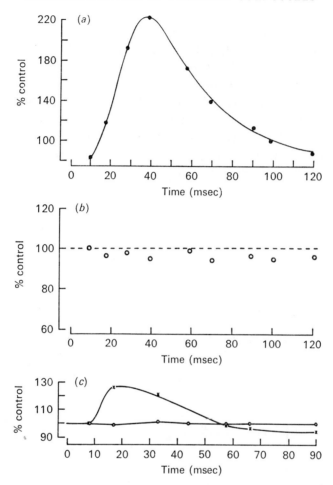

Fig. 4. Time course of changes in excitabilities of afferent fibre groups in the sural and posterior knee joint nerves. (a) shows the changes in the fastest fibres of the antidromic compound action potentials of the sural nerve; (b) shows the changes in the slow fibres (represented by the second elevation in the antidromic compound action potential) of the sural nerve; (c) shows the changes in the first and second elevations of the antidromic potential of the posterior knee joint nerve. Fast fibres, ×—× ; slow fibres, O—O.

tested. Likewise no significant changes in all afferent fibre groups of muscle nerves were detected for conditioning intervals up to 110 msec. This pattern of effect is thus similar to that obtained for the preferential distribution of dorsal root reflexes.

Effects on reflex transmission

Weak vibration consistently failed to affect the size of test monosynaptic reflexes of the muscles of the leg. The results for the conditioned reflex responses of the flexor digitorum longus and triceps surae muscles have been plotted in Fig. 5. Each point on the plots was calculated by comparing the average amplitude of 20–30 test reflex responses, each conditioned by a 30-msec burst of vibration at 500 Hz, to that of an equal number of controls. For intervals (after the onset of vibration) up to a maximum of 140 msec, small changes were observed but these were within the range of normal variations.

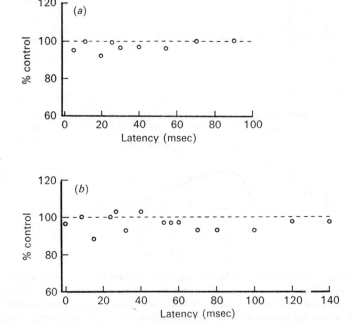

Fig. 5. Effects of vibration on the monosynaptic reflex responses of (*a*) the flexor digitorum longus muscle and (*b*) the triceps surae muscle. Each point on the plot was calculated by comparing the average amplitude of 20–30 test reflex responses, each conditioned by a 30-msec burst of vibration at 500 Hz, to that of an equal number of control reflex responses.

Similar conditioning of the monosynaptic reflex of other hindlimb muscles, including the posterior biceps-semitendinosus, peroneal group, plantaris and tibialis anterior, also failed to reveal any effects. Thus the failure of weak vibration to affect the monosynaptic reflex response was not a local phenomenon.

DISCUSSION

Significance of the dorsal root potential

Reference to Fig. 2 shows that DRPs consistently appeared even with the very weak stimulus of 1 Hz of vibration at the frequency of 1000 Hz. With this stimulus, it was likely that only a very few, possibly just one Pacinian corpuscle, would have been excited. This strongly suggests that the neuronal mechanism generating the DRP needs only a minimal input to gain synaptic transmission. Somewhat similar findings from studies of mechanoreceptors of the cat's foot pad have been reported by Jänig et al. (1968) who found that even a single impulse from a Pacinian corpuscle sufficed to cause a detectable decrease in the electrical threshold of afferent terminals. Furthermore, the fibres depolarised originated from Pacinian corpuscles or other rapidly adapting receptors and the amount of depolarisation was graded according to strength of the stimulus.

The observation that the time course and amplitude of the DRP response was insignificantly altered despite major differences in the vibratory stimulus indicated that once initiated, the process generating the DRP tends to persist for a considerable time and is somewhat refractory to further stimuli from the same input.

A possible neuronal substrate mediating this process is a group of dorsal horn neurones which need for their self-excitation a minimal afferent input from Pacinian corpuscles. Their prolonged action is possibly due to the persistence of transmitters and perhaps the products of transmitter actions, such as K^+ (Curtis & Johnson, 1974). Such a suggestion, however, implies that near their central terminals, afferent fibres from Pacinian corpuscles must give off collaterals to synapse with the spinal neurones vital to the genesis of the DRP. This view is at variance with that of Wall (1973) who stated that afferent fibres from Pacinian corpuscles ascend the dorsal columns without giving off collaterals.

The significance of the DRP is its preferential distribution, as revealed by the antidromic testing of excitability changes, to the rapidly conducting afferent fibres of the sural and the posterior knee joint nerves. (By inference it could also be assumed that the dorsal root reflex discharges occurred in the same afferent groups.) No information on the receptor origin of these afferent fibres was obtained; however, it might be expected from the 'phasic PAD' system of Jänig et al. (1968) that they originated from rapidly adapting mechanoreceptors. Such receptors are known to be present in both the knee (Burgess & Clark, 1969) and the skin (see review by Burgess & Perl, 1973). It is possible that through these mechanisms an input from the Pacinian corpuscles serves to limit the centripetal flow of information from other phasic mechanoreceptors and thereby favours the channelling of more relevant sensory data to the central nervous system.

Significance of the lack of effect of a vibratory stimulus on transmission

By itself, the technique of gauging reflex potency by monosynaptic reflex testing cannot distinguish between effects on the afferent or efferent limb of the reflex pathway. Thus, failure of vibration to affect reflex transmission could have been due either to a genuine lack of effect on both the afferent and efferent components or it could have been due to opposing presynaptic and postsynaptic actions which nullify each other.

However, antidromic testing of excitability changes in the central terminals of muscle afferent fibres showed that all the large fibres were not affected by a vibratory stimulus. It could therefore be concluded that no actions were exerted on the Ia fibres during monosynaptic testing.

By exclusion, it could also be reasonably concluded that no effects were exerted on the motoneurones involved in the reflex. This is in keeping with the observation that fibres from Pacinian corpuscles do not project to the ventral horn of the spinal grey matter.

The author wishes to thank Professor C. S. Seah, Head of the Department of Medicine (III), Singapore General Hospital for reading the manuscripts. He would also like to thank Mrs Louise Yeo for typing the manuscripts, and Mr Henry Lee for photographic assistance.

The materials contained in this article partially fulfilled the requirements of a Ph.D. thesis submitted to Monash University. This work was made possible by a medical postgraduate scholarship from the National Health and Medical Research Council of Australia.

REFERENCES

Burgess, P. R. & Clark, F. J. (1969). Dorsal column projection of fibres from the cat knee joint. *J. Physiol., Lond.* **203**, 301–15.
Burgess, P. R. & Perl, E. R. (1973). Cutaneous mechanoreceptors and nociceptors. In *Handbook of sensory physiology*, ed. A. Iggo, vol. II, pp. 30–78. Berlin: Springer-Verlag.
Curtis, D. A. & Johnson, G. A. R. (1974). Amino acid transmitters in the mammalian central nervous system. *Ergebn. Physiol.* **69**, 98–188.
Hunt, C. C. (1961). On the nature of vibration corpuscles in the hindlimb of the cat. *J. Physiol., Lond.* **155**, 175–86.
Jänig, W., Schmidt, R. F. & Zimmermann, M. (1968). Two specific feedback pathways to the central afferent terminals of phasic and tonic mechanoreceptors. *Expt. Brain Res.* **6**, 116–29.
McIntyre, A. K. (1962). Cortical projection of impulses in the interosseous nerve of the cat's hindlimb. *J. Physiol., Lond.* **163**, 49–60.
McIntyre, A. K. & Proske, U. (1968). Reflex potency of cutaneous afferent fibres. *Aust. J. exp. Biol. med. Sci.* **46**, 19.
Perl, E. R., Whitlock, D. G. & Gentry, J. R. (1962). Cutaneous projections to second order neurones of the dorsal column system. *J. Neurophysiol.* **25**, 337–58.

Petit, D. & Burgess, P. R. (1968). Dorsal column projection of receptors in hairy skin supplied by myelinated fibres. *J. Neurophysiol.* **31**, 849–55.

Silfvenius, H. (1970). Characteristics of receptors and afferent fibres of the forelimb interosseous nerve of the cat. *Acta physiol. scand.* **79**, 6–23.

Wall, P. D. (1958). Excitability changes in afferent fibre terminations and their relation to slow potentials. *J. Physiol., Lond.* **142**, 1–21.

Wall, P. D. (1973). Dorsal horn electrophysiology. In *Handbook of sensory physiology*, ed. A. Iggo, vol. II, pp. 253–70. Berlin: Springer-Verlag.

Some methods for selective activation of muscle afferent fibres

J. J. B. JACK

'The task of discussing central actions of muscle afferent impulses would be greatly simplified if the different classes of receptors were served by separate and distinguishable groups of afferent axons, and if there were completely reliable means of selectively stimulating each receptor-type or its own particular group of afferent fibres.'

[McIntyre, 1974]

In this essay I would like to outline some work which has been concerned with developing methods for selectively stimulating proprioceptive afferents from muscle, with the aim of studying their central actions. The account falls naturally into four parts. In the first stage an analysis was undertaken of the types of receptors represented in the large afferent fibres coming from several cat leg muscles. The next part is an assessment of some methods developed for selective stimulation, by electrical means, of the three main types of large afferent fibres from muscle. Following this is a description of methods available for changing the firing frequency of one type of muscle proprioceptor when the other proprioceptors preserve a constant firing frequency. Finally a brief account is given of the application of some of these methods to the determination of the relative importance of muscle spindle (groups Ia and II) and Golgi tendon organ (group Ib) afferent fibres in stretch and vibration reflexes in the soleus muscle of the decerebrate cat.

LARGE AFFERENT FIBRE COMPOSITION OF SOME CAT LEG MUSCLE NERVES

In a series of experiments by (or with) C. M. L. Coppin, A. K. McIntyre and Catherine MacLennan, cat muscle nerves innervating the semitendinosus, medial gastrocnemius, soleus, peroneus longus and tibialis anterior muscles have been studied. In each case the technique employed was to electrically stimulate the relevant muscle nerve and isolate an individual afferent fibre in a teased dorsal root filament, measure the conduction velocity and electrical excitability of that afferent and also attempt to

Table 1. *The classification and properties of large afferent fibres in the nerves innervating five leg muscles of the cat*

Muscle nerve	Total in sample	Types[1] and number of receptors		Conduction velocity range (m sec^{-1})	Threshold range[2]	% of group I recruited with 5% of group II	% of group II recruited with 100% group I
Semitendinosus[3] (long branch)	167 (11 cats) 98.2% identified	Spindle primary	78	117–74	0.96–1.80	94.5% Ia	
		Tendon organ	51	106–74	1.24–1.90		21%
		Spindle secondary	35	79–28	1.51–5.54	75.5% Ib	
		Unidentified	3	100, 98, 74	1.12, 1.34, 1.67		
Soleus[4]	308 (16 cats) 98.7% identified	Spindle primary	137	120–58.3	0.93–2.58	94% Ia	
		Tendon organ	104	100–61.9	1.03–2.46		35%
		Spindle secondary	61	72.6–21.8	1.40–10.64	86% Ib	
		Paciniform type	2	58.1, 50.9	1.71, 1.73		
		Unidentified	4	99.4, 99.4, 39.2, 31.9	1.37, 1.44, 3.11 3.36		
Peroneus longus[5]	236 (10 cats) 98.3% identified	Spindle primary	113	124.1–69	0.96–2.86	93% Ia	
		Tendon organ	60	124–61.1	1.0–2.63		45%
		Spindle secondary	59	70.5–20	1.83–7.94	87% Ib	
		Unidentified	4	104, 98.9, 97.5, 65.8	1.38, 1.02, 1.25 1.5		
Medial gastrocnemius[6]	110 (8 cats) All identified	Spindle primary	45	118–71	0.99–1.60	95% Ia	
		Tendon organ	41	117–84	1.00–1.71		20%
		Spindle secondary	23	71–31	1.55–6.29	93% Ib	
		Paciniform type	1	69.9	1.68		

Tibialis anterior[7]	339 (8 cats) 96.5 % identified	Spindle primary	117	108.6–72.4	0.82–2.30	87 % Ia
		Tendon organ	60	103.6–54.8	1.1–2.61	
		Spindle secondary	88	71.7–22.6	1.59–11.76	87 % Ib 46 %
		Various mechanoreceptors	62	101.1–34.9	0.98–4.15	
		Uncertain or unidentified	12	110.1–37.8	0.97–3.53	

[1] Spindle fibres have been classified into primary or secondary, not on the basis of the absolute conduction velocity of their afferent fibre (i.e. above or below 72 m sec^{-1}), but on their conduction velocity relative to the fastest fibres in the nerve (i.e. above or below 0.65 of the fastest fibre). The reason for this is that there were large variations in the maximum conduction velocity (see Coppin, Jack & MacLennan, 1970), perhaps related to the fact that many of the animals used were not fully grown.

[2] Thresholds are expressed relative to the lowest threshold fibre in a fraction of the dorsal root. This means that occasionally fibres are isolated which have a lower threshold than any in the monitor fraction.

[3] Coppin, Jack & McIntyre (1969).

[4] Coppin, Jack & MacLennan (1970).

[5] Jack & MacLennan (1971).

[6] Coppin (1973).

[7] MacLennan (1971, 1972).

identify the type of receptor from which it originated. The main questions to which the experiments were addressed were as follows.

(1) What receptor types are represented in the group I and group II afferent fibre conduction velocity range?

(2) What is the range and distribution of electrical thresholds (relative to the lowest threshold) of the afferent fibres coming from the different types of receptor?

Table 1 summarizes some of the results obtained. It may be seen that for the first four muscle nerves listed in the table almost all the fibres isolated (> 98 %) came from an identifiable receptor, the overwhelming majority from spindles or tendon organs, as had been shown earlier for the soleus and medial gastrocnemius muscles (Hunt, 1954). It is tempting to suggest that the failure to identify this small percentage of fibres is due to possible damage to these axons when dissecting the muscle nerve. If this were so, then it appears on the basis of all these results that the only receptors represented in the fibres of conduction velocity in the group I and II range are muscle spindles and Golgi tendon organs, with the exception of occasional 'Paciniform-type' receptors. One qualification to this conclusion is that, in all the muscle nerves listed in Table 1, the group II afferent fibre range has not been sampled as extensively as the group I range. Barker (1962; see also, Boyd & Davey, 1968; Matthews, 1972) has given evidence that there should be roughly as many secondary spindle afferents as primary afferents, whereas in Table 1 the number of group II spindle fibres may be as little as half the number of group Ia fibres.

In any case it is clear that these results cannot be generalized to all muscles, as shown by the work of Hunt & McIntyre (1960) on the nerves to the flexor digitorum longus and flexor hallucis longus and by the data for the nerve to tibialis anterior muscle included in Table 1. This was selected for study by MacLennan because Boyd & Davey (1968) had reported that it contained a large excess of afferent fibres over those expected on the basis of spindle and tendon organ counts in the muscle. In accord with the histological data only 78 % of the large muscle afferents were identified as coming from spindles or tendon organs. MacLennan found that most of the remaining fibres could be activated by mechanical stimuli of various kinds, with some grouping of the response types. Further details of the mechano-receptor types will be found in MacLennan (1971, 1972). For present purposes it is sufficient to note that this muscle would be an inappropriate choice in which to attempt selective activation of any of the three main proprioceptor fibre types by electrical or mechanical means.

SELECTIVE ELECTRICAL ACTIVATION OF GROUP Ia, Ib AND GROUP II FIBRES

Group Ia fibres

The pioneering study of Sumner (1961) on the electrical thresholds of group Ia and Ib afferent fibres in the nerve to medial gastrocnemius muscle, as well as the subsequent data collected in Table 1, give some quantitative justification to the widespread assumption (see, for example, Eccles, 1962) that the fibres of lowest electrical threshold in muscle nerves are the group Ia fibres. However, there turns out to be great variation within the group I range, in the distribution of excitabilities of spindle and tendon organ fibres. Fig. 1 illustrates this in more detail, by showing the percentage of tendon

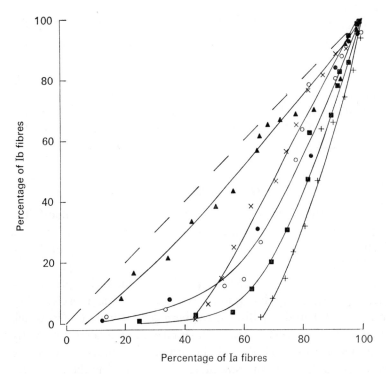

Fig. 1. Data for the muscle nerves listed in Table 1. The abscissa is the percentage of Ia fibres excited by stimuli of various strengths and the ordinate is the percentage of Ib fibres. Thus each point shows the effect of the same stimulus on both Ia and Ib fibres. ▲, peroneus longus; ●, ○, medial gastrocnemius (●, data of Sumner, 1961; ○, data of Coppin, 1973); ■, soleus; ×, tibialis anterior; +, semitendinosus. Note that in the case of the tibialis anterior the recruitment of other large afferent fibres is ignored (see Table 1). The dashed line indicates the recruitment of equal proportions of Ia and Ib fibres.

organ (group Ib) afferents which are recruited by brief electrical shocks to
the muscle nerve for differing percentages of group Ia fibre recruitment.
The best 'threshold separation' is displayed by the long nerve branch to
semitendinous muscle where 60 % of the group Ia afferents can be stimulated without any concomitant group Ib fibre stimulation. In contrast,
there is only a marginal difference in the excitability distributions of group
Ia and Ib fibres for the peroneus longus muscle.

Table 1 also gives some information about the degree of overlap between
the excitabilities of the group I and group II fibres. Although the majority
of group I fibres can be stimulated with minimal activation (i.e., < 5 %)
of the group II fibres, a volley which is just supramaximal for group I will
contain a substantial proportion of the group II population.

The results obtained with electrical stimulation can be summarized fairly
simply. A just suprathreshold shock to a muscle nerve will provide a volley
composed largely or entirely of activity in the group Ia fibres; further
increases in stimulus strength will successively recruit group Ib and group
II fibres, as well as more group Ia fibres, so that great care is necessary in
attributing a central effect to one or other of these fibre types (particularly
if the central effect is not via a monosynaptic pathway, since allowance may
have to be made for spatial summation and perhaps also for the fact that the
central actions of group Ib and group II fibres can only be studied in the
presence of preceding activity of the lower threshold fibres).

Group Ib fibres

The obvious way out of this difficulty is either to block or to modify the
electrical threshold of the different types of fibre. For the group Ib fibres
to be selectively activated by an electrical stimulus it is necessary to elevate
the threshold of all the group Ia fibres in that muscle nerve. Coppin, Jack &
MacLennan (1970) found that this was possible by the technique of low-
amplitude, high-frequency longitudinal vibration of the muscle. They
relied on the careful study of Brown, Engberg & Matthews (1967) which
showed that in the cat soleus muscle such vibration was an almost perfectly
selective means of causing all the spindle primary receptors to fire at the
same frequency as the vibratory stimulus, without altering the firing frequency of the Golgi tendon organs (or spindle secondary endings). Fig. 2
shows the effect on the electrical threshold of a group Ia fibre from the
soleus muscle when it is made to undergo such repetitive activity for
various periods of time. Note that the threshold is considerably elevated
after the vibratory stimulus has ceased (in this experiment the threshold of
the fibre was not measured during the vibration, but in other experiments

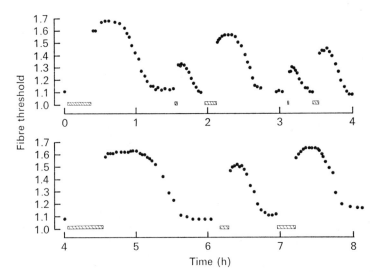

Fig. 2. The effect of repetitive activity on the electrical threshold of a single group Ia fibre in the soleus muscle nerve; longitudinal vibration of 200 Hz was applied to the muscle tendon during the periods marked by cross-hatching. Abscissa, continuous scale of time (h); ordinate, fibre threshold, expressed relative to the initial population threshold.

it has been observed that the threshold rises during the course of the vibration). Systematic studies on a large number of group Ia fibres showed that the magnitude and time course of the threshold increase were primarily dependent on the total number of impulses generated during the vibration period. Coppin *et al.* (1970) found that 240 000 impulses (e.g. 200 imp sec^{-1} for 20 min) was a suitable conditioning because there was a large threshold increase (mean 1.7 times the original value, range 1.4–2.1) and this increase was fairly steadily maintained for nearly 20 min (mean 18, range 10–25 min) after vibration had ceased.

In order to gain some insight into how this would affect the excitability distribution in a muscle nerve, Coppin *et al.* constructed a stimulus growth curve for the large afferent fibres in soleus muscle nerve, on the assumption that all the group Ia fibres had increased their threshold by 1.5 times their normal value. Fig. 3a shows the result, with the normal stimulus growth curves for the three types of afferent fibre (experimentally measured) shown in Fig. 3b. Thus, on the assumption that the vibration shifted the threshold of all the group Ia fibres by this amount without affecting the threshold of either the group Ib or group II fibres, a stimulus of 1.3 times the prevailing group I threshold would activate about 75 % of the group Ib fibres but no group Ia or group II fibres.

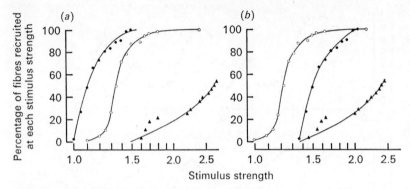

Fig. 3. (*a*) The experimentally determined stimulus growth curves for 308 large afferent fibres from soleus muscle The recruitment of group Ia (●), Ib (○), and II (▲) fibres are shown separately Abscissa, stimulus strength relative to the population threshold (log scale); ordinate, percentage of all the fibres recruited at each stimulus strength (*b*) shows how the stimulus growth curves would be changed if all the Ia fibres increased their electrical threshold by a factor of 1.5 Reproduced, with permission, from Coppin, Jack & MacLennan (1970)

    Two qualifications should be mentioned with respect to the general applicability of this technique. The first is that the proportion of group Ib fibres that can be stimulated without contamination by group Ia fibres will depend on the initial threshold distributions of the group Ia and group Ib fibres in the particular muscle nerve. For example, in the long nerve branch to the semitendinosus only 35 % of the group Ib fibres have thresholds less than 1.5 times the most excitable group Ia fibre and this will be an approximate upper limit to the proportion of group Ib fibres that can be activated alone. The figures for the other muscle nerves are; medial gastrocnemius 85 %, peroneus longus 60 %.

    The second qualification is that in a limited survey of the vibration sensitivity of Golgi tendon organs in the peroneus longus, medial gastrocnemius and tibialis anterior muscles (see MacLennan, 1971) it was found that some tendon organs appear to be relatively more responsive to longitudinal vibration than their fellows in the soleus muscle. Although large amplitudes of longitudinal vibration (> 300 μm) were required to cause the tendon organ to fire one-to-one to the vibration (at 200 Hz), the frequency of firing of some of the tendon organs could be elevated by amplitudes necessary to drive all the spindle primary endings one-to-one. Such an increase in the firing of the Ib afferents would tend to reduce the relative threshold elevation of the group Ia fibres.

Group II fibres

Many techniques have been described in the literature for the selective blocking of large afferent fibres, e.g. compression, asphyxia, cold, 'strong shocks' and steady electrical current. When Richard Roberts and I set out to develop a reliable method we were struck by the fact that most of the methods described had drawbacks due to either (a) insufficient selectivity, (b) difficulty of maintenance or (c) asynchronous firing of fibres, produced by the blocking method. When a block is applied there is commonly a dispersion of the compound action potential, due to slowing in conduction velocity of the fibres, so it seemed essential that any blocking method that was to be adopted required a careful single unit study in order to guarantee that there were no problems with either the selectivity or with asynchronous firing of any of the fibres. Following some unsatisfactory attempts that MacLennan and I had made to block by means of a steady electrical current, we decided to try another method, described not long before by Floyd (1970). This technique was a variation of one used previously by Burke & Ginsborg (1956). Non-polarizable electrodes were applied to a cut nerve trunk, with the cathode on the cut end and the anode on intact nerve. An electrical waveform with a brief rise time (< 200 μsec), but a slow decay (roughly exponential, with a time constant of approximately 1.5 msec) was applied. When this technique was used on a freshly dissected muscle nerve we found it gave very poor selectivity; the group I fibres were blocked only at stimulus strengths which recruited both group II and group III fibres. An accidental observation by Roberts gave us the clue to improvement of the selectivity, because he noted that the group I fibres could be more readily blocked when the nerve had been previously stimulated repetitively by strong electrical stimuli (see Jack & Roberts, 1974). Previous workers (Bishop & Heinbecker, 1935; Laporte & Bessou, 1958, 1959) had reported that 'strong shocks' could block larger afferent fibres and we found that when using weaker and briefer forms of the same conditioning we could depress the excitability of the group I fibres relative to the group II fibres, without actually blocking them. The advantages of the method were: (1) providing the nerve was carefully dissected, it was very easy to set up; (2) once established, the selectivity of the volley remained stable for many hours; (3) the selectivity remained with repetitive stimuli at up to 200 Hz; and (4) we could readily stimulate group I fibres through the same electrode by simply reversing the polarity of the stimulus. Fig. 4 shows the results we obtained with this method in a single unit study of soleus afferent fibres in one animal.

Although it is relatively easy to ensure that no group I fibres are stimu-

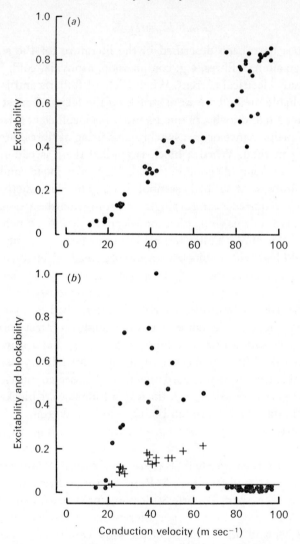

Fig 4. (*a*) The relationship between electrical excitability (the reciprocal of threshold) and conduction velocity for 46 soleus afferent fibres stimulated with rectangular pulses (0.05-msec duration) with the cathode proximal. Excitabilities are expressed relative to the population excitability (○) of 1.0. The nerve had received no
'strong shocks'. (*b*) The relation between excitability (●) and conduction velocity
for 42 of the fibres shown above, when a Floyd stimulus (see text) was employed
and the nerve had been given 'strong shocks' (60 of 1-sec duration and 8 V
strength). The 'strong shocks' were applied about 30 min before the first excitability
measurements were made. It can be seen that a stimulus of about four times the
threshold strength of the most excitable fibre provides a selective group II volley
(i.e. only group II fibres have excitabilities greater than 0.25). The fibres plotted
below the line were not excited by any of the stimulus strengths employed (up to
30 times threshold). The crosses indicate the stimulus strengths at which fibres,
recruited with weaker stimuli, no longer responded to the stimulus ('blockability').

lated by this method it is less certain that some low-threshold group III fibres are not included in the volley. The appropriate stimulus strength can be determined when monitoring afferent fibres in fine dorsal root filaments, but it is more difficult to be certain when the dorsal roots are left intact (see p. 170).

SELECTIVE STIMULATION OF MUSCLE RECEPTORS

Spindle primary endings

Lundberg & Winsbury (1960; see also Stuart, Mosher, Gerlach & Reinking, 1970) described a method for relatively selectively activating muscle spindle primary endings. A brief, small amplitude stretch to the soleus muscle produced a synchronous input restricted to the group Ia afferent fibres. Subsequently, as mentioned above, Brown et al. (1967) showed that in the soleus muscle a very selective increase in the firing frequency of the muscle spindle primary ending could be achieved by carefully choosing the amplitude and frequency of longitudinal vibration of the muscle. On present evidence, the method described by Brown et al. would appear to be the best method for activating the whole population of group Ia fibres in a muscle, although it must be remembered, as with all methods in which the muscle nerve remains intact, that there will be a 'background' input from all those other receptors in the muscle which tend to fire 'spontaneously' (e.g. both Golgi tendon organs and spindle secondary endings which are firing at the resting muscle length). Furthermore, the vibratory stimulus is a powerful excitant of any 'Paciniform-type' receptors in the muscle or nearby tissue (see Table 1). It has already been mentioned that in some muscles Golgi tendon organs may also have their firing frequency elevated by low amplitude ($<$ 200 μm) vibration. These observations were made in passive (i.e., non-contracting) muscle. Brown et al. (1967) noted that when the muscle is contracting the threshold for vibratory activation of Golgi tendon organs is lowered while that of the muscle spindle primary ending is elevated. Although the elevation in vibratory threshold of the primary ending may be offset by concomitant stimulation of γ afferent fibres supplying the spindle (see Fig. 15, Brown et al., 1967), it remains the case that both the selectivity of primary ending activation and the certainty that all primary endings follow vibration one-to-one is much more suspect in the contracting soleus muscle (see p. 173).

Golgi tendon organs

It is well known that Golgi tendon organs are very sensitive to muscle contraction (see Houk & Henneman, 1967) so this is a natural way to activate them. Since muscle spindle endings are classically described as decreasing their firing rate during contraction, this would seem to offer a satisfactory method for their selective activation (e.g. Hunt, 1952), although a systematic single unit study of the behaviour of both tendon organs and spindles has not been made. Unfortunately there is one substantial interpretive difficulty in experiments of this kind when assessing central actions, since it is very difficult to make a quantitative assessment as to whether an observed effect is predominantly the result of increased tendon organ activity or of the concomitant decreased spindle activity (see Houk, Singer & Goldman, 1970).

Lundberg & Winsbury (1960) offered a way out of this difficulty; they described a method whereby a brief stretch was superimposed on the rising phase of a muscle contraction. They reported that if this was done with an initially slack muscle it was 'possible to obtain a synchronous volley in about 75 % of the Ib afferents and with relatively little background discharge'. This method is thus comparable to that described earlier for electrical stimulation of the Ib afferents.

An alternative approach is to use muscle contraction as the stimulus for the tendon organs and to seek means to hold the firing frequency of the spindle endings at the same level as before the start of the contraction. High-frequency vibration of the muscle can, with suitable precautions (see above), hold all the primary endings at a constant frequency but will not combat the tendency of the secondary endings to decrease their firing rate. Roberts and I have made some preliminary attempts to overcome this by adding γ efferent stimulation during the contraction period (see *Spindle secondary endings*). The strength and frequency of γ efferent stimulation is adjusted so that a sample secondary ending (monitored in a dorsal root filament) maintains the same firing frequency during contraction as before.

Spindle secondary endings

These endings, like the primary endings, increase their firing frequency both as a result of an increase in muscle length and static γ efferent stimulation. Since only the primary endings are sensitive to low-amplitude, high-frequency vibration, a selective increase in secondary firing (with a background of steady high-frequency Ia firing) can be obtained by combining vibration either with an increase in muscle length or with static γ efferent stimulation.

Several workers (Matthews, 1969; Westbury, 1972; Kanda & Rymer, 1977) have used an increase in muscle length during vibration as a method of studying the central action of spindle secondary endings.

There are two ways of reliably stimulating a large proportion of the static γ efferent axons innervating a single muscle. The first method, which is direct and elegant (and heroic) is to dissect out functionally single γ efferent axons in the peripheral stump of a cut ventral root and assemble them on a stimulating electrode. This was done successfully by Hunt (1952); he described one experiment in which he isolated 28 γ efferent axons (not classified into static or dynamic, but presumably the bulk of them were static) to the medial gastrocnemius muscle. This represents approximately 16 % of the total γ efferent supply to the muscle.

An alternative is to use the technique of Jack & Roberts (1974) which was described earlier for selective electrical stimulation of group II afferent fibres. We have found that in the ventral root selected it is relatively easy to stimulate at least 50 % of the γ efferent axons, generally with no 'contaminative' stimulation of α motor axons.

THE ROLE OF GROUP Ia, Ib AND II FIBRES IN STRETCH AND VIBRATION REFLEXES OF THE SOLEUS MUSCLE IN THE DECEREBRATE CAT

In 1959, Eccles & Lundberg reported that in the decerebrate cat, stimulation of an extensor muscle nerve of the hindlimb at up to 10 times threshold (i.e., stimulation of all group I and group II fibres) may produce no detectable effect on its own, or synergic motoneurones, other than the monosynaptic excitation due to the group Ia fibres (Eccles & Lundberg, 1959b). The polysynaptic inhibitions observed in the unanaesthetized spinal cat which are produced by the group Ib and group II fibres (Laporte & Lloyd, 1952) were largely or completely absent. If this observation held also for the circumstance of repetitive input from these three types of proprioceptor, then it would be natural to attribute the stretch and vibration reflexes observed in the decerebrate preparation solely to the monosynaptic action of group Ia afferents.

Recently, evidence has been presented that this simple view of the stretch reflex is incorrect. In 1969, Matthews concluded that a substantial fraction of the excitation contributed to motoneurones during stretch and vibration came from the secondary spindle afferents. Since then, further support has been given to this conclusion (McGrath & Matthews, 1973; Rymer & Walsh, 1973; Fromm, Haase & Wolf, 1977; Kanda & Rymer, 1977). There has also been a variety of experiments performed in anaesthetized preparations which indicate that there are both monosynaptic and polysynaptic excitatory actions from group II afferents on extensor moto-neurones (Westbury, 1972; Kirkwood & Sears, 1975; Lundberg, Malm-gren & Schomburg, 1975, 1977; Stauffer *et al.*, 1976; Kato & Fukushima, 1976).

Houk *et al.* (1970) evaluated the 'tension feedback' in the decerebrate cat by studying the increment in tension produced by stimulating a centrally cut ventral root filament in the presence and absence of a tonic stretch reflex. They reported a reduction in the tension increment in the presence of reflex tension, which they assumed to be due to both a reduction in net spindle firing (disfacilitation) and an increase in net tendon organ firing (inhibition). They estimated the importance of these two factors and concluded that the true Ib inhibition could be very significant – in its absence as much as double the reflex tension might have been produced.

A final complication is that Hultborn, Wigström & Wängberg (1975) have presented evidence that the Ia afferents can produce a long-latency, prolonged excitation in the decerebrate animal, indicating that there is also a complex polysynaptic pathway for Ia excitation (see also Kanda, 1972; Fromm & Noth, 1976).

In seeking to develop a method of selectively activating group II fibres by electrical means, Roberts and I had been persuaded by Matthews' (1969, 1973) cogent arguments and were hoping to gain more information about the central pathway(s) responsible for their strong excitatory effect. There seemed to us to be two general ways in which the group II action might be mediated. The first was that there was either a direct excitatory action or a disinhibition when both proprioceptors and motoneurones were not firing; the second possibility was that the effect only became evident in the circumstances of Matthews' experiments, when both proprioceptors and motoneurones were firing. Of course, the two are not mutually exclusive because an effect might be present in the quiescent state but become quantitatively more important when the proprioceptors and/or moto-neurones were active.

We started by looking for the first type of action i.e., in intercollicularly decerebrated cats we cut all the leg muscle nerves, established a group II volley in one of the nerves to the triceps surae muscle and sought changes in the excitability of synergic motoneurones by the Lloyd method of monosynaptic reflex resting. In not one of these experiments (26 cats) was there convincing evidence for an early facilitation with latencies similar to those described for the monosynaptic and polysnaptic excitation reported in anaesthetized preparations (Kirkwood & Sears, 1975; Lundberg *et al.*, 1975, 1977; Stauffer *et al.*, 1976). With a single group II volley the only prominent effect detected in the first 30 msec was an inhibition in 4 preparations; in the other 22 there was no obvious action, confirming the observations of Eccles & Lundberg (1959b). When the inhibition was present, it had a similar time course to that described in spinal preparations using conventional electrical stimulation (Laporte & Lloyd, 1952). With repetitive stimulation (e.g. four group II volleys at 200 Hz) the results were rather different; in 24 preparations inhibition was detected in 11, the other 13 showing no early effect ($<$ 30 msec). In 19 experiments, testing was also carried out at later times and in 12 of these preparations a facilitation of the monosynaptic test reflex was observed.

The properties of this late facilitation were examined. It had a minimal latency of approximately 30 msec and was associated with a decrease in the excitability of the triceps surae group I afferent terminals (tested by the Wall technique, with the cathode in the motor nucleus). The time course of the facilitation was also similar to the time course of an observed inhibition of presynaptic inhibition, the latter effect being produced in the triceps surae group I afferent terminals by repetitive electrical stimulation at group I strength of the nerves to the posterior biceps and semitendinosus muscles. The late facilitation was also abolished by very small doses of Nembutal (5–10 mg kg^{-1} body weight). We therefore concluded that the late facilitation is, at least in large part, caused by inhibition of tonic presynaptic inhibition of the Ia afferent terminals (see also, Rymer & Walsh, 1973).

This facilitation thus seemed to be a good candidate for at least one of the ways in which a group II excitation might be mediated in the decerebrate animal. Furthermore it had the added attraction that in those of our preparations in which it was absent or relatively weak it might be expected to become quantitatively more significant in experiments on vibration reflexes since it was known that vibration of a synergic muscle produced a presynaptic inhibition of the Ia terminals (Barnes & Pompeiano, 1970); thus if the group II afferents inhibited this presynaptic inhibitory path as well, the magnitude of the effect on release of transmitter from the Ia terminals might be large.

One aspect of the results on late facilitation, however, worried us. In some experiments we did not observe the facilitation with a presumed pure group II volley, but it did become readily detectable when the stimulus strength was raised so that it would be certain that group III afferents were activated. In other experiments where it was present with a presumed group II volley it was enhanced by increasing the stimulus strength. This might simply mean that it was a common action of these two components of the flexor reflex afferents (Holmquist & Lundberg, 1961) but, as Matthews

(1972, p. 370) has warned, it seemed important to be particularly careful that secondary spindle afferents were in fact involved.

If one accepts the tentative conclusion (p. 158) that in the nerves to the medial gastrocnemius and soleus muscles there is very little contamination of group II afferents apart from occasional 'Paciniform-type' afferents, then one way to check if secondary spindle afferents are involved is to use carefully controlled electrical stimuli to a muscle nerve, monitoring the conduction velocity of the afferents recruited by recording the incoming volley at the root–cord junction. This was done in nine experiments and the results raised serious doubts about the 'purity' of our group II volley method. In all of these preparations there was a definite late facilitation when stimulus strengths greater than 10 times the group I threshold were used (at this strength group III fibres begin to be recruited, see Eccles & Lundberg, 1959a). In seven of the nine experiments there was no late facilitation at stimulus strengths when the conduction velocity of the slowest afferent fibre stimulated averaged 31 m sec^{-1} (range 23–37 m sec^{-1}), in four there was a weak facilitation when the slowest afferent averaged 28 m sec^{-1} (range 23–35 m sec^{-1}) and it was clearly present when the average conduction velocity was less than 16–23 m sec^{-1} (see also Lund, Lundberg & Vyklický, 1965; Mendell, 1972). These results do not settle the matter conclusively because of the uncertainties of (1) the requirement for spatial summation in the pathway and (2) the likely inadequacies of the sample of receptors whose afferents conduct in this velocity range.

For this reason Roberts and I decided to approach the problem in a quite different way. We adopted techniques similar to those of Matthews (1969) for selective activation of spindle secondary endings. Instead of using the method of vibration plus muscle lengthening (see pp. 166–167) we thought it would be worth trying a combination of vibration and γ efferent stimulation. The advantage of the latter method is that it avoided the criticism (Grillner, 1970, 1973) that changes in active muscle tension might be a result of the length change *per se* rather than in any central effect due to the increased secondary spindle firing.

Results were obtained from a total of 15 animals but I wish here to concentrate mainly on four preparations where we have grounds for believing that the methods we used for selective activation of both secondary spindle endings and Golgi tendon organs were satisfactory. The principles of the methods for the selective activation of these two types of receptor have already been outlined; about half the ventral root supply to the soleus muscle was cut centrally and used to generate a train of either γ efferent or α motor volleys travelling peripherally to the muscle. The muscle was held at constant length and could be submitted to mechanical stimulation in the

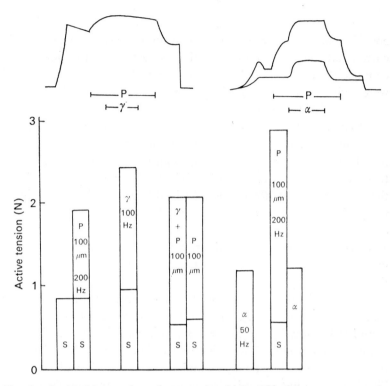

Fig. 5. Results obtained from a decerebrate cat in which 36 % of the α motor supply to the soleus muscle had been cut centrally. The traces above are schematic illustrations of the kind of results obtained. On the left is shown the tension recorded in the soleus muscle in response to stretch, pulses (200 Hz at 100 μm amplitude) and γ efferent stimulation (100 Hz). The timing of the pulses (4 sec) and γ efferent stimulation (2 sec) are shown. On the right the two traces show the results of stretch, pulses and α motor stimulation (2 sec); the lower trace being recorded in the presence of high-frequency stimulation of the ipsilateral peroneal nerve, so that all reflex tension is abolished. The block diagram below shows the results of a sequence of trials, as described in the text. Note that only active tension is plotted (passive tension subtracted). Stretch was from MPL − 9 to MPL − 3 mm, except in the last pair of records, from the same resting length to MPL − 5 mm. In this same pair the additional tension generated by α motor stimulation, on top of the reflex tension generated by stretch and pulses, is plotted to the right to facilitate comparison with the amount of tension generated by the same α volley, when all reflex tension is inhibited. P = pulse; S = stretch.

form of sinusoidal vibration or 'pulses' (see McGrath & Matthews, 1973). The tension developed by the muscle was measured with a conventional transducer and the overall compliance of the system was about 25 μm N⁻¹. Fig. 5 summarizes the results obtained in one of the four experiments. At the muscle length at which the results were obtained (3 or 5 mm less than maximum physiological length, MPL – 3 or MPL – 5 mm) there was a tonic stretch reflex. When the muscle was mechanically stimulated by 200 Hz pulses of 100 μm amplitude an increase in reflex tension occurred, attributable to the selective activation of the spindle primary endings. Repetitive stimulation of the ventral root so that only γ efferents were activated also led to a large increase in reflex tension, as expected, since it would increase the firing frequency of both the spindle primary and secondary receptors. When the γ efferents were stimulated during mechanical stimulation, there was no further increase in tension, although the muscle was capable of generating additional tension (maximum reflex tension available, 13 N). The only simple interpretation of this result is that there is no *net* effect produced in this preparation by an increase in firing of the secondary spindle endings. This result was obtained in all four experiments. In other experiments we have obtained direct evidence in support of the above suggestions, by recording the behaviour of single group Ia and II fibres, isolated from a small, centrally cut, dorsal root filament.

The final pair of results in Fig. 5 is an experiment similar in principle to that of Houk *et al.* (1970). The centrally cut, ventral root filament is now stimulated 'normally', so that α motor fibres are activated. The magnitude of the tension produced by such means is determined when all active reflex tension is inhibited by tetanic stimulation of the ipsilateral peroneal nerve. The increment in tension is also measured when the muscle is contracting reflexly due to both the stretch and the 'pulses'. It is the same magnitude (within the limits of experimental accuracy). The interpretation of this experiment is also straightforward. The 'pulses' should hold the firing of the spindle primary endings constant and since we have previously observed that changes in secondary spindle firing produce no net effect on reflex tension, the lack of reduction of reflex tension with α motor stimulation (i.e. perfect summation of the two tensions) shows that the increase in firing of Golgi tendon organs (which would have been produced by the increased tension of α stimulation) has had no inhibitory effect on the motoneurones.

The same result (zero central effect of an increase in Golgi tendon organ firing) was obtained in one of the other three experiments. In the other two, a small reduction in reflex tension was observed, amounting to 15 % in one and between 5 and 14 % in the other. Thus, the results of these four experiments can be readily summarized. In all of them, an increase in secondary spindle firing produced no net effect centrally and an increase in Golgi tendon organ firing produced, at most, a weak autogenetic inhibition. These conclusions differ markedly from those reached by Matthews (1969),

McGrath & Matthews (1973) and Kanda & Rymer (1977) for the secondary spindle endings, and, to a lesser extent, with those reached by Houk *et al.* (1970) for the Golgi tendon organs. In two of the four experiments it seems reasonable to conclude that the only (known) muscle afferents contributing to stretch and vibration reflexes are those from the spindle primary endings. These experiments do not, of course, determine the extent to which the Ia action is by monosynaptic or polysynaptic paths (Hultborn *et al.*, 1975).

The results of the other 11 of the total of 15 experiments have not been discussed. In all of these, at some point in the experiment, stimulation of the γ efferents did produce an increase in reflex tension when the muscle was vibrated or 'pulsed'. Their interpretation is complicated by the fact that we obtained evidence (of either a direct or indirect kind) in these preparations that the spindle primary endings were not all following the vibratory or 'pulsatile' stimulus in a one-to-one fashion. It is argued elsewhere (Jack & Roberts, 1978) that the most economical interpretation of the results is that there was no significant excitatory effect from the secondary spindle afferents. In these experiments we generally did not attempt to measure the effects of the Golgi tendon organs because it would be expected that the technique we used above would be unreliable, a further increase in muscle tension producing further 'misbehaviour' of the primary endings.

If the above suggestion (that in our preparations the soleus secondary endings produce no net excitation of their own motoneurones) is true, what explanation can be offered for the results of Matthews (1969), McGrath & Matthews (1973) and Kanda & Rymer (1977)? In these publications, the authors relied on the assumption that vibration was a reliable means of causing all primary endings to follow one-to-one, although that has not been our experience. Nevertheless, the design of our experiments tended to favour the tendency for Ia fibres to 'misbehave', because we had cut a substantial portion of the ventral root supply to the muscle and the consequent loss of tonic γ efferent activity would tend to raise the mechanical threshold of these endings. It is possible, therefore, that the type of decerebrate preparation used by these workers differs from ours. The decerebrate preparation is notoriously variable and small differences in the level of the lesion might make a substantial difference in the 'setting' of interneuronal pathways in the spinal cord. Since there is convincing evidence for a variety of alternative pathways from group II afferents which can enhance the excitability of extensor motoneurones by direct or indirect means, it remains a task for the future to find a type of experimental preparation in which it can be unequivocally demonstrated that the excitatory effects dominate the classically described polysynaptic inhibition (if present). The techniques described in this essay may be of some help in such a project.

I would like to thank Dr C. M. L. Coppin, Professor A. K. McIntyre, Dr Catherine R. MacLennan and Dr R. C. Roberts for allowing me to quote unpublished details of their experiments. Dr P. B. C. Matthews and Dr R. C. Roberts gave helpful comments on a draft of the manuscript.

REFERENCES

Barker, D. (1962). The structure and distribution of muscle receptors. In *Symposium on muscle receptors*, ed. D. Barker, pp. 227–40. Hong Kong: Hong Kong Univ. Press.

Barnes, C. D. & Pompeiano, O. (1970). Presynaptic and postsynaptic effects in the monosynaptic reflex pathway to extensor motoneurones following vibration of synergic muscles. *Arch. ital. Biol.* **108**, 259–94.

Bishop, G. H. & Heinbecker, P. (1935). The afferent functions of non-myelinated or 'C' fibres. *Am. J. Physiol.* **114**, 179–93.

Boyd, I. A. & Davey, M. R. (1968). *Composition of peripheral nerves.* Edinburgh: E. & S. Livingstone.

Brown, M. C., Engberg, I. & Matthews, P. B. C. (1967). The relative sensitivity to vibration of muscle receptors of the cat. *J. Physiol., Lond.* **192**, 773–800.

Burke, W. & Ginsborg, B. L. (1956). The electrical properties of the slow muscle fibre membrane. *J. Physiol., Lond.* **132**, 586–98.

Coppin, C. M. L. (1973). A study of the properties of mammalian peripheral nerve fibres. D. Phil. thesis, Univ. Oxford.

Coppin, C. M. L., Jack, J. J. B. & McIntyre, A. K. (1969). Properties of group I afferent fibres from semitendinosus muscle in the cat. *J. Physiol., Lond.* **203**, 45–6P.

Coppin, C. M. L., Jack, J. J. B. & MacLennan, C. R. (1970). A method for the selective activation of tendon organ afferent fibres from the cat soleus muscle. *J. Physiol., Lond.* **219**, 18–20P.

Eccles, J. C. (1962). Central connections of muscle afferent fibres. In *Symposium on muscle receptors*, ed. D. Barker, pp. 81–101. Hong Kong: Hong Kong Univ. Press.

Eccles, R. M. & Lundberg, A. (1959a). Synaptic actions in motoneurones by afferents which may evoke the flexion reflex. *Arch. ital. Biol.* **97**, 199–221.

Eccles, R. M. & Lundberg, A. (1959b). Supraspinal control of interneurones mediating spinal reflexes. *J. Physiol., Lond.* **147**, 565–84.

Floyd, K. (1970). A simple method for demonstrating the mechanical properties of frog slow muscle fibres. *J. Physiol., Lond.* **208**, 47–8P.

Fromm, C., Haase, J. & Wolf, E. (1977). Depression of the recurrent inhibition of extensor motoneurones by the action of group II afferents. *Brain Res.*, **120**, 459–68.

Fromm, C. & Noth, J. (1976). Reflex responses of gamma motoneurones to vibration of the muscle they innervate. *J. Physiol., Lond.* **256**, 117–36.

Grillner, S. (1970). Is the tonic stretch reflex dependent upon group II excitation? *Acta physiol. scand.* **78**, 431–2.

Grillner, S. (1973). A consideration of stretch and vibration data in relation to the tonic stretch reflex. In *Control of posture and locomotion*, ed. R. B. Stein, K. G. Pearson, R. S. Smith & J. B. Redford, pp. 397–405. New York: Plenum.

Holmquist, B. & Lundberg, A. (1961). Differential supraspinal control of synaptic actions evoked by volleys in the flexion reflex afferents in alpha motoneurones. *Acta physiol. scand.* **54** (Suppl. 186), 51 pp.

Houk, J. & Henneman, E. (1967). Responses of Golgi tendon organs to active contractions of the soleus muscle of the cat. *J. Neurophysiol.* **30**, 466–81.

Houk, J. C., Singer, J. J. & Goldman, M. R. (1970). An evaluation of length and force feedback to soleus muscles of decerebrate cats. *J. Neurophysiol.* **33**, 784–811.

Hultborn, H., Wigström, H. & Wängberg, B. (1975). Prolonged activation of soleus motoneurones following a conditioning train in soleus Ia afferents – a case for a reverberating loop? *Neurosci. Lett.* **1**, 147–52.

Hunt, C. C. (1952). The effect of stretch receptors from muscle on the discharge of motoneurones. *J. Physiol., Lond.* **117**, 359–79.

Hunt, C. C. (1954). Relation of function to diameter in afferent fibers of muscle nerves. *J. gen. Physiol.* **38**, 117–31.

Hunt, C. C. & McIntyre, A. K. (1960). Characteristics of responses from receptors from the flexor longus digitorum muscle and the adjoining interosseous region of the cat. *J. Physiol., Lond.* **153**, 74–87.

Jack, J. J. B. & MacLennan, C. R. (1971). The lack of an electrical threshold discrimination between group Ia and group Ib fibres in the nerve to the cat peroneus longus muscle. *J. Physiol., Lond.* **212**, 35–6P.

Jack, J. J. B. & Roberts, R. C. (1974). Selective electrical activation of group II muscle afferent fibres. *J. Physiol., Lond.* **241**, 82–3P.

Jack, J. J. B. & Roberts, R. C. (1978). The role of muscle spindle afferents in stretch and vibration reflexes of the soleus muscle of the decerebrate cat. *Brain Res.* **146**, 366–72.

Kanda, K. (1972). Contribution of polysynaptic pathways to the tonic vibration reflex. *Jap. J. Physiol.* **22**, 367–77.

Kanda, R. & Rymer, W. Z. (1977). An estimate of the secondary spindle receptor afferent contribution to the stretch reflex in extensor muscles of the decerebrate cat. *J. Physiol., Lond.* **264**, 63–87.

Kato, M. & Fukushima, K. (1976). Selective activation of group II muscle afferents and its effects on cat spinal neurones. *Progr. Brain Res.* **44**, 185–96.

Kirkwood, P. A. & Sears, T. A. (1975). Monosynaptic excitation of motoneurones from muscle spindle secondary endings of intercostal and triceps surae muscle in the cat. *J. Physiol., Lond.* **245**, 64–6P.

Laporte, Y. & Bessou, P. (1958). Reflexes ipsilateraux d'origine exclusivement amyélinique chez le chat. *C.r. Séanc. Soc. Biol.* **152**, 161–4.

Laporte, Y. & Bessou, P. (1959). Modification d'excitabilité de motoneurones homonymes provoquées par l'activation physiologique de fibres afférentes d'origine musculaire du groupe II. *J. Physiol., Paris* **51**, 897–908.

Laporte, Y. & Lloyd, D. P. C. (1952). Nature and significance of the reflex connections established by large afferent fibres of muscular origin. *Am. J. Physiol.* **169**, 609–21.

Lund, S., Lundberg, A. & Vyklický, L. (1965). Inhibitory action from the flexor reflex afferents on transmission to Ia afferents. *Acta physiol. scand.* **64**, 345–55.

Lundberg, A., Malmgren, K. & Schomburg, E. D. (1975). Characteristics of the excitatory pathways from group II muscle afferents to alpha motoneurones. *Brain Res.* **88**, 538–42.

Lundberg, A., Malmgren, K. & Schomburg, E. D. (1977). Comments on reflex actions evoked by electrical stimulation of group II muscle afferents. *Brain Res.* **122**, 551–5.

Lundberg, A. & Winsbury, G. (1960). Selective adequate activation of large affe-

rents from muscle spindles and Golgi tendon organs. *Acta physiol. scand.*
49, 155–64.

McGrath, G. J. & Matthews, P. B. C. (1973). Evidence from the use of procaine
nerve block that the spindle group II fibres contribute excitation to the tonic
stretch reflex of the decerebrate cat. *J. Physiol., Lond.* 235, 371–408.

McIntyre, A. K. (1974). Central actions of impulses in muscle afferent fibres. In
Handbook of sensory physiology, vol. III/2, ed. C. C. Hunt, pp. 235–88. Berlin:
Springer-Verlag.

MacLennan, C. R. (1971). Studies on the selective activation of muscle receptor
afferents. D. Phil. thesis, Univ. Oxford.

MacLennan, C. R. (1972). The behaviour of receptors of extramuscular and
muscular origin with afferent fibres contributing to the group I and group II of
the cat tibialis anterior muscle nerve. *J. Physiol., Lond.* 222, 90–1P.

Matthews, P. B. C. (1969). Evidence that the secondary as well as the primary
endings of muscle spindles may be responsible for the tonic stretch reflex of the
decerebrate cat. *J. Physiol., Lond.* 204, 365–93.

Matthews, P. B. C. (1972). *Mammalian muscle receptors and their central actions.*
London: Edward Arnold.

Matthews, P. B. C. (1973). A critique of the hypothesis that the spindle secondary
endings contribute excitation to the stretch reflex. In *Control of posture and
locomotion*, ed. R. B. Stein, K. G. Pearson, R. S. Smith & J. B. Redford, pp. 227–
43. New York: Plenum.

Mendell, L. (1972). Properties and distribution of peripherally evoked presynaptic
hyperpolarization in cat lumbar spinal cord. *J. Physiol., Lond.* 226, 769–92.

Rymer, W. Z. & Walsh, J. V. (1973). Effects of secondary muscle spindle afferent
discharge on extensor motoneurones in the decerebrate cat. In *Control of posture
and locomotion*, ed. R. B. Stein, K. G. Pearson, R. S. Smith and J. B. Redford,
pp. 411–14. New York: Plenum.

Stauffer, E. K., Watt, D. G. D., Taylor, A., Reinking, R. M. & Stuart, D. G.
(1976). Analysis of muscle receptor connections by spike-triggered averaging.
2. Spindle group II afferents. *J. Neurophysiol.* 39, 1393–402.

Stuart, D. G., Mosher, C. G., Gerlach, R. L. & Reinking, R. M. (1970). Selective
activation of Ia afferents by transient muscle stretch. *Exp. Brain Res.* 10, 477–87.

Sumner, A. J. (1961). Properties of IA and IB afferent fibres serving stretch
receptors of the cat's medial gastrocnemius muscle. *Proc. Univ. Otago Med. Sch.*
39, 3–5.

Westbury, D. R. (1972). A study of stretch and vibration reflexes of the cat by
intracellular recording from motoneurones. *J. Physiol., Lond.* 226, 37–56.

Action and excitability in a monosynaptic reflex pathway during and following anoxic insult

D. P. C. LLOYD

Depolarization of neural structures increases excitability but eventually leads to block. Hyperpolarization decreases excitability and also leads to block. Action, as differentiated from excitability in neural structures and prior to establishment of block, is diminished during depolarization and augmented during hyperpolarization, block when established being cathodal in the one instance and anodal in the other.

Antidromic conduction through motoneurone somata suffers cathodal-type block during asphyxia, which rapidly reverses to anodal-type block on readmission of air to the preparation. From general principles one must assume that other intramedullary neural structures, specifically for present discussion the primary afferent projections, behave similarly. In the primary afferent projections acting as agents for transmission the more significant aspect of depolarization would be not increased excitability but rather decreased action, and that of the rebound hyperpolarization not decreased excitability but increased magnitude of action. The phenomenon of meta-tetanic potentiation exemplifies the latter (Lloyd, 1949).

With respect to monosynaptic reflex transmission during and following anoxia, an hypothesis based upon the foregoing would be that presynaptic deficiency is the primary cause of anoxic failure, whilst postsynaptic excitability decrease is the cause of postanoxic depression. Use of brief, high-frequency tetani superimposed upon slow, rhythmic, test reflex stimulation during and following a period of anoxia adds some experimental substance to the hypothesis. The prediction *ex hypothesi* is that presynaptic tetanization would reinforce the anoxic failure and relieve the postanoxic failure of monosynaptic reflex transmission.

Monosynaptic reflex discharges of gastrocnemius motoneurones engendered by infrequent but rhythmic stimulation of the distally severed gastrocnemius nerves were recorded from the first sacral ventral root. The method is essentially that earlier employed in a study of postsynaptic potentials (Lloyd, 1970b) except that nerve stimulation was maximal for the recorded monosynaptic reflex discharge rather than being initially sub-

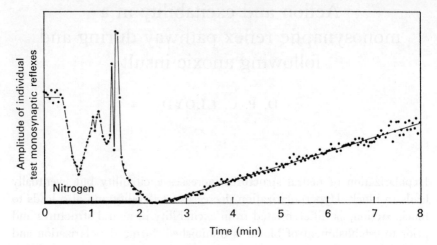

Fig. 1. Ordinate – amplitude of individual test monosynaptic reflexes plotted in arbitrary units against time in minutes on the abscissa. Heavy bar on the abscissa indicates the duration of anoxia.

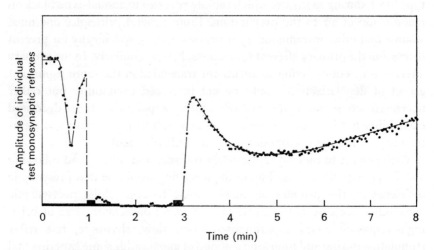

Fig. 2. Ordinate – amplitude of individual test monosynaptic reflexes plotted in arbitrary units against time in minutes. The heavy bar on the abscissa indicates the duration of anoxia and the secondary heavy bars indicate periods of high-frequency tetanization.

liminal. Anoxia was induced by switching intake of the pump necessarily employed for ventilation of the decapitate preparation from air to nitrogen.

To illustrate the experiments the amplitude of each response during and after oxygen deprivation is plotted in Figs. 1 and 2 against time in minutes from the onset of deprivation. About 25 sec after switching the preparation from air to nitrogen there occurs a progressive decline in reflex response, followed by some recovery with convulsive episodes before final decline to extinction. After readmission of air there ensues a slow progressive return of reflex amplitude toward the normal.

In general the course of anoxic and postanoxic change in reflex transmission resembles that of an antidromically induced response of motoneurone somata. Convulsive episodes must be considered largely of nuisance value and extraneous to the central issue of events within the monosynaptic reflex system. But the early decrease seen in reflex response has not been encountered in similar circumstances employing antidromic stimulation of motoneurones. The presumption then is that it is initially a consequence of presynaptic failure. Developing postsynaptic block naturally would add to the final result.

Brief, high-frequency stimulation of primary afferent fibers causes in the normal preparation a rebound hyperpolarization of the stimulated fibers, an increase in impulse amplitude including action at the synapse and a resulting increase in reflex response (Lloyd, 1949). When such a high frequency tetanus falls intercurrent during anoxia (Fig. 2) there is an abrupt drop in reflex response to zero or near zero. As only a small depression would be expected to develop among the motoneurones as a result of the tetanization of the presynaptic fibers (Lloyd, 1949) and as they are in fact hyperexcitable at the time (Lloyd, 1970a), it seems that the intense depression of response at least in its early course must be ascribed to reinforcement of developing cathodal block in the presynaptic terminal regions.

A second brief tetanus falling during the period of postanoxic depression causes the appearance of metatetanic potentiation, entirely normal in character if not in intensity (Lloyd, 1949), written upon and fading into the otherwise anticipated course of postanoxic recovery. In view of this essentially normal presynaptic behavior the weight of evidence points to postanoxic anodal-type depression as the prime cause of postanoxic depression in the monosynaptic reflex.

Occurrence of metatetanic potentiation during postanoxic depression, both phenomena due to anodal-type polarization, provides an exemplar of distinction between action and excitability in the region of synapsis.

REFERENCES

Lloyd, D. P. C. (1949). Post-tetanic potentiation of response in monosynaptic reflex pathways of the spinal cord. *J. gen. Physiol.* **33**, 147–70.

Lloyd, D. P. C. (1953). Influence of asphyxia upon the responses of spinal motoneurons. *J. gen. Physiol.* **36**, 673–702.

Lloyd, D. P. C. (1970a). Early recovery of antidromic conduction through dendrites of spinal motoneurons in the normal, anoxic and postanoxic states. *Proc. nat. Acad. Sci.* **66**, 622–5.

Lloyd, D. P. C. (1970b). Excitatory postsynaptic potential and monosynaptic reflex discharge during anoxic insult. *Proc. nat. Acad. Sci.* **66**, 626–9.

Metabolic and electrical correlates of the clearing of excess potassium in the cortex and spinal cord

G. SOMJEN

It is good to be able to say to Professor McIntyre, on these pages: 'It has been a pleasure and it has been a privilege . . .'. In Archie McIntyre's laboratory one was introduced not just to a technique, but to a style, in the Sherringtonian tradition. To be adopted into the circle, clarity, directness and precision were required in designing experiments and in reporting their results. Multilayered arguments, no matter how brilliant the logic, had to be avoided as long as intermediate steps in long chains of deduction were not demonstrable by experimental evidence. An implacable hostility toward the vague, the woolly and the circumstantial was a basic requirement. And the spinal cord was the preferred object of experimentation, because it lent itself for quantitative study.

In this paper, offered as a tribute to Archie's style, a series of experiments will be described, the goal of which was to relate the movements of ions, particularly of potassium, to electrical events and to the chemical work of the central nervous tissue. The work was performed over several years, but some of the results will be reported in print here for the first time. Some of the experiments to be described were the fruit of collaboration between the laboratory of Dr M. Rosenthal and ours.

Besides the customary techniques of electric recording we had two special tools at our disposal. One was the ion-selective microprobe (Walker, 1971). The familiar capillary micro-electrode gained a new dimension of resolution when it was filled with a liquid ion exchanger. With it the tides of ion activity in extracellular space, long suspected but never witnessed, could be directly observed. The other specialised technique required for this work was the optical method of monitoring the redox state of intra-mitochondrial respiratory enzymes. Based on work by Chance and his colleagues (Chance, Cohen, Jöbsis & Schoener, 1962) and adapted for use on the intact brain by Jöbsis and his colleagues (Jöbsis, O'Connor, Vitale & Vreman, 1971; Rosenthal & Jöbsis, 1971; Jöbsis, Keizer, LaManna & Rosenthal, 1977), the method allows measurements to be made on intact tissue, without the need of homogenising, explanting, slicing or otherwise disrupting the organ. At first we trained the reflectance fluorometer on the surface of the cerebral cortex. More recently, encouraged by Dr Y. Yamada of Loma Linda, we tried the double beam spectrophotometer (Jöbsis *et al.*, 1977) on the spinal cord, and discovered to our delight that more than enough of the light penetrated through the dorsal white matter to yield a high-quality signal of the redox state of cytochrome a,a_3 in the dorsal horns. Moreover, by proper selection of wavelengths it is possible to obtain a relative measure of the amount of blood in the tissue, and of the oxygenation of the haemoglobin therein contained. It is especially fitting that we can print our first recordings from spinal cord in this volume, honouring Archie McIntyre.

In the following pages we shall first pose the general problem of ion homoeostasis in the central nervous system, and then offer a solution, albeit a partial one.

THE RELEASE OF POTASSIUM FROM NEURONES INTO EXTRACELLULAR SPACE

Three quarters of a century ago Overton (1902) proposed that during excitation electric current is carried by an exchange of intracellular potassium for extracellular sodium. How his theory became accepted (almost) universally has been recounted on numerous occasions, and needs no retelling here. In much of the early theoretical work concerning the generation and the conduction of impulses it had been assumed that excitable cells are confined containers of ions, surrounded by a well-stirred extracellular sea of infinite size. Ion activity in the extracellular medium could thus be assumed to remain constant at all times. That this may not always be so, was first disclosed by Frankenhaeuser & Hodgkin (1956) who inferred, from indirect evidence, that K^+ ions tend to accumulate near the outside of nerve membranes after the firing of an impulse. With the advent of the electron microscopic study of the brain the suspicion grew that a special problem may exist in central nervous tissue, where little space appeared to remain free between neurones and neuroglial cells. Estimates of the size of extracellular space in central gray matter have again grown more generous recently, but the problem did not thereby disappear.

Using reasonable estimates for geometric relationships and of quantities of ions moving, Lebovitz (1970) calculated that significant interactions of membrane potential and excitability may occur between adjacent cells, as a consequence of the release of ions during activity.

Besides theoretical calculations indirect estimates of extracellular potassium activity ($[K^+]_o$) were attempted by the recording of the membrane potential of neuroglial cells. In coldblooded nervous systems, Kuffler and his associates (Kuffler & Nicholls, 1966) showed that the membrane potential of glial cells remains closer to the equilibrium potential for potassium than that of nerve cells. It appears therefore that the glial membrane is more selective for potassium than is the neuronal membrane. In agreement with these observations on coldblooded animals it was found that the presumed glial cells ('idle cells') of the mammalian gray matter usually had high and stable resting membrane potentials (e.g. Phillips, 1956; Krnjević & Schwartz, 1967). It also turned out, however, that during activity in surrounding neurones, glial cells were subject to depolarising shifts of membrane potential (Karahashi & Goldring, 1966; Grossman & Hampton, 1968; Somjen, 1969, 1970; reviewed by Somjen, 1973, 1975). Because of the similarity of behaviour of glial cells in mammalian and poikilotherm nervous systems, it was reasonable to assume that the depolarisation of mammalian glia was caused by K^+. The possibility that other agents were acting on the glial membrane was, however, not ruled out; Krnjević & Schwartz (1967), for example, found that acetylcholine and gamma aminobutyric acid caused depolarisation of some but not all glial cells.

With the introduction of the ion-selective micro-electrodes all possible doubt was removed that activation of groups of neurones in the central gray matter is often associated with appreciable increase of $[K^+]_o$ (Krnjević & Morris, 1972; Vyklický, Syková, Kříž & Ujec, 1972; Vyskočil, Kříž & Bureš, 1972; Prince, Lux & Neher, 1973).

The responses of $[K^+]_o$ are not artefacts caused by massive electrical stimulation. They can be observed in the visual cortex when the visual system is 'adequately' stimulated (Kelly & Van Essen, 1974; Singer & Lux, 1975), and in the spinal response to somatic input (Fig. 1b). One can also regularly observe small spontaneous elevations of $[K^+]_o$ with each cortical spindle burst (Fig. 1a), without deliberately stimulating the animal at all. It is clear from Fig. 1 that, although smaller than the maximal responses evoked by electrical stimulation, these physiological variations are of a similar order of magnitude.

Accepting that the fluctuations of $[K^+]_o$ in the central nervous system are a manifestation of normal neuronal activity, we must ask, from where is the potassium released and to where and how is it removed. Possible sources are axons, while they conduct impulses, axon terminals, cell bodies, and the dendritic expanse, or any combination of these elements. Currents of action may contribute more than synaptic currents, or the reverse may be true. Masses of small neurones may make multiple small contributions or there may be a few large donors.

(a)

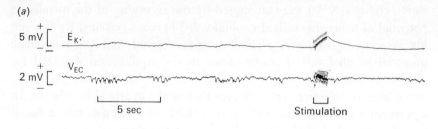

(b)

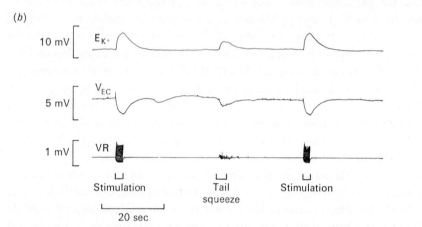

Fig. 1. Responses of extracellular potassium and potential in the cerebral cortex (a) and the spinal cord (b). E_{K+}, voltage recorded differentially between potassium-selective and 'reference' barrels of a double-barrelled electrode; V_{EC}, extracellular potential, recorded direct coupled and single-ended from 'reference' barrel; VR, electrical activity of spinal ventral root. Cortical recording (a) 750 μm beneath the surface; spontaneous activity consists of spindle bursts; at stimulus mark, electric stimulation of cortical surface at 15 V, 15 Hz, 0.5-msec pulse train occurred. Spinal Cord (b) stimulated either by 20-Hz pulse train applied to the dorsal root, or by squeezing the tail. (Unpublished experiment of G. Cordingley.)

Myelinated fibres do not seem to add significantly to $[K^+]_o$, even during high-frequency firing. In the dorsal white matter of the spinal cord little or no response of $[K^+]_o$ is seen, unless stimulus intensity is raised to a level activating C fibres as well (Lothman & Somjen, 1975). This is not surprising, since myelinated fibres exchange ions only at nodes of Ranvier. Cell bodies also seem to contribute little, because antidromic stimulation of ventral roots, to the highest frequency which the somata of motoneurones can follow, evokes little or no change of $[K^+]_o$ in the ventral horn (Somjen & Lothman, 1974). This leaves the ingredients of neuropil, the expanses of dendritic trees and the terminal unmyelinated arborisation of axons as the probable main sources of excess potassium during activity. That impulse conduction in unmyelinated axons raises $[K^+]_o$ in their environment was made probable already by old observations of Frankenhaeuser & Hodgkin (1956). That synaptic currents may be an additional major source is suggested by recent experiments of C. J. Karwoski and L. M. Proenza (unpublished). They treated isolated retinas of necturus with tetrodotoxin. Receptors, horizontal cells and bipolars in this tissue do not fire action potentials, and therefore can operate normally in the presence of this toxin. When the potassium responses evoked by light flashes in tetrodotoxin-treated retinas were compared to normal controls, the quantity of K^+ released into extracellular space in the inner layers of the retina was reduced somewhat, but even after blocking impulse traffic, sizeable K^+ responses were evoked. It is clear that in the necturus retina synaptic currents are a major contributor to the flash-evoked $[K^+]_o$ responses, and the same may well be true in the case of synaptic currents in the mammalian nervous system.

POSSIBLE EFFECTS OF RELEASED POTASSIUM

The electrical signals most closely related to elevated $[K^+]_o$ are the stimulus-evoked sustained shifts of extracellular electric potential. So close is the correlation between the log of $[K^+]_o$ and extracellular sustained potential (SP) shifts, (Somjen & Lothman, 1974; Lothman & Somjen, 1975), that d.c. coupled recordings could become the poor man's potassium gauge. The main generator of negative SP shifts appears to be depolarisation of glial cells even though neurones presumably contribute to all electric recordings made in the central nervous system. The onset and offset of extracellular SP shifts differ from the course of glial depolarisation probably, because of the synaptic activity in neurones, but the extracellular voltage changes which are sustained for a second or more are very closely correlated with the behaviour of the membrane potential of neuroglial cells, and not at all with that of neurones (Somjen, 1969, 1970, 1973; Castellucci & Goldring, 1970). The glial membrane potential follows closely the Nernst function for potassium at body temperature in the mammalian spinal cord (Lothman & Somjen, 1975), just as it does in leech ganglia (Kuffler & Nicholls, 1966). It would seem therefore that there is no reason to assume that glial membrane potential is influenced by agents other than $[K^+]_o$, at least under the conditions of these experiments.

The membrane potential of neurones must undoubtedly also be affected by changes of $[K^+]_o$. Potassium could affect postsynaptic excitability by direct action on the membrane potential. It could also influence presynaptic function. According to Cooke & Quastel (1973) it has a specific effect on the release of transmitter at the neuromuscular junction which is greater than could be expected from depolarisation of the motor nerve terminal. Whether such an effect operates in the central nervous system and within the physiological range of concentrations remains to be seen.

Even if it had no 'specific' effect on transmitter release in the central nervous system of the kind suggested by Cooke & Quastel's (1973) experiments on the neuromuscular junction, $[K^+]_o$ could affect the membrane potential of the presynaptic terminals. The lower the membrane potential of a presynaptic ending, the greater the spontaneous release of transmitter. At the same time and more importantly, depolarisation could curtail the amplitude of incoming orthodromic impulses and hence the release of transmitter. This effect would be entirely impartial, affecting excitatory and inhibitory, axo-dendritic, axo-somatic and axo-axonic junctions alike.

Such effects of potassium would amount to an unregulated diffuse feedback, built into all tissues where excitable cells are densely packed. Under normal conditions, central neurones appear, however, to be protected against those fluctuations of $[K^+]_o$, to which they regularly are exposed. In the first place, within the physiological range of 3 to 5 mM, the neuronal membrane potential is relatively insensitive to variations of $[K^+]_o$ (Kuffler & Nicholls, 1966). One might speculate that the 'resting' baseline of $[K^+]_o$ in the normal environment of central neurones is maintained around 3 mM (compared to the 4.5 mM found in plasma and most other body fluids) in order to limit the fluctuations of $[K^+]_o$ within this innocuous range. Additionally, neuronal membranes, on the postsynaptic side at least, are under the dominating influence of transmitter substances which, by their ability to regulate ionic permeability, override the relatively weak effects of variations of $[K^+]_o$. Consequently no correlation whatsoever is detected in simultaneous recordings of extracellular potassium and neuronal membrane potential (Lothman & Somjen, 1975).

The above considerations may not apply when the normal bounds of physiological activity are broken. High levels of potassium have repeatedly been implicated in theories of epileptogenesis (e.g. Zuckerman & Glaser, 1968; Fertziger & Ranck, 1970; Pollen & Trachtenberg, 1970). By a combination of pre- and postsynaptic action excess K^+ could, as we have seen, partially isolate neurone assemblies from synaptic input and favour autochthonous activity and non-synaptic interactions between neighbours. Recently-available experimental data do not, however, support a causal role for excess potassium in the genesis of epileptiform discharges, either ictal or interictal, either in the cortex or in the spinal cord, at the foci of glial scars or induced by penicillin, nor indeed by the blocker of membrane-bound, Na-K-activated ATPase, digitoxigenin (Glötzner, 1973; Lothman et al., 1975; Lothman & Somjen, 1976; Pedley, Fisher, Futamachi &

Prince, 1976; Pedley, Fisher & Prince, 1976; Cordingley & Somjen, 1978; Somjen *et al.*, 1978). While not a *cause* or trigger of epileptiform seizures, it may nevertheless be a *modifier*, shaping the course of the discharges. This proposition is very difficult to prove or refute, but the level of $[K^+]_o$ around 8 to 12 mM during tonic seizure activity compared with the normal 3 to 5 mM certainly allows for such a role.

Some protective mechanism seems to guard the central nervous system against unbridled rise of $[K^+]_o$ even during seizures, for a ceiling of 10 to 12 mM is apparently respected during paroxysmal activity in the neocortex and hippocampus and around 8 to 10 mM in the spinal cord. The limiting value is transgressed only during spreading depression, when $[K^+]_o$ rises to levels of 30 to 80 mM (Vyskočil *et al.*, 1972; Hubschmann, Grossman, Mehta & Abramson, 1973; Mayevsky, Zeuthen & Chance, 1974; Lothman *et al.*, 1975). Several years before the advent of K^+-selective micro-electrodes, Grafstein (1956) proposed that spreading depression occurred when the rate of release of K^+ into extracellular fluid exceeded the maximal rate of its removal. Measurements made with K^+-sensitive micro-electrodes are compatible with this theory, but the observation of spreading depression propagating through a tetrodotoxin-treated area of cortex (Sugaya, Takato & Noda, 1971) is, at first sight, hard to reconcile with it. Additional factors, such as glutamate (Van Harreveld & Fifkova, 1970, 1973), may play a part, either in conjunction with or independently from K^+.

A physiological role for K^+ was suggested by some investigators using the K^+-sensitive micro-electrode (Krnjević & Morris, 1972; Vyklický *et al.*, 1972). When it appeared that the largest responses of $[K^+]_o$ to afferent stimulation occurred in the regions of the spinal cord richest in primary afferent terminals, the inference was that K^+ may be the agent transmitting the negative dorsal root potential (DRP) and hence presynaptic inhibition. Not only the K^+ released by primary afferents, but also that coming from postsynaptic membrane with which they make contact could feed back onto presynaptic terminals. There are, however, at least three compelling reasons, which make it extremely unlikely that the rise of $[K^+]_o$ is the agent of the main negative wave of the DRP (termed DR V by Lloyd & McIntyre, 1949). First, the DR V rises and falls much faster than the level of $[K^+]_o$ in the dorsal horn or anywhere else in the spinal gray matter. True, the measured response of $[K^+]_o$ may be retarded since K^+ must diffuse from its source to the micro-electrode. That the error of measurement is not significant is indicated, however, by the exact match between the time course of glial depolarisation and the electrode potential (Lothman & Somjen, 1975). From the supposed sources of K^+, neuroglial cells are not farther removed than are primary afferent terminals. Moreover, the cus-

tomary point of recording DRPs is several electrotonic lengths distant from
the site of primary afferent depolarisation, whereas glial potentials are
measured near to the very site of the action. The second argument follows
from the lack of correlation between the magnitudes of DR V and $[K^+]_o$
responses (Somjen & Lothman, 1974). Third, while picrotoxin suppresses
the DR V, it enhances the responses of $[K^+]_o$; by contrast, barbiturates,
which in moderate doses enhance and prolong the DR V, depress the
response of $[K^+]_o$ at all dose levels. (For both sides of the argument concerning $[K^+]_o$ and DRPs, see the contributions by Vyklický, and by Somjen, in Ryall & Kelly, 1978.) It should be added that, while the 'classical'
DR V is presumably caused by some agent other than potassium, an additional depolarisation of a different character and much smaller amplitude
may occur in primary afferents (see also ten Bruggencate, Lux & Liebl,
1974; Lothman & Somjen, 1975; Vyklický, Syková & Mellerová,
1976).

Under ordinary conditions, assuming that the mechanism of its clearing
is operating normally, the distribution of $[K^+]_o$ responses gives a good,
albeit not a linear, measure of total neuronal activity. In evoked potentials
related to synaptic activity, currents flowing in opposite directions cancel,
but in the responses of $[K^+]_o$ (and the consequent glial depolarisation) the
contributions from excitatory and inhibitory, pre- and postsynaptic responses, from synaptic currents, and from currents of action, are all
summed. The spatial distribution of maximal amplitudes of $[K^+]_o$ responses
therefore maps the distribution of neural activation. Different spatial profiles are obtained by stimulation of the ventroposterolateral (VPL) nucleus
and of the cortical surface (Fig. 2; and Cordingley & Somjen, 1978). In the
spinal cord, when a cutaneous nerve is stimulated, the maximal responses
occur in the dorsal horn. When a muscle nerve is stimulated, the responses
are smaller and the maximum is at a deeper point. After the administration
of a convulsant dose of penicillin, responses are enhanced, and the maximal
responses move into a substantially more ventral location (Lothman &
Somjen, 1976). The spatial profile of SP shifts invariably mirrors that of the
$[K^+]_o$ responses.

Responses of $[K^+]_o$ in the spinal cord resemble those of the cerebral
cortex. There is, however, a consistent quantitative difference between the
two tissues. For a given increase of $[K^+]_o$ the sustained shift of potential is
considerably larger in the spinal cord than in the cortex (Fig. 3; also Somjen, 1975; Cordingley & Somjen, 1978).

The reason for the difference is not immediately obvious. Joyner &
Somjen (1974; see Somjen, 1973) have analysed, with the aid of a formal
model, the factors determining the distribution of voltage between intra-

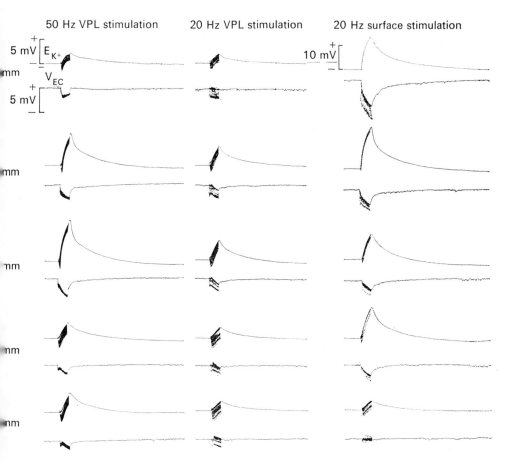

Fig. 2. Responses of potassium and potential recorded at varying depths in the somatic receiving area of the cerebral cortex. E_{K+}, voltage recorded differentially between potassium-selective and 'reference' barrels of a double-barrelled electrode; V_{EC}, extracellular potential, recorded direct coupled and single-ended from the 'reference' barrel. Note that maximal responses occur deeper in the tissue when the ventroposterolateral (VPL) nucleus is stimulated, than with surface stimulation. (Unpublished experiment of G. Cordingley.)

and extracellular compartments of a network of electrotonically coupled cells. In such a system, the higher the membrane resistance in relation to the resistance which links cells, the farther the voltage changes, both intra- and extracellular, spread. With the higher membrane resistance (i.e. larger space constant), while the voltage changes are spread farther, at the centre of the array the maximal effect is attenuated. In other words, the current drawn by adjacent 'passive' cells shunts some of the voltage at the site of the activity, and voltage gradients become shallower. Increasing the

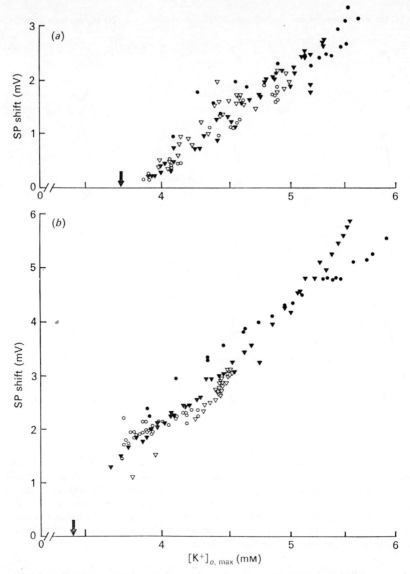

Fig. 3. Correlation of the amplitude of evoked sustained potential (SP) shifts and the maximal levels of responses of extracellular potassium activity ($[K^+]_{o, max}$) in the cerebral cortex (a) and the spinal cord (b). Cortical responses evoked by stimulation of the ventroposterolateral (VPL) nucleus. Spinal responses, measured in the dorsal horn, evoked by stimulation of the dorsal root. Each graph is from one recording site, the different symbols denoting the variation of different stimulus parameters. Vertical arrows mark unstimulated baseline level of $[K^+]_o$. (Unpublished experiment of G. Cordingley.)

coupling resistance between cells increases intracellular but attenuates extracellular responses. Varying the extracellular resistance, or the resistance from gray matter through the remainder of the body to ground, has more effect on extracellular potential than on intracellular potential.

There are two reasons which suggest that the space constant of the glial electronic net in 'real' nervous systems is short. When the spatial profile of K^+ responses is compared to the spatial profile of SP shifts, the two are, as already mentioned, very close to being mirror images in both cord and cortex. In other words, voltage changes do not spread far outside the region where an electromotive force acts on the glial membrane. Second, the membrane potential change follows closely the Nernst function (Lothman & Somjen, 1975), suggesting that not much of the electromotive force is shunted by current drawn as 'passive' adjacent cells (confirming the relatively short space constant). The parameters which could best explain the differences between cord and cortex are extracellular and tissue-to-ground resistances. These are probably higher in the cord and lower in the cortex (cf. Fig. 11, Somjen, 1973).

THE CLEARING OF POTASSIUM FROM EXTRACELLULAR FLUID

There has been some uncertainty about the mechanism of clearing potassium from extracellular fluid. In principle the spilled ions have several ways to be removed. They could diffuse through extracellular clefts from areas of high to low activity. They could be removed by way of the blood stream, if they could enter capillaries. They could be taken up by glial cells. And finally they could be pumped back into the neurones which released them. Eventually neurones must of course regain what they have lost, or nervous function would cease. At issue, however, is whether K^+ ions are recaptured so fast that they have no time to move anywhere else, or whether neurones have to retrieve them after they have been displaced.

Available evidence indicates that, while K^+ added to the extracellular fluid of central nervous tissue from an exogenous source dissipate by diffusion, K^+ ions released by activated neurones are cleared by active transport. Lux & Neher (1973) injected K^+ ions into cortex by micro-iontophoresis and found that their movement from iontophoretic electrode to recording electrode could be described by diffusion equations, using a diffusion constant one sixth of that determined for physiological saline. Fisher, Pedley & Prince (1976) let artificial cerebrospinal fluid of a high K^+ content flow over the surface of the cortex, and measured the rise and fall of $[K^+]_o$ at varying depths under the surface. They found a barrier to diffusion at the pial surface, but for movement of K^+ within the gray matter they derived a diffusion constant very similar in magnitude to that found by Lux & Neher (1973).

In a study of the clearing of K^+ released from neurones into extracellular fluid in the spinal cord and cortex, Cordingley compared the actual rates of decay of $[K^+]_o$ transients with those expected for dissipation by diffusion from a volume similar in size to that determined experimentally (Cordingley & Somjen, 1978). The observed rate of clearing $[K^+]_o$ was more than 100 times faster than that calculated for diffusion, using the diffusion constants derived for the cortex by Lux & Neher (1973) and by Fisher et al. (1976). Even if the constant for free diffusion in saline were substituted, the experimentally determined half-decay times were 30 times shorter than the calculated ones. Krnjević & Morris (1975a), who made a similar comparison of experimental and theoretical data, found a smaller discrepancy. They assumed, however, that K^+ is released from multiple point sources, whereas the experimentally determined profiles of spatial distribution suggest a solid volume source, at least for the particular experimental conditions in which the measurements were made.

Additional evidence supporting the role of active transport in the clearing of K^+ is found in experiments in which the effects of temperature (Lewis & Schuette, 1975b) and of digitalis derivatives (Krnjević & Morris, 1975b; Cordingley & Somjen, 1978) were investigated.

It seems then, that K^+ from an external source moves through gray matter by diffusion, whereas K^+ released from endogenous sources is recaptured by neurones by an active mechanism. The explanation of this difference may be that whenever neurones are activated, membrane-bound ATPase is stimulated by a rise in both $[K^+]_o$ and $[Na^+]_i$, but the $[Na^+]_i$ content of neurones remains unchanged when K^+ is administered from an external source.

In recent years much has been written about the possible role of neuroglia in regulating extracellular potassium activity in central nervous tissue. This topic is discussed in some detail by Varon & Somjen (1978), and only the main conclusions will be reiterated here.

While most of the K^+ ions released from active neurones are forthwith returned to the cells which lost them, some seem to leak away, and are recaptured later. Two facts indicate this. The first is the existence of extracellular potential gradients, apparently related to depolarisation of neuroglia. Such extracellular potential gradients imply current flowing through glial processes (see above), and since the glial membrane is selectively permeable to K^+, the current presumably is carried mostly by this ion. Another indication of movement by diffusion is found in the frequent but not invariable subsidance of $[K^+]_o$ below the 'resting' baseline some seconds after cessation of stimulation. Such undershooting of $[K^+]_o$ is best explained by assuming that, in restoring K^+ content in cytoplasm to the pre-stimulation baseline, neurones create a temporary shortage in the extracellular medium, because some of the K^+ that had previously been released had diffused away. This movement from regions of high to regions of normal activity may, in part, take place by way of the neuroglial net.

In line with the reasoning of the previous paragraph, post-activation undershooting of $[K^+]_o$ was no longer observed in preparations treated with digitoxigenin, one of the inhibitors of membrane-bound ATPase. More difficult to understand is the similar suppression of $[K^+]_o$ undershoots by diphenylhydantoin, which is a drug believed to stimulate membrane ATPase (LaManna et al., 1977). We found, however, other reasons as well for doubting that phenytonin stimulates K^+ transport in central nervous tissue.

Glial cells, and perhaps also the endothelial cells lining the capillaries of the brain, may have a subsidiary yet active role in the homoeostatic regulation of the extracellular activity of $[K^+]_o$. Reasons for suspecting this role are, however, rather indirect and, besides, not pertinent to the main subject of this article (cf. Varon & Somjen, 1978).

METABOLIC CORRELATES OF THE MOVEMENTS OF POTASSIUM IN CORTEX AND SPINAL CORD

(This section is based, in part, on work in progress by J. LaManna, M. Younts, M. Rosenthal and G. Somjen.)

There is a precise correlation between the responses of $[K^+]_o$, evoked by stimulation of the cortical surface or of a thalamic afferent projection nucleus, and the oxidation of intramitochondrial NADH, monitored by reflectance fluorometry of the cerebral cortex (Lewis & Schuette, 1975a; Lothman et al., 1975). In agreement with what was discussed in the previous pages, this correlation indicates that a major fraction of the energy expended by the cortical tissue is related to the active transport of ions. The normally consistent relationship is broken, however, during convulsive activity and during spreading depression, when the demand for metabolic energy appears to exceed greatly that which is usual, for equivalent levels of $[K^+]_o$, in non-convulsing tissue. (See Lothman et al., 1975; discussed in detail by Somjen et al., 1978.)

In the unstimulated cerebral cortex in situ, cytochrome a, a_3 is in a partially reduced state and responds to stimulation with a transient oxidation, the magnitude of which is a function of the electrical response of the tissue (Jöbsis et al., 1975; Rosenthal et al., 1976). This is in contrast with the

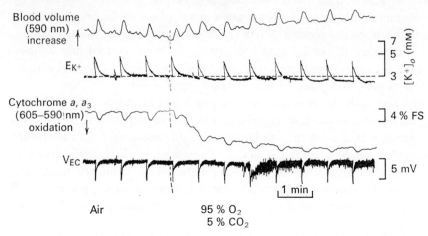

Fig. 4. Metabolic, electric and ionic responses of the spinal cord evoked by brief trains of stimuli applied to a dorsal root of a decapitate cat. Double beam reflectance spectrophotometry (Jöbsis *et al.*, 1977) was used to measure oxidation/reduction level of cytochrome a,a_3. The sample wavelength (605 nm) is an absorption maximum for cytochrome a,a_3. The reference wave length (590 nm) is one at which optical density changes of haemoglobin, related to oxygenation, equal those occurring at the sample wavelength (equibestic point). A signal proportional to reflection at the reference wavelength (590 nm) indicates relative blood volume in the tissue. Upward movement of the uppermost trace indicates increase of blood volume; downward movement of the third trace indicates oxidation of cytochrome a,a_3 (calibration in percentage of full scale, FS). Measurements of potassium and of extracellular potential similar to Figs. 1 and 2. Calibration (mM scale) converts E_{K^+} into $[K^+]_o$; at the vertical broken lines inhaled gas was changed from room air to 95 % O_2, 5 % CO_2: note substantial oxidation of cytochrome a,a_3, accompanied by vasodilation, and a decrease of 'baseline' potassium. There is an episode of convulsive activity about 3 min after changing the gas mixture; this was unusual for 5 % CO_2, but was regularly seen with more severe hypercapnia. (Unpublished experiment by J. C. LaManna, M. Rosenthal, G. Somjen and M. Younts.) E_{K^+}, voltage recorded differentially between potassium-selective and 'reference' barrels of a double-barrelled electrode; V_{EC}, extracellular potential, recorded coupled and single-ended from 'reference' barrel.

behaviour of mitochondria isolated *in vitro*, in which cytochromes a and a_3 are almost completely oxidised, and can be reduced only by severe hypoxia, (PO_2 < 0.01 kPa) (Chance & Williams, 1955). Addition of ADP or of a substrate does not cause a significant change in the redox state of cytochrome a or a_3 in isolated mitochondria. Obviously, they could not become more oxidised, since they already are nearly fully oxidised 'at rest'.

In the spinal cord the responses of cytochrome a,a_3 are qualitatively very similar to those in the cortex (Figs. 4 and 5). There are, however, several important quantitative differences. In the unstimulated spinal cord cyto-

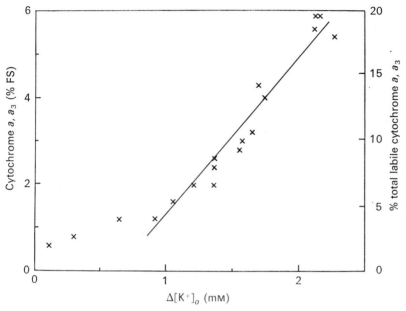

Fig. 5. The correlation of the responses of cytochrome a,a_3 and of extracellular potassium activity in spinal gray matter, evoked by varying intensities and varying frequencies of stimulation of the dorsal root of the L7 spinal segment. Left-hand ordinate scale in percentage of total optical signal ('full scale', FS), right-hand ordinate scale in percentage of available labile signal (i.e. optical signal as a fraction of the amplitude of the change measured when breathing gas is changed from 95 % O_2 + 5 % CO_2, to 100 % N_2). Plotted from recordings similar to those of Fig. 4. (Unpublished experiment by J. C. LaManna, M. Rosenthal, G. Somjen and M. Younts.)

chrome a,a_3 is 60 % reduced, as gauged by the fraction of 'available labile optical signal' (see legend to Fig. 5; Jöbsis *et al.*, 1977). From this relatively more reduced level, oxidative responses can be evoked by stimulation of an afferent pathway, which are of greater amplitude and hence better resolved with available instrumentation than the ones customarily seen in the cortex. Furthermore, the method of making oxygen available seems to differ in cord and in cortex. In the cord but not in the cortex, afferent stimulation evokes well-marked increases in the relative blood volume monitored by the reflectance spectrophotometer (Figs. 4 and 6).

Increasing the oxygenation of either the cortex or of the spinal cord causes a slight but reproducible oxidation of cytochrome a,a_3 and, usually, a decline of the unstimulated baseline level of $[K^+]_o$. Both effects are more marked and more regularly seen when the animal is given 95 % O_2 and 5 % CO_2 to breathe (Fig. 4), than when it is changed from room air to 100% O_2. Clearly the vasomotor effect of carbon dioxide is very effectively aiding

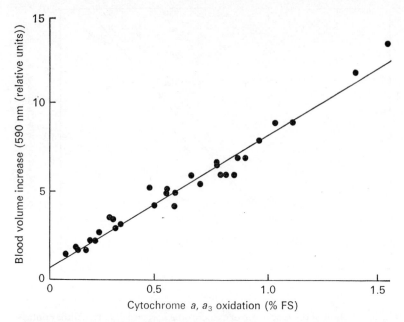

Fig. 6. The correlation of the responses of relative blood volume and of the oxidation of cytochrome a,a_3 evoked by varying trains of stimuli applied to the L7 dorsal root of the spinal cord of a decapitate cat. From records similar to those of Fig. 4. (Unpublished experiment by J. C. LaManna, M. Rosenthal, G. Somjen and M. Younts.) FS = full scale.

the oxygenation of the tissue. It is rather interesting that making more oxygen available brings about a decrease of $[K^+]_o$ in unstimulated gray matter, for it suggests that when breathing room air at or near sea level, central gray matter is somewhat 'hypoxic'.

In contrast with the cerebral cortex, spreading depression has never been observed in the spinal cord. It would be tempting to relate this immunity to the relatively greater oxidation reponses evoked by afferent stimulation in the spinal cord. It could be that a more active transport prevents accumulation of potassium in the cord under circumstances where the clearing mechanism in the cortex would fail. There are two reasons suggesting that this explanation is probably not the correct one. First, the actual rate of clearing $[K^+]_o$ from spinal extracellular fluid is not significantly faster than in the cortex. Second, even when the spinal cord is deprived of oxygen, so that active transport fails, spreading depression still does not occur. Fig. 7 illustrates the changes seen when a high spinal preparation was given 100 % N_2 to breathe. The oxidative responses of cytochrome a,a_3 evoked by intermittent stimulation ceased within a minute of nitrogen breathing, as

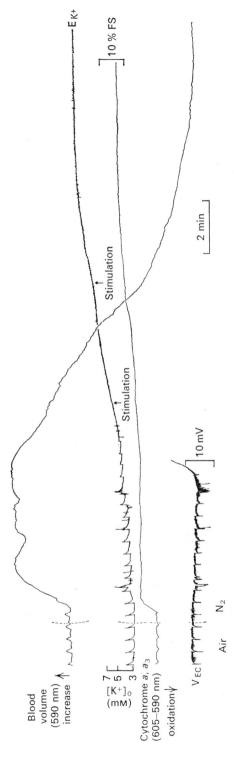

Fig. 7. The agonal anoxic changes of blood volume, cytochrome a,a_3 oxidation/reduction level, potassium activity, and electric potential in the L7 segment of the spinal cord of a decapitate cat exposed to 100 % N_2 in inhalation (see mark). Recording conditions similar to those of Fig. 4. Responses evoked by regularly recurring trains of stimuli applied to the dorsal root of the L7 segment. (Unpublished experiment by J. C. LaManna, M. Rosenthal, G. Somjen and M. Younts.) E_{K^+}, voltage recorded differentially between potassium-selective and 'reference' barrels of a double-barrelled electrode; V_{EC}, extracellular potential, recorded direct coupled and single-ended from 'reference' barrels.

cytochrome a,a_3 became abruptly more reduced. The relative blood volume increased at first with vasodilatation, then diminished as the circulation failed. The responses of $[K^+]_o$ and the associated negative potential shifts continued for a considerable period after cessation of the metabolic responses, albeit of progressively diminishing amplitude. The dissipation of the $[K^+]_o$ responses was very much slower than in the control state. After failure of active transport the kinetics of the last few $[K^+]_o$ transients seem to correspond to the rate at which K^+ can move through the gray matter by diffusion. The unstimulated 'baseline' level of $[K^+]_o$ rose very gradually during the anoxic period, and it did not come to a steady state for over an hour (not illustrated). Cytochrome a,a_3 reached its final reduced level also very slowly. Yet among all these agonal events there did not occur the explosive outpouring of K^+, which is the hallmark of spreading cortical depression, and the inevitable consequence of cortical anoxia.

It seems therefore that the spinal cord is protected from spreading depression not by a more efficient ion transport mechanism, but by some other unknown feature. It may be that some component of a positive feedback loop, required for the explosive depression seen in cortex, is missing from spinal tissue.

Unlike in the healthy brain, in neoplastic brain tissue cytochrome a,a_3 appears to be very nearly fully oxidised, similar to the state of isolated mitochondria (Sylvia, Rosenthal, Bigner & Somjen, 1977). Although our data are very preliminary, so far this seems to be the state of both astrocytomas and neuroblastomas induced by the inoculation of cultured cell lines in rats, whether the implant is placed into the brain or into subcutaneous tissue. One possible interpretation, offered by F. F. Jöbsis (personal communication) is that neoplastic tissue, and also isolated mitochondria, lack a controlling factor which in normal brain cells regulates the flow of electrons in the respiratory chain.

Work reported here for the first time was supported by grants NS 11933, NS 10384 and NS 13319 of the USPHS.

REFERENCES

ten Bruggencate, G., Lux, H. D. & Liebl, L. (1974). Possible relationship between extracellular potassium activity and presynaptic inhibition in the spinal cord of the cat. *Pflügers Arch. ges. Physiol.* **349**, 301–17.

Castellucci, V. F. & Goldring, S. (1970). Contribution to steady potential shifts of slow depolarization in cells presumed to be glia. *Electroenceph. clin. Neurophysiol.* **28**, 109–18.

Chance, B., Cohen, P., Jöbsis, F. & Schoener, B. (1962). Intracellular oxidation-reduction states *in vivo*. *Science* **137**, 499–508.

Chance, B. & Williams, G. R. (1955). Respiratory enzymes in oxydative phosphorylation. *J. biol. Chem.* **217**, 383–427.

Cooke, J. D. & Quastel, D. M. J. (1973). The specific effect of potassium on transmitter release by motor nerve terminals and its inhibition by calcium. *J. Physiol. Lond.* **228**, 435–58.

Cordingley, G. E. & Somjen, G. G. (1978). The clearing of excess potassium from extracellular space in spinal cord and cerebral cortex. *Brain Res.* (In Press.)

Fertziger, A. P. & Ranck, J. B. (1970). Potassium accumulation in interstitial space during epileptiform seizures. *Exp. Neurol.* **26**, 571–85.

Fisher, R. S., Pedley, T. A. & Prince, D. A. (1976). Kinetics of potassium movement in normal cortex. *Brain Res.* **101**, 223–37.

Frankenhaeuser, B. & Hodgkin, A. L. (1956). The after-effects of impulses in the giant nerve fibers of Loligo. *J. Physiol., Lond.* **131**, 341–76.

Glötzner, F. L. (1973). Membrane properties of neuroglia in epileptogenic gliasis. *Brain Res.* **55**, 159–71.

Grafstein, B. (1956). Mechanism of spreading cortical depression. *J. Neurophysiol.* **19**, 154–71.

Grossman, R. G. & Hampton, T. (1968). Depolarization of cortical glial cells during electrocortical activity. *Brain Res.* **11**, 316–24.

Hubschmann, O., Grossman, R. G., Mehta, P. & Abramson, M. (1973). Spreading depression of electrocortical activity studied with K^+ ion specific microelectrodes. *Am. EEG Soc. 27th A. Meet.* Abstr. 18, 25.

Jöbsis, F. F., Keizer, J. H., LaManna, J. C. & Rosenthal, M. (1977). *In vivo* reflectance spectrophotometry of cytochrome a, a_3 in the cerebral cortex of the cat. *J. appl. Physiol.* **43**, 858–72.

Jöbsis, F. F., O'Connor, M., Vitale, A. & Vreman, H. (1971). Intracellular redox changes in functioning cerebral cortex. I. Metabolic effects of epileptiform activity. *J. Neurophysiol.* **34**, 735–49.

Jöbsis, F. F., Rosenthal, M., LaManna, J., Lothman, E., Cordingley, G. & Somjen, G. (1975). Metabolic activity in epileptic seizures. In *Brain work, Alfred Benzon symposium VIII*, ed. D. Ingvar & N. Lassen, pp. 185–96. Copenhagen: Munksgaard.

Joyner, R. & Somjen, G. (1974). A model simulating the contribution of glia to sustained potential shifts of central nervous tissue. *Electroenceph. clin. Neurophysiol.* **37**, 417.

Karahashi, Y. & Goldring, S. (1966). Intracellular potentials from 'idle' cells in cerebral cortex of cat. *Electroenceph. clin. Neurophysiol.* **20**, 600–7.

Kelly, J. P. & Van Essen, D. C. (1974). Cell structure and function in the visual cortex of the cat. *J. Physiol., Lond.* **238**, 515–47.

Krnjević, K. & Morris, M. E. (1972). Extracellular K^+ activity and slow potential changes in spinal cord and medulla. *Can. J. Physiol. Pharmac.* **50**, 1214–17.

Krnjević, K. & Morris, M. E. (1975a). Factors determining the decay of K^+ potentials and focal potentials in the central nervous system. *Can. J. Physiol. Pharmac.* **53**, 923–34.

Krnjević, K. & Morris, M. E. (1975b). Strophantidine effects on extracellular K^+ and electrogenic pumping in the cuneate nucleus. *J. Physiol. Lond.* **250**, 36–7P.

Krnjević, K. & Schwartz, S. (1967). Some properties of unresponsive cells in the cerebral cortex. *Exp. Brain Res.* **3**, 306–19.

Kuffler, S. W. & Nicholls, J. G. (1966). The physiology of neuroglial cells. *Ergebn. Physiol.* **57**, 1–90.

LaManna, J., Lothman, E., Rosenthal, M., Somjen, G. & Younts, M. (1978). Phenytoin, Electric, Ionic, and Metabolic Responses in Cortex and Spinal Cord. *Epilepsia* **18**, 327–9.

Lebovitz, R. M. (1970). A theoretical examination of ionic interactions between neural and non-neural elements. *Biophysical J.* **10**, 423–44.

Lewis, D. V. & Schuette, W. H. (1975a). NADH fluorescence and $[K^+]_o$ changes during hippocampal electrical stimulation. *J. Neurophysiol.* **38**, 405–17.

Lewis, D. V. & Schuette, W. H. (1975b). Temperature dependence of potassium clearance in the central nervous system. *Brain Res.* **99**, 175–8.

Lloyd, D. P. C. & McIntyre, A. K. (1949). On the origins of dorsal root potentials. *J. gen. Physiol.* **32**, 409–43.

Lothman, E., LaManna, J., Cordingley, G., Rosenthal, M. & Somjen, G. (1975). Responses of electrical potential, potassium levels, and oxidative metabolic activity of the cerebral neocortex of cats. *Brain Res.* **88**, 15–36.

Lothman, E. & Somjen, G. G. (1975). Extracellular potassium activity, intracellular and extracellular potential responses in the spinal cord. *J. Physiol., Lond.* **252**, 115–36.

Lothman, E. & Somjen, G. G. (1976). Functions of primary afferents, and responses of extracellular K^+ during spinal epileptiform seizures. *Electroenceph. clin. Neurophysiol.* **41**, 253–67.

Lux, H. D. & Neher, E. (1973). The equilibrium time course of $[K^+]_o$ in cat cortex. *Exp. Brain Res.* **17**, 190–205.

Mayevsky, A., Zeuthen, T. & Chance, B. (1974). Measurements of extracellular potassium, ECoG and pyridine nucleotide levels during cortical spreading depression in rats. *Brain Res.* **76**, 347–9.

Overton, E. (1902). Beiträge zur allgemeinen Muskel und Nervenphysiologic. II. Mitteilung. *Pflügers Arch. ges. Physiol.* **92**, 346–86.

Pedley, T. A., Fisher, R. S. & Prince, D. A. (1976). Focal gliosis and potassium movement in mammalian cortex. *Exp. Neurol.* **50**, 346–61.

Pedley, T. A., Fisher, R. S., Futamachi, K. J. & Prince, D. A. (1976). Regulation of extracellular potassium concentration in epileptogenesis. *Fedn Proc. Fedn Am. Socs exp. Biol.* **35**, 1254–9.

Phillips, C. G. (1956). Intracellular records from Betz cells in the cat. *Q. J. exp. Physiol.* **41**, 58–69.

Pollen, D. A. & Trachtenberg, M. C. (1970). Neuroglia: gliosis and focal epilepsy. *Science* **167**, 1252–3.

Prince, D. A., Lux, H. D. & Neher, E. (1973). Measurement of extracellular potassium activity in cat cortex. *Brain Res.* **50**, 489–95.

Rosenthal, M. & Jöbsis, F. F. (1971). Intracellular redox changes in functioning cerebral cortex. II. Effects of direct cortical stimulation. *J. Neurophysiol.* **34**, 750–61.

Rosenthal, M., LaManna, J. C., Jöbsis, F. F., Levasseur, J. E., Kontos, H. A. & Patterson, J. L. (1976). Effects of respiratory gases on cytochrome A in intact cerebral cortex: is there a critical P_{O_2}? *Brain Res.* **108**, 143–54.

Ryall, R. W. & Kelly, J. S. (1978). *Iontophoresis and transmitter mechanisms in the mammalian central nervous system.* Amsterdam: Elsevier/North-Holland. (In Press.)

Singer, W. & Lux, H. D. (1975). Extracellular potassium gradients and visual receptive fields in the cat striate cortex. *Brain Res.* **96**, 378–83.

Somjen, G. G. (1969). Sustained evoked potential changes of the spinal cord. *Brain Res.* **12**, 268–72.

Somjen, G. (1970). Evoked sustained focal potentials and membrane potential of neurons and of unresponsive cells of the spinal cord. *J. Neurophysiol.* **33**, 562–82.

Somjen, G. G. (1973). Electrogenesis of sustained potentials. In *Progressive neurobiology*, ed. G. A. Kerkut & J. W. Phillis, vol. 1, pp. 199–237. London: Pergamon Press.

Somjen, G. G. (1975). Electrophysiology of neuroglia. *A. Rev. Physiol.* **37**, 163–90.

Somjen, G. (1978). A comment of the effect of potassium on dorsal root potentials. In *Iontophoresis and transmitter mechanisms in the mammalian central nervous system*, ed. R. W. Ryall & J. S. Kelly. Amsterdam: Elsevier/North-Holland. (In Press.)

Somjen, G. G. & Lothman, E. W. (1974). Potassium, sustained focal potential shifts, and dorsal root potential of the mammalian spinal cord. *Brain Res.* **69**, 151–7.

Somjen, G., Lothman, E., Dunn, P., Dunaway, T. & Cordingley, G. E. (1978). Microphysiology of spinal seizures. In *Abnormal neuronal discharges*, ed. N. Chalazonitis. Yew Nork: Raven Press. (In Press.)

Sugaya, E., Takato, M. & Noda, Y. (1971). Spreading depression under the effects of tetrodotoxin. *J. physiol. Soc. Jap.* **33**, 591–2.

Sylvia, A., Rosenthal, M., Bigner, D. & Somjen, G. (1978). Vascular response and level of oxidation of *cytochrome a, a₃* in astrocytomas *in situ. 7th A. Meet. Soc. Neurosci.* Abstr. 214, 71.

Van Harreveld, A. & Fifkova, E. (1970). Glutamate release from the retina during spreading depression. *J. Neurobiol.* **2**, 13–29.

Van Harreveld, A. & Fifkova, E. (1973). Mechanism involved in spreading depression. *J. Neurobiol.* **4**, 375–88.

Varon, S. & Somjen, G. (1978). Neuron–glia interactions. Neurosciences Research Program Bulletin, Report of a Work Session. (In Press.)

Vyklický, L. (1978). Transient changes in extracellular potassium and presynaptic inhibition. In *Iontophoresis and transmitter mechanisms in the mammalian central nervous system*, ed. R. W. Ryall & J. S. Kelly, Amsterdam: Elsevier/North-Holland. (In Press.)

Vyklický, L., Syková, E., Kříž, N. & Ujec, E. (1972). Post-stimulation changes of extracellular potassium concentration in the spinal cord of the rat. *Brain Res.* **45**, 608–11.

Vyklický, L., Syková, E. & Mellerová, B. (1976). Depolarization of primary afferents in the frog spinal cord under high Mg^{2+} concentrations. *Brain Res.* **117**, 153–6.

Vyskočil, F., Kříž, N. & Bureš, J. (1972). Potassium-selective microelectrodes used for measuring the extracellular brain potassium during spreading depression and anoxic depolarization in rats. *Brain Res.* **39**, 255–9.

Walker, J. L. (1971). Ion specific liquid ion exchanger microelectrodes. *Anal. Chem.* **43**, 89–93A.

Zuckermann, E. C. & Glaser, G. H. (1968). Hippocampal epileptic activity induced by localized ventricular perfusion with high-potassium cerebrospinal fluid. *Exp. Neurol.* **20**, 87–110.

Dendritic spines and synaptic potency

W. RALL

My interest in synaptic transmission and in modifications of synaptic potency dates back to my apprenticeship with Professors Eccles and McIntyre at the University of Otago, in Dunedin. At that time we discussed post-tetanic potentiation and disuse atrophy; we made distinctions between homosynaptic and heterosynaptic effects and between presynaptic and postsynaptic mechanisms (Eccles & Rall, 1951a,b; Eccles & McIntyre, 1953). We did not then distinguish between dendritic and somatic locations of synapses, nor did we consider dendritic spines.

Recent interest in dendritic spines has followed from the electron microscopic demonstration of synaptic contacts upon spine heads (Gray, 1959), and from subsequent evidence that some cortical neurones receive nearly all of their synaptic contacts upon their dendritic spines (Colonnier, 1968; Scheibel & Scheibel, 1968; Diamond, Gray & Yasargil, 1970). Also, it has recently been reported, for several neurone types, that dendritic spines with long and thin spine stems occur more frequently on the more distal dendritic branches of small diameter, while stubby dendritic spines occur more frequently on dendritic trunks and the more proximal dendritic branches of larger diameter. This was explicitly noted for granule cells in the dentate gyrus (Laatsch & Cowan, 1966), and for pyramidal cells of neocortex (Jones & Powell, 1969; Peters & Kaiserman-Abramof, 1970). This interesting correlation begs for functional interpretation. At first glance, it seems strange to compound the synaptic attenuation resulting from distal dendritic location with the synaptic attenuation resulting from high spine stem resistance. However, it will be seen below, that a possible functional interpretation can be provided by a theoretical consideration of the ratio of spine stem resistance to dendritic input resistance.

That the high resistance of a thin spine stem would attenuate the effect upon the postsynaptic neurone of synaptic input delivered to the spine head was probably first pointed out by Chang (1952). He concluded that the summation of many such inputs would be needed for effective excitation of such neurones. But what is the evolutionary survival value of such synaptic input attenuation? We proposed (abstracts of W. Rall & J. Rinzel in 1971 for the XXV International Physiological Congress and for the First Annual Meeting of the Society for Neuroscience) that fine adjustments of spine stem resistances of many spines could provide one way for an organism to adjust the relative weights of the many synaptic inputs to each neurone, thus providing a possible contribution to plasticity and learning. Maximum synaptic potency would be sacrificed in exchange for fine control.

Here, circumstances permit only a brief summary of the biophysical argument for steady-state conditions; different details have been presented briefly by Rall (1974) and by Jack, Noble & Tsien (1975).

SPINE HEAD POTENTIAL

Let V_{SH} represent the potential inside the spine head, relative to its resting value. Kirchhoff's law for the three currents flowing out of the spine head, during a steady synaptic input conductance, can be expressed as

$$V_{SH}/R_{SH}+(V_{SH}-V_{EQ})/R_{SYN}+V_{SH}/(R_{SS}+R_{BI}) = 0 \qquad (1)$$

where the first term represents leakage current across the resting resistance, R_{SH}, of the spine head membrane, while the second term represents synaptic current across the opened synaptic resistance, R_{SYN}, driven by the difference between V_{SH} and the synaptic equilibrium potential, V_{EQ}; the third term represents current that flows out through the spine stem resistance, R_{SS}, and onward through the branch input resistance, R_{BI}, of the dendrite. By noting the rearrangement

$$V_{EQ}/R_{SYN} = V_{SH}(1/R_{SYN}+1/(R_{SS}+R_{BI})+1/R_{SH})$$

one can express the ratio V_{EQ}/V_{SH} in terms of resistance ratios, as follows

$$V_{EQ}/V_{SH} = 1+(R_{SYN}/R_{SH})[1+R_{SH}/(R_{SS}+R_{BI})] \qquad (2)$$

Here, one can confirm that an extremely large synaptic input would make R_{SYN} approach zero, and that this would make V_{SH} approach V_{EQ}. Also, when R_{SH} is much greater than $(R_{SS}+R_{BI})$, which is usually the case, then equation (2) simplifies to the approximation

$$V_{EQ}/V_{SH} \approx 1+R_{SYN}/(R_{SS}+R_{BI}) \qquad (3)$$

For example, when R_{SYN} equals one tenth of $(R_{SS}+R_{BI})$ this implies that $V_{EQ}/V_{SH} \cong 1.1$, or that V_{SH} is depolarized to within 10 % of its limiting value, V_{EQ}; however, if R_{SYN} were equal to $(R_{SS}+R_{BI})$ the value of V_{SH} would be reduced to half of V_{EQ}.

ATTENUATION FROM V_{SH} TO V_{BI}

Let V_{BI} represent the intracellular potential (relative to its resting value) at the branch input site, where the spine stem is attached to the dendrite. To obtain the steady attenuation from V_{SH} to V_{BI}, we note that the steady spine stem current can be expressed as

$$I_{SS} = (V_{SH} - V_{BI})/R_{SS} \qquad (4a)$$

$$= V_{BI}/R_{BI} \qquad (4b)$$

$$= V_{SH}/(R_{SS} + R_{BI}) \qquad (4c)$$

where (4a) is simply Ohms law for current flow in the spine stem (I_{SS}) (stem length is usually less than 0.1 λ, implying that leakage across the stem membrane can be neglected); epuation (4b) represents the definition of the input resistance, R_{BI}, while equation (4c) expresses the same current in terms of the full voltage drop across these two resistances in series. From (4b and 4c) it follows that the attenuation factor can be expressed as

$$V_{SH}/V_{BI} = 1 + R_{SS}/R_{BI} \qquad (5)$$

Thus when the ratio R_{SS}/R_{BI} has values of 0.01, 0.1, 1, 10 and 100, it follows that this attenuation factor, V_{SH}/V_{BI}, has the values 1.01, 1.1, 2, 11 and 101. From this it is clear that a further increase of spine stem resistance would produce greater attenuation, but that any reduction of spine stem resistance to less than one hundredth of the branch input resistance value would negligibly reduce the already negligible attenuation from V_{SH} to V_{BI}.

ATTENUATION FROM V_{BI} TO SOMA

Both branch input resistance and steady attenuation from a branch input site to the soma of a dendritic neurone model have been treated and discussed at length by Rall & Rinzel (1973). When the input site is not restricted to the terminal of a dendritic branch, the attenuation factor can be expressed as

$$V_{BI}/V_{SOMA} = (R_{BI}/R_N)(\cosh L)/\cosh(L - X_i) \qquad (6)$$

where R_N represents the input resistance of the neurone when measured at the soma, V_{SOMA} represents the potential at the soma, while L and X_i are electrotonic distances from the soma to the dendritic terminal and to the input site, respectively; this expression and the general expression for R_{BI}/R_N can be found as equations (A-12 and A-11) in the appendix of Rall & Rinzel (1973).

For an illustrative example, we set $L = 1.5$, and compare a mid-dendritic location where $X_i = L/2$ and $R_{BI}/R_N = 10$, with a distal dendritic location where $X_i = L$ and a small-diameter branch of high branching order (see Table 1 of Rall & Rinzel, 1973) makes $R_{BI}/R_N = 100$. Then, for the mid-dendritic input site, equation (6) implies that V_{BI}/V_{SOMA} equals $(10)(2.35)/(1.3)$, or about 18, while the distal input site implies a value of 235 for this attenuation factor.

DEPENDENCE OF ATTENUATION UPON R_{SS}

To obtain the total attenuation, from V_{EQ} to V_{SOMA}, one would multiply the successive factors, V_{EQ}/V_{SH}, V_{SH}/V_{BI} and V_{BI}/V_{SOMA} as defined by equations (2), (5) and (6). Because only equations (2) and (5) depend upon R_{SS}, we consider here the combination of these two effects. Suppose that $R_{SH}/R_{BI} = 10^2$; then equation (2) implies that

$$V_{EQ}/V_{SH} = 1 + (R_{SYN}/R_{SH})[1 + 100/(1 + R_{SS}/R_{BI})] \qquad (7)$$

For a synaptic intensity corresponding to $R_{SYN}/R_{SH} = 10^{-2}$, it can be seen that R_{SS}/R_{BI} values of 0.01, 0.1, 1, 10 and 100 would imply V_{EQ}/V_{SH} values of 2, 1.92, 1.51, 1.1 and 1.02, respectively. When these values are multiplied by the corresponding V_{SH}/V_{BI} values (cited above, in the sentence following equation (5)), the resulting V_{EQ}/V_{BI} ratios are found to be approximately 2, 2.1, 3, 12 and 103, respectively. Here it can be seen that reduction of R_{SS} to less than one tenth of R_{BI} yields almost negligible gain in the value of V_{BI}.

For a tenfold increase of synaptic intensity, corresponding to $R_{SYN}/R_{SH} = 10^{-3}$, the effect of R_{SS} upon V_{EQ}/V_{SH} is smaller (the values are 1.1, 1.09, 1.05, 1.01 and 1.00, respectively) and the resulting V_{EQ}/V_{BI} ratios are found to be approximately 1.1, 1.2, 2.1, 11 and 101, respectively. Again, it can be seen that reduction of R_{SS} to less than one tenth of R_{BI} yields almost negligible gain in the value of B_{BI}.

Thus, for a given synaptic input intensity and a given dendritic input location, it is clear that the effect of spine stem resistance upon synaptic potency depends upon the ratio R_{SS}/R_{BI}. If the spine stem does serve as a locus for fine adjustments of synaptic potency, one might expect R_{SS}/R_{BI} values between 0.1 and about 5 to represent an optimal range, because R_{SS}/R_{BI} values below 0.1 all correspond to maximum potency and values above 10 all correspond to more than tenfold attenuation of potency.

TENTATIVE APPLICATION

For the variety of spine dimensions reported by Peters & Kaiserman-Abramof (1970), one can make tentative estimates of resistance values. The area of the spine head membrane seems to range from 0.3 to 3 μm^2 except for mushroom-shaped spine heads, which are larger. The membrane resistivity is not known but could range between 1000 and 6000 Ωcm^2. These values imply a range approximately from 10^{10} to 10^{12} Ω for R_{SH}.

For the classical thin spine stem, the diameter seems to range from 0.05 to 0.3 μm, while the length may range from 0.5 to 3 μm. Treating the stem as a cylinder with a volume resistivity of about 63 Ωcm, one obtains a range of resistance values from about 4×10^6 to 10^9 Ω for R_{SS}; however, a stubby spine might correspond to a value as low as 10^5 Ω. Thus it is useful to estimate that the long, thin spines at distal dendritic locations would have R_{SS} values of 10^8 to 10^9 Ω, while more moderate spines at mid-dendritic locations would have smaller R_{SS} values of about 10^7 to 10^8 Ω, and the stubby spines at proximal dendritic locations would have still smaller R_{SS} values of about 10^5 to 10^6 Ω.

If we take $R_N = 10^7$ Ω as a representative pyramidal cell input resistance, measured at the soma (see p. 672 of Rall & Rinzel, 1973, for references), then using the previously mentioned R_{BI}/R_N ratios of 10 and 100 for mid- and distal dendritic input locations, we obtain estimates of 10^8 and 10^9 Ω for the corresponding R_{BI} values.

Although these are all rough estimates, they do permit us to make at least preliminary order of magnitude estimates of the ratio R_{SS}/R_{BI} at the different input locations. For the distal dendritic location we have the range, 10^8 to 10^9, divided by 10^9, or a range of 0.1 to 1, which corresponds to the theoretically designated optimal range for fine adjustments of synaptic potency. Again, for the mid-dendritic location, we have the range, 10^7 to 10^8, divided by 10^8, yielding the same range of 0.1 to 1. However, the estimates for proximal locations give the range, 10^5 to 10^6, divided by 10^7, yielding a range of 0.01 to 0.1, which corresponds to maximal potency with negligible adjustability.

DISCUSSION

Although the numerical estimates of the previous section are subject to revision as more comprehensive data become available, the combination of these estimates with the earlier attenuation analysis supports the plausibility of the proposed hypothesis. In other words, if changes in spine stem resistance are used to provide fine adjustments of relative synaptic potency, then one would expect the larger stem resistance values of long, thin stems

208 W. RALL

to be matched with the higher input resistance values at distal dendritic locations; one would expect a scatter of R_{SS}/R_{BI} values, but would hope to find most of them in the optimal operating range of about 0.1 to 5. On the other hand, synapses being kept at maximum potency would have R_{SS}/R_{BI} values less than 0.1 and would benefit from proximal dendritic locations.

It may be briefly noted that transient response functions (Rinzel & Rall, 1974) have been coupled computationally to the transient synaptic activation of a spine head; although voltage attenuation factors are then greater, the distribution of charge corresponds to steady-state voltage (see pp. 779–84 of Rinzel & Rall, 1974) and it should not be surprising that the dependence upon R_{SS}/R_{BI} values was found to be similar to that for the steady state.

Although the focus here has been upon spine stem resistance, there is no intention of excluding the possibilities that synaptic potency could be modified by changes at the synapse, or by changes involving the incompletely understood spine apparatus, or by changes in the dendritic branch or the dendritic tree (Rall, 1962) to which the spine is attached.

Returning briefly to the opening reference to distinctions between homosynaptic and heterosynaptic effects, it seems useful to note that changes at the spine stem locus, as well as at a dendritic branch or a dendritic tree, lie somewhere between the purely homosynaptic character of a presynaptic change and the purely heterosynaptic character of a change at the spike trigger zone of the postsynaptic neurone. If a change occurs in one dendritic tree as distinguished from other trees of the same neurone then the change is upstream from the convergence zone; it is not fully heterosynaptic because it does not change the weights of synapses on the other trees; yet it is not purely homosynaptic, because it can effect several pathways which have a significant proportion of their synapses on this tree; such overlap could add to the efficacy of a conditioning process (Rall, 1962). In contrast, when the change is localized to one dendritic branch, a small number of synapses share a correlated change in potency, and when the change is localized to a dendritic spine it becomes in most respects homosynaptic, although the locus is postsynaptic.

REFERENCES

Chang, H. T. (1952). Cortical neurons with particular reference to the apical dendrites. *Cold Spring Harb. Symp. quant. Biol.* **17**, 189–202.

Colonnier, M. (1968). Synaptic patterns on different cell types in the different laminae of the cat visual cortex. An electron microscope study. *Brain Res.* **9**, 268–87.

Diamond, J., Gray, E. G. & Yasargil, G. M. (1970). The function of the dendritic spine: an hypothesis. In *Excitatory synaptic mechanisms*, ed. P. Anderson & J. K. S. Jansen, pp. 213–22. Oslo: Universitetsforlaget.

Eccles, J. C. & McIntyre, A. K. (1953). The effects of disuse and of activity on mammalian spinal reflexes. *J. Physiol., Lond.* **121**, 492–516.

Eccles, J. C. & Rall, W. (1951*a*). Effects induced in a monosynaptic reflex path by its activation. *J. Neurophysiol.* **14**, 353–76.

Eccles, J. C. & Rall, W. (1951*b*). Repetitive monosynaptic activation of moto-neurones. *Proc. R. Soc., B* **138**, 475–98.

Gray, E. G. (1959). Axo-somatic and axo-dendritic synapses of the cerebral cortex: an electron microscopic study. *J. Anat.* **93**, 420–33.

Jack, J. J. B., Noble, D. & Tsien, R. W. (1975). *Electric current flow in excitable cells.* Oxford: Clarendon Press.

Jones, E. G. & Powell, T. P. S. (1969). Morphological variations in the dendritic spines of the neocortex. *J. Cell. Sci.* **5**, 509–29.

Laatsch, R. H. & Cowan, W. M. (1966). Electron microscopic studies of the dentate gyrus of the rat. I, Normal structure with special reference to synaptic organization. *J. comp. Neurol.* **128**, 359–96.

Peters, A. & Kaiserman-Abramof, I. R. (1970). The small pyramidal neuron of the rat cerebral cortex. The perikaryon, dendrites and spines. *Am. J. Anat.* **127**, 321–56.

Rall, W. (1962). Electrophysiology of a dendritic neuron model. *Biophys. J.* **2**, 145–67.

Rall, W. (1974). Dendritic spines, synaptic potency and neuronal plasticity. In *Cellular mechanisms subserving changes in neuronal activity*, ed. C. D. Woody, K. A. Brown, T. J. Crow, Jr & J. D. Knispel, pp. 13–21. Brain Information Service Research Report No. 3. Los Angeles: Univ. California.

Rall, W. & Rinzel, J. (1973). Branch input resistance and steady attenuation for input to one branch of a dendritic neuron model. *Biophys. J.* **13**, 648–88.

Rinzel, J. & Rall, W. (1974). Transient response in a dendritic neuron model for current injected at once branch. *Biophys. J.* **14**, 759–90.

Scheibel, M. E. & Scheibel, A. B. (1968). On the nature of dendritic spines – report of a workshop. *Commun. behav. Biol.* **1A**, 231–65.

Renshaw cell activity in the human spinal cord

J. L. VEALE

THE F RESPONSE

This was first described by Magladery & McDougal (1950) as a reflex response to peripheral nerve stimulation, recordable in the small peripheral muscles of the hand and foot. Using two stimulation sites, they considered that the afferent pathway was significantly slower than the efferent pathway, and so concluded that the F response was a polysynaptic reflex elicited by afferent nerve stimulation. However, Dawson & Merton (1956) challenged the existence of any difference in conduction velocity in the up-going and down-going pathways, asserting them to be identical. McLeod & Wray (1966) demonstrated F responses in the hand muscles of deafferented baboons. Mayer & Mawdsley (1965) found that, during recovery from spinal anaesthesia, the F response returns first, along with spontaneous movement, whereas the H reflex returns some hours later, along with sensation. The recurrent nature of the response has also been supported by Miglietta (1973). He reports four cases of traumatic spinal cord injury with severe spasticity. Neurosurgery was undertaken to relieve this, the operation being a version of longitudinal myelotomy originally introduced by Bischof (1951). The operation eliminated all reflex activity: clonus, flexor and extensor spasms, tendon reflexes and H reflexes all disappeared postoperatively. F responses, however, continued to be obtained postoperatively, and with the accepted characteristics.

It must be noted that Coggeshall & Ito (1977) report the existence of sensory fibres in the ventral roots of L7 and S1 in the cat. They conclude that the cell bodies of these fibres are in the dorsal root ganglia, and favour the conclusion that the central axons enter the spinal cord in the ventral root. This work has yet to be confirmed in the primate or man. If it is, then the appropriate control for the current view on the nature of the F response will be its persistence following extirpation of the dorsal root ganglia (rather than following severance of the dorsal roots, or longitudinal myelotomy).

It has, however, become the accepted view that, in a relaxed limb, the F response arises because of 'back-firing' within the motoneurone pool secondary to the stimulation of motor axons. It is accordingly not considered to be a reflex in the conventional sense of arising secondary to stimulation of physiological afferent pathways. It is thus becoming usual to refer to the F response, rather than to the F reflex.

Apart from its intrinsic interest, the F response can be used for both clinical and theoretical purposes. Thus King & Ashby (1976) have used it to estimate conduction velocities in proximal segments of motor nerves, and Shahani & Young (1976a), observing that vibration applied to the muscle suppresses the F response, interpreted this as indicating a postsynaptic action of vibration on motoneurone excitability.

The F response occurs with a latency comparable to the H reflex, but is clearly distinguishable on a basis of its behaviour with increasing strength of stimulation. First, there is a characteristic change-over in amplitude between the direct motor (M) response and H reflex (due to collision in the motor axons) which is not seen with the F response. Second, the F response does not appear until the stimulus strength is rather high (as judged by the motor response). Third, the F wave may continue to increase in amplitude after the M response has become maximal. These are accepted properties.

The state of relaxation of the subject is important. Upton, McComas & Sica (1971) have shown the importance of voluntary activity in the muscle on its responses. They show that it is possible to produce responses with H reflex characteristics during voluntary effort, and even the formulation of the 'desire' to move is sufficient to induce a response. They called this response V_1, and considered that its latency was 2–3 msec shorter than that of the F response. Sica, Sanz & Colombi (1976) have shown that voluntary contraction of other muscles can potentiate the F response.

The F response exhibits a marked variability, much more than the H reflex. Furness & Jessop (1976) have developed a method of display (using a LINC-8 computer) which enables sequential trials to be displayed on a single diagram. Their Fig. 1 clearly shows the variability of the F response as compared with the M response. This variability, however, is referring to the response in the muscle as a whole, that is, to the surface electromyogram. Trontelj (1973a) has investigated the timing of single muscle fibres in the F response. The direct M responses had a standard deviation of their latencies of 59 μsec, attributed to variability of neuromuscular transmission and of time and place of action potential generation. The standard deviation of the latencies of the single fibres contributing to the F response was only somewhat larger: 73 μsec. The variability of H reflex responses was markedly larger (Trontelj, 1973b).

Thorne (1965) investigated the F response, using single motor unit recordings in the muscle. In all 11 units reported by him, if a unit discharged during the F wave, then it had also discharged during the M response, a finding confirmed by Trontelj (1973*a*). However, Veale & Hewson (1973), undertaking a recording technique similar to Thorne, did find two units which did discharge from time to time during the F wave without having discharged in the M response. The all-or-nothing nature of the units could be established. Furthermore, the unit did also occur (following a single stimulation) in both the M and F waves, which is a necessary finding to exclude it being a long-latency motor axon reflex, as described by Fullerton & Gilliatt (1965). The crucial event (a unit discharging in the F response but not in the M wave) was quite infrequent (17 successes in 370 trials), but this is sufficient (along with its consistent timing) to exclude it as being a chance discharge.

It perhaps should be stated that the two units were selected from a larger group of units. That is, the actual incidence of such units (if confirmed) is no doubt small.

In these experiments, normal subjects were used. It remains a reasonable objection that these represent very infrequent genuine reflexes, i.e responses secondary to afferent stimulation. This, however, is unlikely on timing grounds: it would be remarkable if the latencies would be the same when generated by antidromic motor back-firing or orthodromic afferent stimulation.

This finding, if accepted, influences consideration of the central mechanism for generation of the F response. Veale & Hewson (1973) discuss three possibilities:

(1) Antidromic action potentials in motor axons, having reached the cell body, cause sufficient depolarisation (or other changes) to generate an action potential in the original axons (presumably at first in the initial segment). This Veale & Hewson discard: not only is the mechanism unclear, but it does not invoke any interaction between the motoneurones. If correct, it would give no reason why F responses should be more easily elicited with high stimulus strengths than with low. In any case it is inconsistent with the units which discharge in the F response but not in the M wave.

(2) Interaction between motoneurones could occur via collaterals to Renshaw cells. The antidromic action potentials in motor axons invade the collaterals and excite Renshaw cells. Renshaw himself (1941) described excitatory effects upon reflexes for about a third of the time in this situation. Subsequent work (Wilson & Burgess, 1962; Hultborn, Jankowska & Lindström, 1971*a,b,c*) show that such facilitation arises because of disinhibition. Such a mechanism for generation of the F response is unlikely. Effects following Renshaw cell stimulation spread to other muscle groups

(Renshaw, 1941; Eccles, Fatt & Koketsu, 1954; Wilson, 1959). The F response remains restricted to the muscle group stimulated, which is not characteristic of recurrent Renshaw effects. Furthermore, facilitation via Renshaw cells persists over several milliseconds, whereas an F response responds with quite precise timing (Trontelj, 1973a). For both these reasons, generation of the F response by Renshaw disinhibition is unlikely.

(3) The third possibility is concerned with dendro-dendritic interaction within the motoneurone pool, and is favoured by Veale & Hewson. The antidromic action potentials invade the dendrites and can collectively induce sufficient depolarisation in a few motoneurones (dendrites or soma) to initiate an action potential that would be orthodromically transmitted to the muscle. Two mechanisms are possible: (a) dendro-dendritic synapses, which have been described in various parts of the central nervous system, though not yet (and not excluded) in motoneurone pools of the spinal cord; (b) electrical interaction. This has been described by Nelson (1966) in the cat. He obtained brief facilitation of monosynaptic reflexes following shocks to large ventral root segments, and considered that this arose from electrical interaction between dendrites. Grinnell (1966), working on the frog spinal cord, also obtained evidence of electrical interaction between the dendrites of a motoneurone pool. Such work has been extended by Magherini, Precht & Schwindt, 1976a,b, who favour the view that the effects are electrical, not chemical.

Such an electrical (i.e. electrotonic) interaction between dendrites can explain many of the features of the F response. The necessity to stimulate a large number of axons simultaneously follows at once, since only this will generate sufficient electrical activity to influence adjacent motoneurones. The limitation of an F response to the muscle stimulated is also explained: motoneurone pools do not have significant overlap within the spinal cord. The meagre nature of the F response would depend not only on the generally weak level of the electrical interaction, but also on the fact that this mechanism has a self-cancelling aspect: an action potential may be generated in an axon but suffer collision with an antidromic action potential from the original stimulus. Furthermore, many action potentials will arrive more or less simultaneously, and interactions will fail because of the absolute refractory period. All this makes the response a chancy affair, but, should it occur, its timing should be quite precise and consistent, as it is.

Thus the hypothesis that F responses arise owing to electrical interactions between dendrites of the motoneurone pool explains many of the features of the F response, and has support from animal experiments.

THE H REFLEX

The H, or Hoffmann, reflex (Hoffmann, 1922; Magladery & McDougal, 1950; Paillard, 1955) is a monosynaptic reflex contraction elicited in the triceps surae following the electrical stimulation of the posterior tibial nerve in the popliteal fossa. The response is generally displayed electrically: a surface electromyogram is recorded. Two responses can be seen: (*a*) a direct M (for muscle) response, with a brief (5–10 msec) latency due to direct stimulation of α motor axons, and (*b*) a reflex H response due to stimulation of Ia afferent fibres. The induced volley passes centripetally, relays monosynaptically in the spinal cord, and generates a centrifugal volley in the α motor axons. This passes to the muscle, eliciting a compound muscle action potential with a latency of 30–40 msec.

There is a characteristic pattern of response of the M and H waves in the electromyogram with increasing strength of electrical stimulation. The M response, once its threshold is reached, grows with increasing strength of stimulus to a maximum, when all α axons are stimulated. Under suitable stimulus conditions (Hugon, 1973; Veale, Rees & Mark, 1973*b*), an H reflex is elicited with a threshold strength below that of the M response. It will grow with increasing strength up to a limit, but then will undergo a characteristic extinction, decreasing in size while the M response grows. It is accepted that this extinction occurs due to collision in the motor axons: the orthodromic, or efferent, volley in the α axons (that would otherwise directly elicit the H response) collides with the antidromic volley generated in the α axons. This pattern of recruitment of M and H responses, and the extinction of the H response at high intensities of stimulation while the M response continues to grow, is the essential feature needed to describe a reflex as an H reflex.

Hugon (1973) gives an account of current methodology for undertaking H reflex studies in man and includes specimen examples of recruitment curves of the M and H responses that can be obtained.

The H reflex has become almost the standard investigation for the clinical neurophysiology of the spinal cord, and it would require a lengthy review to cover all aspects of its alterations in spasticity, Parkinsonism, the Jendrassik manoeuvre, age and so on. There are supraspinal influences (Táboříková & Sax, 1969; Táboříková, 1973) of considerable complexity.

Trontelj (1973*b*) has undertaken investigations of the H reflex, using single fibre electromyography for recording responses. He has identified the variation in latencies of motor units participating in the H reflex with changes of stimulus strength, and with the Jendrassik manoeuvre.

Two accounts have been published, using quite different lines of attack, but both using H reflexes to display effects considered secondary to stimulating Renshaw cells.

The first method is due to Veale, Rees and Mark (Veale *et al.*, 1973*a*,*b*; Veale & Rees, 1973). They began by consideration of the strength–duration curves in the ulnar nerve as obtained by stimulation with external electrodes. Growth curves of electromyograms and neurograms were examined, along with the results of collision experiments. It was concluded that sensory axons were more sensitive than motor axons if the pulse duration was 1 msec or more, whereas the reverse was true for short-duration pulses (200 μsec or less).

Using these unfortunately rather small differences in sensitivities, they investigated the H reflex. Pulses of 1-msec duration were used to elicit a test reflex. Short-duration pulses were used as a conditioning stimulus to elicit a motor response with no H reflex response. Care was taken with this conditioning stimulus to ensure that no H reflex response occurred when the triceps surae was at rest, and also that no H reflex response occurred when the triceps surae was voluntarily contracted (which can facilitate subliminal Ia stimulation). These two stimuli (test and conditioning) were delivered (with various timings) through the same electrode. It perhaps should be stressed that the electrode placement was critical, and the strengths of stimuli had to be kept always very weak. It could take up to an hour of trial and error to find a satisfactory, stable stimulation site, and in a substantial minority of subjects no such site could be found. There is, of course, no physiological reason for such sites to occur.

In successful experiments, then, an H reflex could be elicited by a test stimulus, and preceded by a conditioning stimulus which stimulated motor axons only. It was then possible to demonstrate: (*a*) that there was no commonality between the portion of the motoneurone pool involved with the H reflex and the portion which was antidromically invaded by the conditioning (motor) stimulus; (*b*) that conditioning with the motor stimulus gave a pattern of inhibition (for 2–3 msec) followed by facilitation (peaking at 5–8 msec) of the H reflex.

Since there was no commonality between the test and conditioned portions of the motoneurone pool, it was concluded that the effect is transferred via axon collaterals of the motor axons stimulated by the conditioning volley. Thus the effects are secondary to the stimulation of Renshaw cells.

Finally, (*c*) in spasticity the initial inhibition is missing: the conditioning curve is all facilitation, augmented and prolonged. This raises a possibility concerning the mechanism of the exaggerated tendon jerk in spastic states. Any volley in Ia afferents arriving at the spinal cord as a result of a tendon tap will have some degree of temporal dispersion, especially when compared with a volley generated electrically. The first action potential can activate α motoneurones and, through the axon collaterals, excite the Renshaw cells. If these are biased to strong facilitation, then the remainder of the Ia volley will arrive at an α motoneurone pool which has been facilitated, and hence an exaggerated tendon jerk will follow. Such a mechanism is, of course, not in conflict with the usual α or γ type of spasticity, but is an additional possibility that could serve to exaggerate the α or γ type. This remains to be quantified.

The second method of demonstrating effects secondary to stimulation of Renshaw cells is due to Pierrot-Deseilligny and his colleagues (Pierrot-Deseilligny & Bussel, 1975; Pierrot-Deseilligny, Bussel, Held & Katz, 1976). A conditioning shock, submaximal (and thus able to elicit an H reflex), is delivered to the posterior tibial nerve. This is then followed (say, some 15 msec later) by a maximal shock to the posterior tibial nerve, sufficient to stimulate all α axons and Ia afferents. The H reflex elicited by the conditioning shock is, of course, extinguished. But the motoneurones that were excited by the conditioning shock can now be re-excited by the Ia volley generated by the second maximal shock, and an H reflex is generated. The size of this second H reflex can be expressed as a fraction of the H reflex elicited by the conditioning shock alone. (This fraction necessarily can never exceed unity.) Variations in this fraction with variations in strength of the conditioning shock are studied. As the strength is increased, the fraction falls below unity. Two possible mechanisms are considered: (a) the effect is secondary to Ib and II afferent stimulation; (b) the effect is secondary to Renshaw cell stimulation. The latter is favoured on the basis of changes in the ratio secondary to stimulation of the radial nerve.

TREMOR

A completely relaxed limb is without electromyographic activity. Under these circumstances, it is possible to record vibrations in the fingers which are undoubtedly of cardiac origin (Brumlik & Yap, 1970). Its persistence with vascular occlusion, and its general correlation with cardiac activity, warrant the term 'ballistocardiac oscillations', or perhaps 'ballistocardiac tremor'. What is not warranted is the view that such a mechanism underlies physiological tremor, as Brumlik & Yap claim. Marsden, Meadows, Lange & Watson (1969) show convincingly the distinction, and that physiological tremor arises secondary to skeletal muscular activity and is, in general, of much larger amplitude. Of course, ballistocardiac oscillations remain as an important source of interference for the investigator of tremor of skeletal muscular origin.

Here we take physiological tremor to be the naturally-occurring, fine, rapid tremor that occurs during voluntary activity, with a predominant frequency in the 8–12-Hz range. Shahani & Young (1976*b*) discuss some of the newer information on drug actions in tremor and their use in the classification of clinical tremors.

The view has been maintained for many years that physiological tremor (8–12 Hz) originates in the stretch reflex arc. Lippold presents the case in Chapter 5 of his book (Lippold, 1973), which may be consulted for further details and references.

Broadly, the view is based on the following experimental evidence:

(1) The servo-loop of the stretch reflex arc is opened in tabes dorsalis, and such patients show both a general reduction in tremor at all frequencies and, more specifically, no peak in the 8–12-Hz range.

(2) Parameters of the servo-loop of the stretch reflex arc can be altered and induce appropriate alterations in physiological tremor. The gain of the loop can be decreased by ischaemia induced by a sphygmomanometer cuff inflated to 27 kPa. Such a procedure in the anaesthetised cat yields a sustained, high-frequency discharge of the spindles, which would imply loss of sensitivity to stretch. This is consistent with the observed result in man: a generalised decrease in the amplitude of tremor without any particular effect on any frequency range.

The total delay in the servo-loop can be changed by a number of procedures. In each case there are the expected alterations in the frequency of tremor. The procedures include cooling the limb, alterations in length, and fatigue. Cooling causes progressive slowing of tremor. There are consistent differences in tremor if a muscle is lengthening under load, as compared with shortening under load. Fatigue increases the amplitude of tremor, but is not selective in increasing the amplitude in the 8–12-Hz range. All these observations can be interpreted in terms of a servo-loop hypothesis.

Lippold (1973) considers that the main delay is in the time for a muscle twitch to develop tension (about 150 msec) and is not especially dependent upon conduction time. Thus tremor frequency can be expected to be more or less independent of relative distance of the muscle from the central nervous systems, as it is.

(3) Perturbations of the servo-loop. A 30-msec, step-wise displacement of the extended finger will induce 5–20 sinusoidal oscillations at the same frequency of physiological tremor. The response is consistent and phase-locked to the displacement. The frequency of the induced oscillations slows with cooling and quickens with warming the limb, which is consistent with the servo-loop hypothesis. Similarly, ischaemia, fatigue and some other procedures have the expected effects on the induced oscillations.

Furthermore, compound muscle action potentials recorded during such procedures show similar frequency responses, excluding the effects as being mechanical (in the muscle) rather than neurological in origin.

Perturbations can also be given (of much diminished amplitude), varying the moment of delivery with regard to the ongoing physiological tremor and to electrically induced contractions. Results consistent with the servo-loop hypothesis are obtained, it being possible (depending upon the phase relationship) to exaggerate or diminish the response to the displacement.

However, there is evidence that is in direct conflict with the hypothesis, and some of the results are challenged. Marsden *et al.* (1967) report of a patient who, following elective arm deafferentiation, had a clear 9.5-Hz peak of tremor in both the normal and deafferented limb. These authors also found tremor in the physiological range in patients with tabes dorsalis. Marsden, Meadows & Lange (1970) found variations in amplitude but not in the dominant frequency of tremor in patients with thyrotoxicosis and myxoedema, in spite of such patients having significant changes in their speeds of muscular contraction. Milner-Brown, Stein & Yemm (1973) have measured (using an averaging technique) the developed twitch in a single motor unit following its electromyogram. The delay between the electromyogram of a motor unit and the peak of tension was more than 50 msec, rather than the 150 msec taken by Lippold (1973). Thus the intramuscular delay in the servo-loop is not as dominant as assumed by Lippold.

Fatigue has effects on tremor which are clearly unrelated to any servoloop hypothesis for tremor. Furness, Jessop & Lippold (1977) have shown that intense, brief contractions can increase finger tremor (over all frequencies) for periods of up to 4 h. Comparable intense muscular contractions induced by peripheral nerve stimulation do not enhance tremor. If the intense effort is attempted during nerve block (with no ensuing contractions), then, following the nerve block, an increase in tremor is observed. Furness *et al.* (1977) conclude that the mechanism is of central origin.

Agarwal & Gottlieb (1977) have undertaken a detailed analysis of the responses to sinusoidal torque applied to the ankle joint. They observed resonance in the compliance in the 5–8-Hz range, and consider this observation to be difficult to reconcile with a servo-loop hypothesis generating tremor in the 8–12-Hz range.

In a recent and important paper, Elble & Randall (1976) recorded not only the mechanical response (third finger) and the surface electromyogram, but also motor unit activity. Extensive computer analysis of the frequency components of the tremor and of the rectified surface electromyogram was undertaken, as well as interspike-interval analysis, and coherence studies. In general, finger tremor (under a rather large constant load) showed a pronounced peak in the 8–12-Hz range, as did the rectified electromyogram, and the coherency was strongly positive.

Attention was then turned to examining the relationship between motor unit discharge and the surface-rectified electromyogram (under load conditions). About half the motor units studied contributed to the 8–12-Hz range of the surface electromyogram (as shown by coherency analysis) even though this was not the average frequency of discharge of the units. Such units showed transient sequences of double discharges: short intervals alternating with long. Such motor units were more easily recordable from subjects with prominent 8–12-Hz tremor. The analysis was taken further by undertaking phase and coherency analysis on six pairs of motor units, recorded simultaneously. An example is given of six pairs of motor units with differing mean frequencies of discharge, but each with 8–12-Hz spectral peaks. A strong correlation was found in this frequency range.

Elble & Randall (1976) consider that there are two major objections to the servo-loop hypothesis: the mean frequency of tremor should fall with increased load, but instead it rises. The mean frequency should be a function of reflex arc delay, but it is not. They quote Milner-Brown et al. (1973) that the intramuscular delay is much less than assumed by Lippold (1973). They also report (in the Discussion) a failure to find variations in tremor frequency with limb temperature, again in direct opposition to Lippold's findings.

Elble & Randall's (1976) suggested explanation for tremor involves a cycle of recurrent inhibition followed by rebound-excitation through the Renshaw cells. Such a sequence could generate an 8–12-Hz modulation of α motoneurone discharge and is consistent with much of the evidence, such as persistence with deafferentation, and independence from reflex arc length. They envisage the cycle of inhibition–excitation generating the double discharge pattern, especially at high levels of discharge. If the claim of Veale et al. (1973b) is correct, and Renshaw cells play a part in the genesis of spasticity, then changes in tremor in spastic states should be expected.

REFERENCES

Agarwal, G. C. & Gottlieb, G. L. (1977). Oscillation of the human ankle joint in response to applied sinusoidal torque on the foot. *J. Physiol., Lond.* **268**, 151–76.

Bischof, W. (1951). Die longitudinale Myelotomie. *Zentbl. Neurochir.* **11**, 79–88.

Brumlik, J. & Yap, C. B. (1970). *Normal tremor.* Springfield, Illinois: Charles C. Thomas.

Coggeshall, R. E. & Ito, H. (1977). Sensory fibres in ventral roots L7 and S1 in the cat. *J. Physiol., Lond.* **267**, 215–35.

Dawson, G. D. & Merton, P. A. (1956). Recurrent discharges from motoneurones.

In *XXe Congres International Physiology*, pp. 221–2. Brussels, Resumes des Communications.

Eccles, J. C., Fatt, P. & Koketsu, K. (1954). Cholinergic and inhibitory synapses in a pathway from motor-axon collaterals to motoneurones. *J. Physiol., Lond.* **126**, 524–62.

Elble, R. J. & Randall, J. E. (1976). Motor-unit activity responsible for 8- to 12-Hz component of human physiological finger tremor. *J. Neurophysiol.* **39**, 370–83.

Fullerton, Pamela M. & Gilliatt, R. W. (1965). Axon reflexes in human motor nerve fibres. *J. Neurol. Neurosurg. Psychiat.* **28**, 1–11.

Furness, P. & Jessop, Jennifer (1976). A rapid method of F-wave analysis, using a three-dimensional plotting technique. *J. Physiol., Lond.* **258**, 44–5.

Furness, P., Jessop, Jennifer & Lippold, O. C. J. (1977). Long-lasting increases in the tremor of human hand muscles following brief, strong effort. *J. Physiol., Lond.* **265**, 821–31.

Grinnell, A. D. (1966). A study of the interaction between motoneurones in the frog spinal cord. *J. Physiol., Lond.* **182**, 612–48.

Hoffman, P. (1922). *Untersuchungern über die Eigenreflexe (Sehnenreflexe) menschlicher Muskeln*. Berlin: Springer-Verlag.

Hugon, M. (1973). Methodology of the Hoffmann relex in man. In *New developments in electromyography and clinical neurophysiology*, ed. J. E. Desmedt, vol. 3, pp. 277–93. Basel: Karger.

Hultborn, H., Jankowska, Elżbieta & Lindström, S. (1971a). Recurrent inhibition from motor axon collaterals of transmission in the Ia inhibitory pathway to motoneurones. *J. Physiol., Lond.* **215**, 591–612.

Hultborn, H., Jankowska, Elżbieta & Lindström, S. (1971b). Recurrent inhibition of interneurones monosynaptically activated from group Ia afferents. *J. Physiol., Lond.* **215**, 613–36.

Hultborn, H., Jankowska, Elżbieta & Lindström, S. (1971c). Relative contribution from different nerves to recurrent depression of Ia IPSPs in motoneurones. *J. Physiol., Lond.* **215**, 637–64.

King, D. & Ashby, P. (1976). Conduction velocity in the proximal segments of a motor nerve in the Guillain–Barré syndrome. *J. Neurol. Neurosurg. Psychiat.* **39**, 538–44.

Lippold, O. C. J. (1973). *The origin of the alpha rhythm*. London: Churchill Livingstone.

Magherini, P. C., Precht, W. & Schwindt, P. C. (1976a). Electrical properties of frog motoneurones in the *in situ* spinal cord. *J. Neurophysiol.* **39**, 459–73.

Magherini, P. C., Precht, W. & Schwindt, P. C. (1976b). Evidence for electrotonic coupling between frog motoneurones in the *in situ* spinal cord. *J. Neurophysiol.* **39**, 474–83.

Magladery, J. W. & McDougal, D. B., Jr (1950). Electrophysiological studies on nerve and reflex activity in normal man: I. Identification of certain reflexes in the electromyogram and the conduction velocity of peripheral nerve fibres. *Bull. Johns Hopkins Hosp.* **86**, 265–90.

Marsden, C. D., Meadows, J. C. & Lange, G. W. (1970). Effect of speed of muscle contraction on physiological tremor in normal subjects and in patients with thyrotoxicosis and myxoedema. *J. Neurol. Neurosurg. Psychiat.* **33**, 776–82.

Marsden, C. D., Meadows, J. C., Lange, G. W. & Watson, R. S. (1967). Effect of deafferentation on human physiological tremor. *Lancet* ii, 700–2.

Marsden, C. D., Meadows, J. C., Lange, G. W. & Watson, R. S. (1969). The role of the ballistocardiac impulse in the genesis of physiological tremor. *Brain* **92**, 647–62.

Mayer, R. F. & Mawdsley, C. (1965). Studies in man and cat of the significance of the H wave. *J. Neurol. Neurosurg. Psychiat.* **28**, 201–11.

McLeod, J. G. & Wray, Shirley, H. (1966). An experimental study of the F wave in the baboon. *J. Neurol. Neurosurg. Psychiat.* **29**, 196–200.

Miglietta, O. E. (1973). The F response after transverse myelotomy. In *New developments in electromyography and clinical neurophysiology*, ed. J. E. Desmedt, vol. 2, pp. 323–27. Basel: Karger.

Milner-Brown, H. S., Stein, R. B. & Yemm, R. (1973). The contractile properties of human motor units during voluntary isometric contractions. *J. Physiol., Lond.* **228**, 285–306.

Nelson, P. G. (1966). Interaction between spinal motoneurones of the cat. *J. Neurophysiol.* **29**, 275–87.

Paillard, J. (1955). *Réflexes et régulations d'origine proprioceptive chez l'homme*. Paris: Arnette.

Pierrot-Deseilligny, E. & Bussel, B. (1975). Evidence for recurrent inhibition by motoneurons in human subjects. *Brain Res.* **88**, 105–8.

Pierrot-Deseilligny, E., Bussel, B., Held, J. P. & Katz, R. (1976). Excitability of human motoneurones after discharge in a conditioning reflex. *Electroenceph. clin. Neurophysiol.* **40**, 279–87.

Renshaw, B. (1941). Influence of discharge of motoneurones upon the excitation of neighbouring motoneurons. *J. Neurophysiol.* **4**, 167–83.

Shahani, B. T. & Young, R. R. (1976a). Effect of vibration on the F response. In *The motor system: neurophysiology and muscle mechanisms*, ed. M. Shahani, pp. 189–95. Amsterdam: Elsevier.

Shahani, B. T. & Young, R. R. (1976b). Physiological and pharmacological aids in the differential diagnosis of tremor. *J. Neurol. Neurosurg. Psychiat.* **39**, 772–83.

Sica, R. E. P., Sanz, Olga, P. & Colombi, A. (1976). Potentiation of the F wave by remote voluntary muscle contraction in man. *Electromyog. and clin. Neurophysiol.* **16**, 623–35.

Táboříková, Helena (1973). Supraspinal influences of H-reflexes. In *New developments in electromyography and clinical neurophysiology*, ed. J. E. Desmedt, vol. 3, pp. 328–35. Basel: Karger.

Táboříková, Helena & Sax, D. S. (1969). Conditioning of H-reflexes by a preceding subthreshold H-reflex stimulus. *Brain*, **92**, 203–12.

Thorne, J. (1965). Central responses to electrical activation of the peripheral nerves supplying the intrinsic hand muscles. *J. Neurol. Neurosurg. Psychiat.* **28**, 482–95.

Trontelj, J. V. (1973a). A study of the F response by single fibre electromyography. In *New developments in electromyography and clinical neurophysiology*, ed. J. E. Desmedt, vol. 3, pp. 318–22. Basel: Karger.

Trontelj, J. V. (1973b). A study of the H-reflex by single fibre EMG. *J. Neurol. Neurosurg. Psychiat.* **36**, 951–9.

Upton, A. R. M., McComas, A. J. & Sica, R. E. P. (1971). Potentiation of 'late' responses evoked in muscles during effort. *J. Neurol. Neurosurg. Psychiat.* **34**, 699–711.

Veale, J. L. & Hewson, N. D. (1973). Unit analysis of the F wave. *Proc. Aust. Ass. Neurol.* **10**, 129–37.

Veale, J. L. & Rees, Sandra (1973). Renshaw cell activity in man. *J. Neurol. Neurosurg. Psychiat.* **36**, 674–83.

Veale, J. L., Mark, R. F. & Rees, Sandra (1973a). Differential sensitivity of motor and sensory fibres in human ulnar nerve. *J. Neurol. Neurosurg. Psychiat.* **36**, 75–86.

Veale, J. L., Rees, S. & Mark, R. F. (1973b). Renshaw cell activity in normal and spastic man. In *New developments in electromyography and clinical neurophysiology*, ed. J. E. Desmedt, vol. 3, pp. 523–37. Basel: Karger.

Wilson, V. J. (1959). Recurrent facilitation of spinal reflexes. *J. gen. Physiol.* **42**, 703–13.

Wilson, V. J. & Burgess, P. R. (1962). Disinhibition in the cat spinal cord. *J. Neurophysiol.* **25**, 392–404.

Techniques for examining the arterial baroreceptor reflexes in the conscious state

J. LUDBROOK

A priori, it seems likely that blood pressure- and blood volume-sensed reflexes are of special importance in man, both because of his unusual ability to undergo rapid changes of posture in the course of his normal daily activities, and because of his unique susceptibility to the disease process of arterial hypertension. Yet it has not been possible to study even the arterial baroreceptor reflex system with any great precision in conscious man or even in conscious animals.

In anaesthetized animals of various species the stimulus–response characteristics of the carotid sinus baroreceptor reflex have been intensively studied, using some variant of the Moissejeff (1927) method in which transmural pressure is varied in the vascularly isolated, perfused, carotid sinus (with or without denervation of other baroreceptors, in order to create a classical open-loop preparation). However, it has become clear that in animals (Korner, 1971, 1978; Kirchheim, 1976) and in man (Bristow *et al.*, 1969*b*; Vatner, Franklin & Braunwald, 1971) general anaesthesia causes more or less profound alterations in the baroreceptor reflexes, and especially in their interactions with neural inputs that originate from, or have synaptic relays in, supramedullary brain centres. For this reason a variety of methods has been devised for reproducible stimulation or destimulation of carotid sinus baroreceptors in unanaesthetized animals. These all suffer from one or other defect or limitation. Carotid sinus nerve stimulation (Vatner, Franklin, Van Citters & Braunwald, 1970) mimics increase, but not decrease, in carotid sinus transmural pressure (CSTMP). Common carotid artery occlusion (Kirchheim, 1969; Higgins, Vatner, Eckberg & Braunwald, 1972) by means of an implanted inflatable cuff allows deductions to be made about the effects of reduction in CSTMP, but the magnitude of the reduction is uncertain and the effect is an on–off one. The information gained from techniques in which the whole population of arterial baroreceptors is stimulated or destimulated by the inflation of cuffs around the descending thoracic aorta (Scher & Young, 1970), or of cuffs around the aorta and vena cava (Korner, Shaw, West & Oliver, 1972); or in which arterial pressure is varied by the injection of vasoconstrictor drugs (Alexander & DeCuir, 1966; Vatner, Boettcher, Heyndrickx & McRitchie, 1975), have the grave limitation that baroreflex sensitivity can be gauged only with respect to heart rate or R–R heart interval. Observations of cardiovascular variables made before and after selective denervation of arterial baroreceptors provides information by subtraction (Heymans & Neil, 1958; Korner, 1965; Chalmers, Korner & White, 1967), but in an all-or-none and unidirectional fashion. The ideal technique for use in conscious animals would be a means for inducing graded variations in arterial transmural pressure that are bi-directional and quantifiable. Such a method has been described for the carotid sinus baroreceptor reflex in dogs (Shubrooks, 1972) in the form of a variable-pressure neck collar, though it does not appear to have been applied to conscious animals.

When it comes to making observations of the characteristics of the arterial baroreceptor reflexes in conscious man the technical problems are even greater. There have been some opportunities to examine the cardiovascular responses to chronically implanted carotid sinus nerve stimulators (Carlsten *et al.*, 1958; Epstein *et al.*, 1969). Common carotid artery compression under carefully controlled circumstances has been used to lower CSTMP (Roddie & Shepherd, 1957), but the technique suffers from the same disadvantages as do implanted inflatable carotid artery cuffs in animals. The tachycardia in the hypotensive period following a Valsalva manoeuvre has been used as an index of arterial baroreflex sensitivity (Wade *et al.*, 1970), but interpretation of the heart rate response to the complex stimulus is not easy (see Korner, Tonkin & Uther, 1976). Vasoactive drugs that supposedly have no cardiac or central nervous action (angiotensin, and more recently phenylephrine; amyl nitrite, and more recently glyceryl trinitrate) have been used to produce transient changes in arterial pressure, beat-to-beat change in heart rate or in R–R heart interval being used as the index of arterial baroreceptor reflex sensitivity (Robinson, Epstein, Beiser & Braunwald, 1966; Smyth, Sleight & Pickering, 1969). However, the magnitude of this beat-to-beat response is less for falling than for rising arterial pressure (Pickering, Gribbin & Sleight, 1972), so that Korner, West, Shaw & Uther (1974) have preferred to measure the 'steady-state' R–R heart interval in response to brief plateau-changes in arterial pressure induced by drugs.

Thus neither in conscious man, nor in conscious animals, has it been possible to inscribe the response of the controlled variable, arterial pressure, over the full range of the appropriate stimulus to the arterial baroreceptors, change in transmural pressure to the carotid sinus or aortic arch. Indeed the closest approximation to this has been achieved in man, by enclosing the neck within a sealed chamber inside which the pressure can be varied and thus cause inverse changes in CSTMP. Such a technique in which suction is applied to the neck was first described by Ernsting & Parry (1957), and employed systematically to examine the responses of arterial pressure, heart rate, cardiac output and forearm blood flow by Bevegård & Shepherd (1966). A simplified form of neck chamber has been more recently used by Eckberg, Cavanaugh, Mark & Abboud (1975) to study the heart rate responses. Thron, Brechmann, Wagner & Keller (1967) enclosed the head and neck in a chamber, so that either negative or positive pressure could be applied, but this necessitates the trained subject breathing through a mouthpiece, and having his external auditory canals sealed off. Ludbrook, Mancia, Ferrari & Zanchetti (1977b) have recently found it possible to apply both negative and positive pressures to the neck of up to ± 8 kPa with the head outside the chamber, so that the technique is applicable to untrained normal subjects and patients. We have also been able to show by direct measurement (Ludbrook et al., 1977b) that while the pneumatic pressure changes are not transmitted perfectly to the outside of the carotid sinus, the transmission is predictable enough to allow calculation of the actual change in CSTMP. Eckberg (1976) found that the inertial time delay of negative pressure transmission to the region of the carotid sinus was short. Eckberg et al. (1975) had shown previously that neck suction does not interfere with chemoreceptor activity, and we have shown that application of a positive pressure of up to 8 kPa does not appear to alter chemoreceptor activity, nor do applied pressures of ± 8 kPa alter cerebral blood flow (Ludbrook et al., 1977b).

Several aspects of carotid baroreceptor reflex function have been studied by this technique. The magnitude of the immediate bradycardiac response to sudden increase in CSTMP seems to depend on the time relation of the stimulus to the electrical events of the cardiac cycle (Eckberg, 1976). In normal subjects, for equivalent changes in CSTMP the immediate brady-cardic and hypotensive response to sudden increase in CSTMP is greater than the tachycardic and hypertensive response to sudden reduction (Thron et al., 1967; Ludbrook et al., 1977a; Mancia et al., 1977a,b; see also Fig. 1). However, if the altered CSTMP is sustained for 1–2 min, the situation reverses itself: the response to a sustained reduction in CSTMP is greater than that to a sustained increase (Thron et al., 1967; Bjurstedt, Rosen-hamer & Tydèn, 1975; Ludbrook et al., 1977a; Mancia et al., 1977a; see also Fig. 1). On the other hand, from a study of normotensive and hyper-tensive subjects it appears that the greater the resting level of blood pres-sure, the greater the depressor response to sustained increase in CSTMP, and the less the sustained pressor response to the opposite stimulus

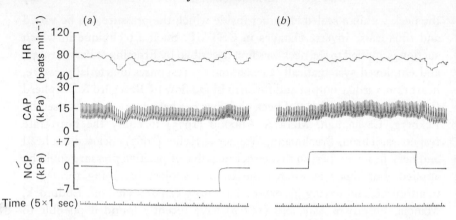

Fig. 1. Effect of alteration in carotid sinus transmural pressure (CSTMP) by means of a variable-pressure neck chamber on beat-to-beat heart rate (HR) and full-wave carotid arterial pressure (CAP) in a normal, seated subject. (a) Increase in CSTMP; (b) reduction in CSTMP. Bottom trace indicates changes in neck chamber pressure (NCP) that produce corresponding initial changes in CSTMP of +4.4 kPa (a) and −4.5 kPa (b).

(Mancia et al., 1977c and unpublished observations). This suggests that in normal human subjects the set point for blood pressure is towards the lower inflection of the sigmoid stimulus–response curve for blood pressure versus CSTMP, while in hypertensive subjects the set point is at the upper inflection. In other words, normotensive subjects may be better adapted to resist fall in blood pressure, hypertensives to resist further rise.

In normal subjects it appears that the depressor response of the carotid baroreceptor reflex is brought about mainly by fall in peripheral resistance (Bjurstedt et al., 1975), and a comparable fall in vascular resistance in fore-arm muscle and forearm skin has been described (Bevegård & Shepherd, 1966; Beiser et al., 1970). On the other hand, increase in cardiac output makes an important contribution to the pressor response of the carotid baroreceptor reflex (Bjurstedt et al., 1975), though there is a suggestion that splanchnic vasoconstriction may also occur (Mark, Eckberg & Abboud, 1975). These phenomena are consistent with the observation that while the pressor response of the carotid baroreceptor reflex is accompanied by a parallel rise in heart rate, the depressor response is not necessarily accom-panied by a sustained fall in heart rate (Bjurstedt et al., 1975; Mancia et al., 1977a; Ludbrook et al., 1977a; see also Fig. 1). This does create some doubt that changes in heart rate or R–R interval (vide supra) can be used as a reliable index of the capacity of the arterial baroreceptor reflexes to effect homeostasis of blood pressure.

Comparison of the sensitivity or gain of the carotid baroreceptor reflex with respect to arterial pressure has been made between normotensive and hypertensive subjects, with somewhat confusing results. Wagner & Thelen (1968) and Wagner, Wackerbauer & Hilger (1968) found no difference. As described earlier, we (Mancia *et al.*, 1977*c* and unpublished observations) have found that the more severe the hypertension, the smaller the gain to reduction in CSTMP and the greater the gain to increase in CSTMP. Neither of these sets of results, in which baroreflex control of blood pressure was examined directly, accord with the reports of marked reduction, in hypertensives, of gain for heart interval, tested by vasoactive drug injection by Bristow *et al.* (1969*a*) and Korner *et al.* (1974).

Postural change seems to have little effect on the capacity of the carotid baroreflex to control blood pressure (Wagner & Thelen, 1968; Eckberg *et al.*, 1976); though there is just a suggestion that foot-down tilting (Bjurstedt *et al.*, 1975) or lower-body suction (Bevegård, Castenfors & Linblad, 1977) may cause an increase in reflex gain with respect to increase in CSTMP. The state of physical fitness does not seem to affect the reflex (Stegemann, Busert & Brock, 1974).

Interesting results have been obtained by studying the carotid baroreceptor reflex during exercise. On a basis of the beat-to-beat heart interval response to the rising blood pressure after phenylephrine injection, Cunningham *et al.* (1972) had suggested that during exercise the sensitivity of the arterial baroreceptor reflexes is greatly diminished, thus permitting the simultaneous rise in blood pressure and heart rate that is observed to occur. However, using the neck chamber technique Mancia *et al.* (1977*b*) found that while the transient increase of heart interval that follows step-increase in CSTMP was greatly reduced during isometric handgrip exercise, the response to step-decrease in CSTMP was scarcely altered. A recent and more exhaustive study (Ludbrook *et al.*, 1977*a*) has shown that carotid baroreflex gain with respect to blood pressure control is entirely unaffected by isometric handgrip. This allows the hypothesis to be advanced that the systemic cardiovascular changes of exercise are initiated by elevation of the central set point for blood pressure in the central nervous system, so that arterial pressure (and heart rate) rise towards this new set point.

Nevertheless, for several reasons the variable-pressure neck chamber technique has distinct limitations. The transmission of pneumatic pressure to the carotid sinus is much better predicted within the individual than between individuals, so that the technique is likely to give more precise results for within-individual rather than between-individual comparisons of reflex gain. In human volunteers considerations not merely of ethics but also of safety severely limit the study of circumstances that might interact with the carotid baroreceptor reflex to alter its characteristics. CSTMP can

230   J. LUDBROOK

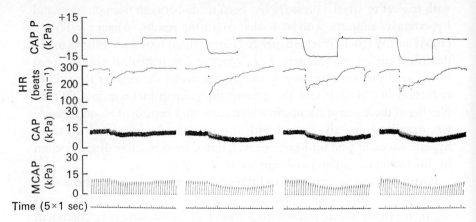

Fig. 2. Effects of reduction in pressure within a capsule chronically implanted around the carotid bifurcation of a NZ White rabbit (causing equal and opposite increase in carotid sinus transmural pressure) on heart rate and arterial blood pressure. Four sets of observations made in the conscious state.

CAP P = change in pressure in the capsule; HR = beat-to-beat heart rate; CAP = full wave arterial pressure at the level of the carotid sinus, via the central ear artery; MCAP = integrated (mean) arterial pressure.

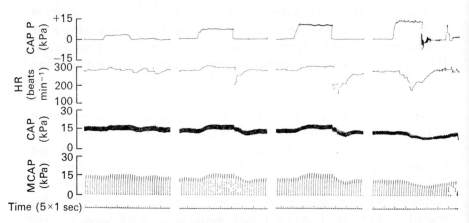

Fig. 3. Effects of increase in pressure within a capsule chronically implanted around the carotid bifurcation of a NZ White rabbit (causing equal and opposite decrease in carotid sinus transmural pressure) on heart rate and arterial blood pressure. Four sets of observations made in the conscious state. CAP P = change in pressure in the capsule; HR = beat-to-beat heart rate; CAP = full wave arterial pressure at the level of the carotid sinus, via the central ear artery; MCAP = integrated (mean) arterial pressure. Note the paradoxical effect when pressure in the capsule exceeds systolic carotid arterial pressure.

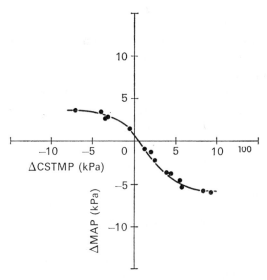

Fig. 4. Plot of change in mean steady-state arterial pressure (ΔMAP) against measured steady-state change in carotid sinus transmural pressure (ΔCSTMP) in a conscious NZ White rabbit. ΔCSTMP caused by altering pressure in a capsule around one carotid bifurcation, the opposite sinus nerve and both aortic nerves having been divided. Line of best fit ($r = 0.985$, $n = 14$) calculated by logit transformation of y axis.

be altered over too narrow a range to allow the upper and lower limits of responses of the cardiovascular variables to be attained. The aortic baroreceptor reflex would be expected partly to buffer the effects of change in CSTMP, even though its effectiveness in man is questionable (Holton & Wood, 1965; Mancia et al., 1977a). The attenuation of the initial depressor response, and the continuing evolution of the pressor response (Fig. 1), suggest that even in trained volunteers activity of the neck muscles may be superimposing an effect of isometric exercise.

For these several reasons we have recently set out to devise a method of altering CSTMP in a quantifiable fashion in the conscious animal. This has recently been achieved in rabbits (J. Ludbrook and G. G. Jamieson, unpublished). It has been possible to chronically implant a capsule around the carotid bifurcation, which on the Guyton (1963) principle communicates with the tissue fluid, and which is accessible to percutaneous puncture so that the pressure outside the carotid sinus can be measurably varied over a range of ± 16 kPa. The opposite carotid sinus nerve, and the aortic nerves, can be divided in order to eliminate the buffer effects of other arterial baroreflexes. It has thus far been possible to construct repeatedly and with great precision the full sigmoid curve relating arterial pressure change to

change in CSTMP in the conscious rabbit (Figs. 2, 3, 4). It is noteworthy
that the response of heart rate does not correlate well with that of the controlled variable, arterial pressure. This heightens the doubts that deductions about the arterial baroreceptor reflex that are based on heart rate
responses will fairly reflect the characteristics of the reflex so far as blood
pressure control is concerned. Several aspects of this technique remain to be
evaluated – for instance, whether alteration of pressure within the capsule
affects the carotid chemoreceptors – but it offers promise as a method for
precise study of the carotid baroreceptor reflex that can be adapted to any
species of laboratory animal.

I am most grateful to my colleagues, Professor G. Mancia of the Istituto di Ricerche
Cardiovascolari dell'Università di Milano and Mr G. G. Jamieson of my own
Department of Surgery, for the original work in this review, and to the Università
di Milano of Italy, the National Heart Foundation of Australia and the University
of Adelaide for supporting it.
    However, the greatest debt I owe is to Archie McIntyre, who, while I was a
B.Med.Sc. student at the University of Otago, inculcated in me what rudimentary
insight I possess into central nervous system reflexes.

REFERENCES

Alexander, N. & DeCuir, M. (1966). Loss of baroreflex bradycardia in renal hyper-
    tensive rabbits. *Circulation Res.* **19**, 18–25.
Beiser, G. D., Zelis, R., Epstein, S. E., Mason, D. T. & Braunwald, E. (1970). The
    role of skin and muscle resistance vessels in reflexes mediated by the baroreceptor
    system. *J. clin. Invest.* **49**, 225–31.
Bevegård, S., Castenfors, J. & Linblad, L. E. (1977). Blood pressure and heart rate
    regulating capacity of the carotid sinus during changes in blood volume distri-
    bution in man. *Acta. physiol. scand.* **94**, 300–12.
Bevegård, S. & Shepherd, J. T. (1966). Circulatory effects of stimulating the carotid
    arterial stretch receptors in man at rest and during exercise. *J. clin. Invest.* **45**,
    132–42.
Bjurstedt, H., Rosenhamer, G. & Tydèn, G. (1975). Cardiovascular responses to
    changes in carotid sinus transmural pressure in man. *Acta physiol. scand.* **94**,
    497–505.
Bristow, J. D., Honour, A. J., Pickering, G. W., Sleight, P. & Smyth, H. S. (1969a).
    Diminished baroreflex sensitivity in high blood pressure. *Circulation* **39**, 48–54.
Bristow, J. D., Prys-Roberts, C., Fisher, A., Pickering, T. G. & Sleight, P. (1969b).
    Effects of anesthesia on baroreflex control of heart rate in man. *Anesthesiology* **31**,
    422–8.
Carlsten, A., Folkow, B., Grimby, G., Hamberger, C. A. & Thulesius, O. (1958).
    Cardiovascular effects of direct stimulation of the carotid sinus nerve in man.
    Acta physiol. scand. **44**, 138–45.
Chalmers, J. P., Korner, P. I. & White, S. W. (1967). Effects of haemorrhage on
    the distribution of the peripheral blood flow in the rabbit. *J. Physiol., Lond.* **192**,
    561–74.
Cunningham, D. J. C., Petersen, E. S., Peto, R., Pickering, T. G. & Sleight, P.

(1972). Comparison of the effect of different types of exercise on the baroreflex regulation of heart rate. *Acta physiol. scand.* **86**, 444–55.

Eckberg, D. L. (1976). Temporal response patterns of the human sinus node to brief carotid baroreceptor stimuli. *J. Physiol., Lond.* **258**, 769–82.

Eckberg, D. L., Abboud, F. M. & Mark, A. L. (1976). Modulation of carotid baroreflex responsiveness in man: effects of posture and propranolol. *J. appl. Physiol.* **41**, 383–7.

Eckberg, D. L., Cavanaugh, M. S., Mark, A. L. & Abboud, F. M. (1975). A simplified neck suction device for activation of carotid baroreceptors. *J. Lab. clin. Med.* **85**, 167–73.

Epstein, S. E., Beiser, G. D., Goldstein, R. E., Stampfer, M., Wechsler, A. S., Glick, G. & Braunwald, E. (1969). Circulatory effects of electrical stimulation of the carotid sinus nerves in man. *Circulation* **40**, 269–76.

Ernsting, J. & Parry, D. J. (1957). Some observations on the effects of stimulating the stretch receptors in the carotid artery of man. *J. Physiol., Lond.* **137**, 45P.

Guyton, A. C. (1963). A concept of negative interstitial pressure based on pressures in implanted perforated capsules. *Circulation Res.* **12**, 399–414.

Heymans, C. & Neil, E. (1958). *Reflexogenic areas of the cardiovascular system.* London: Churchill.

Higgins, C. B., Vatner, S. F., Eckberg, D. L. & Braunwald, E. (1972). Alterations in the baroreceptor reflex in conscious dogs with heart failure. *J. clin. Invest.* **51**, 715–24.

Holton, P. & Wood, J. B. (1965). The effects of bilateral removal of the carotid bodies and denervation of the carotid sinuses in two human subjects. *J. Physiol., Lond.* **181**, 365–78.

Kirchheim, H. (1969). Effect of common carotid occlusion on arterial blood pressure and on kidney blood flow in unanaesthetised dogs. *Pflügers Arch. ges. Physiol.* **306**, 119–34.

Kirchheim, H. (1976). Systemic arterial baroreceptor reflexes. *Physiol. Rev.* **56**, 100–76.

Korner, P. I. (1965). The effect of section of the carotid sinus and aortic nerves on the cardiac output of the rabbit. *J. Physiol., Lond.* **180**, 266–78.

Korner, P. I. (1971). Integrative neural cardiovascular control. *Physiol. Rev.* **51**, 312–67.

Korner, P. I. (1978). Central nervous control of the heart and circulation. In *Handbook of physiology: circulation,* 2nd edn Washington: Am. Physiol. Soc. (In Press.)

Korner, P. I., Shaw, J., West, M. J. & Oliver, J. R. (1972). Central nervous system control of baroreceptor reflexes in the rabbit. *Circulation Res.* **31**, 637–52.

Korner, P. I., Tonkin, A. M. & Uther, J. B. (1976). Reflex and mechanical circulatory effects of graded Valsalva maneuvers in normal man. *J. appl. Physiol.* **40**, 434–40.

Korner, P. I., West, M. J., Shaw, J. & Uther, J. B. (1974). 'Steady-state' properties of the baroreceptor–heart rate reflex in essential hypertension in man. *Clin. exp. Pharmac. Physiol.* **1**, 65–76.

Ludbrook, J., Faris, I. B., Iannos, J., Jamieson, G. G. & Russell, J. W. (1977a). Lack of effect of isometric handgrip exercise on the blood pressure response to carotid baroreceptor stimulation and destimulation in man. *Proc. Aust. physiol. pharmac. Soc.* **8** (No. 2), 199.

Ludbrook, J., Mancia, G., Ferrari, A. & Zanchetti, A. (1977b). The variable pressure neck chamber method for studying the carotid baroreflex in man. *Clin. Sci. mol. Med.* **53**, 165–72.

Mancia, G., Ferrari, A., Gregorini, L., Valentini, R., Ludbrook, J. & Zanchetti, A.

(1977*a*). Circulatory reflexes from carotid and extracarotid baroreceptor areas in man. *Circulation Res.* **41**, 309–15.

Mancia, G., Iannos, J., Jamieson, G. G., Lawrence, R. H., Sharman, P. R. & Ludbrook, J. (1977*b*). The effect of isometric handgrip exercise on the carotid sinus baroreceptor reflex in man. *Clin. Sci. molec. Med.* **54**, 33–7.

Mancia, G., Ludbrook, J., Ferrari, A., Gregorini, L., Valentini, R. & Zanchetti, A. (1977*c*). Carotid baroreceptor reflex in normotensive and hypertensive subjects. *Clin. Sci. molec. Med.* **51**, 343–5*s*.

Mark, A. L., Eckberg, D. L. & Abboud, F. M. (1975). Selective contribution of cardiopulmonary and carotid baroreceptors to forearm and splanchnic vasoconstrictor responses during venous pooling in man. *Physiologist, Wash.* **18**, 305.

Moissejeff, E. (1927). Zur Kenntnis des Carotissinusreflexes. *Z. ges. exp. Med.* **53**, 696–704.

Pickering, T. G., Gribbin, B. & Sleight, P. (1972). Comparison of the reflex heart rate response to rising and falling arterial pressure in man. *Cardiovascular Res.* **6**, 277–83.

Robinson, B. F., Epstein, S. E., Beiser, G. D. & Braunwald, E. (1966). Control of heart rate by the autonomic nervous system; studies in man on the interrelation between baroreceptor mechanisms and exercise. *Circulation Res.* **19**, 400–11.

Roddie, I. C. & Shepherd, J. T. (1957). The effects of carotid artery compression in man with special reference to changes in vascular resistance in the limbs. *J. Physiol., Lond.* **139**, 377–84.

Scher, A. M. & Young, A. C. (1970). Reflex control of heart rate in the unanesthetised dog. *Am. J. Physiol.* **218**, 780–9.

Shubrooks, S. J. (1972). Carotid sinus counterpressure as a baroreceptor stimulus in the intact dog. *J. appl. Physiol.* **32**, 12–19.

Smyth, H. S., Sleight, P. & Pickering, G. W. (1969). Reflex regulation of arterial pressure during sleep in man: a quantitative method of assessing baroreflex sensitivity. *Circulation Res.* **24**, 109–21.

Stegemann, J., Busert, A. & Brock, D. (1974). Influence of fitness on the blood pressure control system in man. *Aerospace Med.* **45**, 45–8.

Thron, H. L., Brechmann, W., Wagner, J. & Keller, K. (1967). Quantitative untersuchungen über die Bedeutung der Gefässdehnungsreceptoren in Rahmen der Kreislaufhomoistase beim wachen Menschen. *Pflügers Arch. ges. Physiol.* **293**, 68–99.

Vatner, S. F., Boettcher, D. H., Heyndrickx, G. R. & McRitchie, R. J. (1975). Reduced baroreflex sensitivity with volume loading in conscious dogs. *Circulation Res.* **37**, 236–42.

Vatner, S. F., Franklin, D., Braunwald, E. (1971). Effects of anesthesia and sleep on circulatory response to carotid sinus nerve stimulation. *Am. J. Physiol.* **220**, 1249–55.

Vatner, S. F., Franklin, D., Van Citters, R. L. & Braunwald, E. (1970). Effects of carotid sinus nerve stimulation on blood flow distribution in conscious dogs at rest and during exercise. *Circulation Res.* **27**, 495–503.

Wade, J. G., Larson, C. P., Hickey, R. F., Ehrenfeld, W. K. & Severinghaus, J. W. (1970). Effect of carotid endarterectomy on carotid chemoreceptor and baroreceptor functions in man. *New Engl. J. Med.* **282**, 823–9.

Wagner, J. & Thelen, M. (1968). Über den Einfluss der Carotis-Sinusreceptoren auf arteriellen Blutdruck und Herzfrequenz beim Hypertoniker im Liegen und Stehen. *Verh. d. Ges. inn. Med.* **74**, 703–7.

Wagner, J., Wackerbauer, J. & Hilger, H. H. (1968). Arterielles Blutdruck- und Herzfrequenzverhalten bei Hypertonikern unter Änderung des transmuralen Druckes im Karotissinusbereich. *Z. Kreislaufforsch.* **57**, 703–12.

Neural factors in the
control of breathing during exercise

J. D. SINCLAIR

A neurophysiologist would begin to study the control of breathing by studying the characteristics of the cervical and thoracic motoneurones which determine the ventilatory activity of the diaphragm and the intercostal muscles. Andersen & Sears (1970) used this approach to produce noteworthy evidence concerning the role of the apneustic centre. However, most physiologists interested in the control of breathing maintain the traditional preoccupation with central pathways: with the basic organization of hindbrain centres and with peripheral and central factors influencing these centres (Mitchell & Berger, 1975). The difference of approach may well result not from the difference of discipline but simply from a difference in estimates of where the major findings are likely to be produced by current techniques.

Respiratory physiologists see the control of breathing during exercise as an outstanding example of a precisely governed system, and here is where their excitement is centred. Alveolar ventilation is so closely related to oxygen consumption that arterial levels of oxygen do not change when metabolism increases to five or ten times the resting level. This well-established phenomenon, reviewed by Wasserman & Whipp (1975), is further supported by careful studies of gas exchange in healthy people undertaken by Harris and his colleagues at Green Lane Hospital, Auckland (Harris, Seelye & Whitlock, 1976; Bradley, Harris, Seelye & Whitlock, 1976). They confirm the increase of physiological deadspace in exercise and demonstrate the reduction of shunting of venous blood through lung spaces. This sort of study has been central in respiratory physiology for 25 years.

The accuracy of our information on gas exchange in exercise stands in sharp contrast with the imprecision of information on respiratory control, which is only now receiving widespread attention. Details of neural mechanisms, chemical influences and central rhythmicity affecting the normal human subject are just beginning to emerge.

Neurogenic theories of the control of exercise ventilation have been explored by those hypothesizing an origin in the 'command' stimulus and others seeking major excitatory input from muscle spindles or other limb receptors. The standing of the fast, neurogenic element outlined by Dejours (1964) has advanced so little, however, that Guz (1975) has concluded that its existence and meaning cannot be solved with present methods.

Beaver & Wasserman (1970) concluded that the command signal was a learned response; they found no consistent change in ventilation at the onset and end of short periods of exercise through a range of workloads. There was no consistency between the rapid changes at onset and termination, nor was there a significant difference according to workload. Yet Asmussen (1973), in naive subjects, demonstrated a fast component of ventilatory response which increased with the bicycle work performed, but which did not occur in the presence of hyperoxygenation and hypocapnia, did not occur in isometric work, and increased with the frequency of movements associated with the work. It is difficult to believe that a learned response would be suppressed by removal of chemical drive in a normal subject. Further support for the existence of a significant 'command' signal to ventilation was produced by the ingenious experiments of Goodwin, McCloskey & Mitchell (1971). They applied a vibrator to tendons to confuse their subjects as to the amount of physical work they were undertaking. The exercise was isometric, at 50% of the subjects' capacities. Ventilation increased in a way that was closer to the subjects' interpretation of the involved work than to the actual physical work done.

As for the influence of chemical factors, orthodox studies of human responses to hypoxia and hypercapnia continue to produce useful information; a short period in which bilateral excision of the carotid bodies was undertaken for the relief of asthma produced some valuable subjects for human experiments; the precise role of the carotid bodies has been established in a variety of animals; and the nature and role of central chemoreceptors is being clarified with increasing satisfaction.

Some of the most interesting current developments in respiratory physiology concern the organization of respiratory rhythm in the pontomedullary centres. Anatomical and physiological studies have greatly clarified the nature and role of apneustic and pneumotaxic centres, and of the origin of rhythmicity. The significance of vagal input to the hindbrain from inflation receptors of airways and lung has been demonstrated convincingly: the results have allowed the presentation of a specific model for the mechanisms by which chemical drive and pulmonary inflation modulate a basic respiratory pattern (Clark & von Euler, 1972). The special factors relevant to exercise respiration should now be restudied in the light of this model. But it is likely to be a long time before there is clear evidence as to the influence of the forebrain and midbrain on patterns demonstrable in hindbrain controls of anaesthetized animals.

In the present article, major emphasis will be given to recent experiments on patterns of respiration and on chemosensitivity during exercise. Other aspects of exercise ventilation will be reviewed briefly.

GENERAL METHODOLOGICAL PROBLEMS

Animal studies

The desirability of designing exercise studies on unanaesthetized animals is at least equalled by the technical difficulties of their achievement. Precise analysis of neural patterns requires monitoring of arterial or at least alveolar gas levels; measurement of respiratory frequency and tidal volume, preferably with functional residual volume so that inflation input is known; the capacity to produce local ablation or stimulation in the central nervous system, or at least to separate hindbrain, midbrain and forebrain influences; and a certainty of upper and lower motoneurone output, especially where mechanical changes in lung or thorax might reduce the value of ventilation as an index of neural function. It is difficult to achieve all these in a single species.

There is increasing use of the rat in respiratory studies, and this animal is proving valuable for studies of patterns of frequency and tidal volume; further, it is readily enough trained to exercise on a simple treadmill or by swimming. Unfortunately it is too small for quantitative studies of blood gas levels, and difficult for studies of central pathways dependent on transection of the brain. In the dog, which is readily trained to exercise, surgical procedures and repeated arterial sampling are feasible but the measurement of tidal volume without tracheostomy is not. Large animals such as the sheep and llama are satisfactory for resting studies but could prove cumbersome on the treadmill. Perhaps the goat offers the best combination of size and educability; its friskiness is an obvious limitation.

Specific components of neural control generally must be studied in the anaesthetized animal. Inputs from the exercising limb have been variously stimulated. Kao, Lahiri, Wang & Mei (1967) produced muscle contraction by stimulation of the sciatic nerve. They also produced respiratory stimulation without an increase in oxygen consumption by rhythmic squeezing of the calf muscles at a rate of 120 min^{-1}. They argued that passive movement of the limbs was not comparable with active exercise and that the associated increases in ventilation were only proportional to increases of oxygen uptake. Barron & Coote (1973) studied the role of joint receptors in the decerebrate cat after partially denervating the limb to remove other neural inputs. Muscle spindle inputs have been stimulated by vibration (Hodgson & Matthews, 1968) and by the intra-arterial administration of succinyl-choline (Gautier, Lacaisse & Dejours, 1969).

Studies of basic respiratory patterns in the unanaesthetized animal have been much assisted by the development of the whole body plethysmograph which makes possible measurements of tidal volume and frequency in the unrestrained animal; adequate routine training and familiarization allow surprisingly complex studies, such as those of Tenney & Ou (1977), in which chemical responses were compared in normal, decorticated and decerebrate cats. These are steady-state studies, tidal volume being calculated from the pressure changes in a closed box, resulting from changes of temperature and water vapour pressure of the inspired air.

Human studies

Current studies involve modest modifications of the classical approach. Forms of exercise are usually based on the bicycle ergometer or treadmill. For specific purposes it may be desirable to compare the effects of leg exercise with arm exercise, or isometric exercise with isotonic exercise (Hanna, Hill & Sinclair, 1975) but the respiratory physiologists have not gone as far as the exercise physiologists in comparing the effect of exercise in one trained leg with the other leg as an untrained control (Saltin *et al.*, 1976).

Steady-state procedures for measuring exercise response are being supplemented with measurements of responses to transient change; these vary from the rebreathing method of following progressive changes in the ventilation:pCO_2 relationship, popularized by Read (1967), to the two-breath changes used by Cunningham *et al.* (1977). Beaver & Wasserman (1970) compared 'on' and 'off' respiratory transients, while Asmussen (1973) studied the transient 'on' effect in the presence of varying chemical stimulation.

Measurement

In respiratory studies, it has long been assumed that alveolar ventilation represents neural inspiratory drive. This assumption depends on constancy of mechanical properties of lungs and thorax, whereas these are affected by carbon dioxide (altering airway resistance) hypoxia (altering functional residual volume, FRC) exercise (altering FRC) and anaesthesia (altering lung distension). Recent methods of measuring neural drive involve integration of phrenic nerve activity, integration of electrical activity of the diaphragm, or the measurement of airway pressure during very brief occlusion; but these seem unlikely to be practical in exercise studies.

PONTO-MEDULLARY ORGANIZATION

Specific neural inputs associated with exercise can be considered now against increasing appreciation of the structural organization of the classical pontine and medullary 'centres'.

The medulla has been shown to contain three distinct groups of neurones involved in respiratory control. There is a dorsal group, associated with the nucleus of the tractus solitarius and responding rhythmically to lung inflation or to vagal stimulation. These cells apparently activate others in the medulla and almost certainly act as the first central synapse in the pathway of the Hering–Breuer reflex (von Euler, Hayward, Martilla & Wyman, 1973). Then, among ventral neurones, Merrill (1970), from experiments using antidromic stimulation, claims that there are two groups, one associated with the nucleus ambiguus and one with the nucleus retroambigualis. Respiratory neurones in the area of the nucleus ambiguus are thus placed in close proximity to the motor nucleus of the IXth and Xth cranial nerves, serving accessory muscles of respiration. The second ventral group was found by Merrill to consist of neurones of both inspiratory and expiratory characteristics, sited as a column in the nucleus retroambigualis and extending as far caudally as the first cervical segment. Many are found to project to the thoracic cord. Their detailed function has not yet been demonstrated.

The 'apneustic centre' can be taken from the work of Andersen & Sears (1970) to represent the site of origin of reticulo-spinal fibres which connect to respiratory motoneurones of the spinal cord. The sustained contraction of the external intercostal muscles, produced by stimulation medially in the reticular formation of the medulla, resulted from stimulation of both α and γ motoneurones; at the same time there was inhibition of the motoneurones supplying the internal intercostal muscles. Thus, the apneustic centre no longer appears to be primarily involved in the establishment of respiratory rhythm, but to affect the fine control of respiratory movement.

The 'pneumotaxic centre', on the other hand, is now regarded as the site of the oscillatory network establishing respiratory rhythm (Bertrand, Hugelin & Vibert, 1974). The neurones are sited dorsally, laterally and rostrally in the pons in the area of the nucleus parabrachialis. There are clear groups of inspiratory neurones, expiratory neurones and neurones whose pattern of excitation spans the full respiratory cycle. Stimulation at various sites can either advance or terminate phrenic inspiratory activity. It appears that these are the neurones which determine the basic inspiratory–expiratory activity of the respiratory neurones of the medulla.

The chemosensitive function of the medulla remains something of a mystery (Feldberg, 1976). There has been no evidence, however, that the respiratory neurones of the pons and medulla show a special sensitivity to carbon dioxide or H⁺. A few specific studies suggest that they share the general neuronal depression by carbon dioxide (Marino & Lamb, 1975; Mitchell & Berger, 1975). The presumptive chemoreceptor area is on the ventrolateral surface of the medulla where ventilatory responses to the direct application of carbon dioxide and H⁺ were demonstrated by Mitchell, Loeschcke, Massion & Severinghaus (1963), confirming suspicions from earlier experiments of Winterstein, Leusen and Loeschcke, reviewed by Mitchell *et al.* Often, however, the pH or application of perfusates was not adequately controlled; carefully controlled perfusion had been shown to give reliable results (Feldberg & Malcolm 1959).

Mitchell *et al.* (1963) showed that carbon dioxide, H⁺, nicotine and acetylcholine stimulated respiration by an action on the ventrolateral medulla, but not in the fourth ventricle. The effects of perfusion were confirmed by direct application of agents on pledgets 2–5 mm² in area. The latency in effect of elevated pCO_2 in a perfusing medium chemically like the cerebrospinal fluid was 3 ± 1 sec; the latency of response to acids other than carbonic was 8 ± 4 sec. Respiration was depressed by perfusion with procaine. Because of the speed of ventilatory responses, the investigators concluded that the chemosensory area was superficial; and was therefore separate from the major respiratory centres of the pons and medulla.

Many subsequent experiments support the conclusions of Mitchell *et al.* (1963), e.g. in terms of the interrelationships of gaseous and non-gaseous acids, of H⁺ and K⁺ and other ions, the effects of hyperventilation, altitude, metabolism, etc. The evidence is reviewed by Leusen (1972). Trouth, Loeschcke & Berndt (1973) have identified likely chemosensitive cells in the superficial 0.2 mm of the medulla.

It should still be remembered that there has been one major contradictory finding. Pappenheimer, Fencl, Heisey & Held (1965) perfused the ventriculo-cisternal system of unanaesthetized goats through indwelling cannulae, and were able to study the ventilatory responses to changes of the cerebrospinal fluid bicarbonate, carbon dioxide and H⁺. Responses were much greater than in the anaesthetized animal. Ventilation was not a single function of cerebrospinal fluid pH but of pH at a position, by Pappenheimer's calculations, 70% of the distance, measured as a bicarbonate gradient, from the cerebrospinal fluid to the blood. The reason to remember this anomalous result is that it comes from one of the few experiments conducted on central chemoreceptor function in the intact and unanaesthetized animal.

PATTERNS OF FREQUENCY AND TIDAL VOLUME

Anaesthetized animals

The neural mechanisms serving increases of frequency and tidal volume have been greatly clarified by the work of von Euler and his associates (Clark & von Euler, 1972; Bradley, von Euler, Marttila & Roos, 1975). They have incorporated into a relatively simple model the three basic elements of respiratory rhythm – the bulbopontine oscillator, the vagally conveyed pulmonary inflation input, and the chemical drive. Limited as the model may be, it remains the basis onto which exercise influences need to be superimposed.

In their most important study, Clark & von Euler (1972) demonstrated that in the cat, anaesthetized with pentobarbitone, there was a constant relationship of tidal volume (V_T) to the duration of inspiration (T_I). At increasing levels of ventilation, tidal volume increased and inspiratory duration decreased. Inspiratory activity was cut off when tidal volume, or experimentally applied pulmonary inflation, reached a threshold level; and this threshold fell with the passage of time. There was thus established a hyperbolic relationship between tidal volume and inspiratory time, of the form: $(VT - Vo) T_I = C$ where Vo and C are constants. Clark & von Euler did not find T_I dependent on rate of increase of V_T, only on volume. The duration of expiration was dependent on the duration of inspiration. And the relationship of V_T to T_I was the same whether increased ventilation resulted from hypercapnia or from experimentally produced inflation.

When pulmonary inflation input was excluded by the process of bilateral vagotomy, respiratory frequency no longer increased with respiratory drive; so that in the absence of reductions of T_I, increased ventilation resulted only from increases of V_T. Elevation of body temperature, by contrast, specifically reduced T_I and thus increased respiratory frequency.

Temporarily ignoring other parts of the experiments, it is convenient to take this model of Clark & von Euler's and to consider the role of the three basic components.

The bulbopontine oscillator has intrinsic rhythmicity; its organization therefore sets a basic inspiration–expiration process; it is directly influenced by temperature so that increased temperature produces increased frequency of oscillation; it responds to increased chemical drive with increased inspiratory firing but not by change of frequency; and its inspiratory phase is cut off by the input of inflation receptors at a threshold that falls with increasing duration of inspiration.

Chemical drive to breathing, whatever the chemicals, receptors, and pathways involved, produces increases of inspiratory firing rate and therefore an inclination to increased tidal volume.

Pulmonary inflation input cuts off inspiratory activity of the central mechanism; but because of the falling central threshold, a greater tidal volume is required to produce cut-off early in inspiration, and conversely, a smaller tidal volume produces cut-off if inspiration is slow.

This model has a number of implications for studies of a more complex nature, for example in the unanaesthetized animal or in the human subject.

(1) Hypoxia and hypercapnia have the same effect on the respiratory pattern at any given level of ventilation.

(2) Despite the diversity of their neural and respiratory effects, hypoxia and hypercapnia act on the respiratory pattern in one way only, as does pyrexia, though by a different mechanism.

(3) Expiratory time is dependent only on inspiratory time.

(4) Optimization of ventilatory work as an ideal of its elastic and viscous components must be programmed in central characteristics rather than achieved through feedback controls adjusting tidal volume and frequency.

Before considering patterns in more complex situations, further experiments of Clark and von Euler (1972) should be noted. When they anaesthetized cats with urethane, to produce rapid shallow respiration, they demonstrated a phase of respiration which lacked inflation cut-off, in which tidal volume increased without an increase of frequency. They called this pattern Range 1, and proposed that here ventilation was so shallow that vagal cut-off was not attained. As ventilation increased under carbon dioxide stimulation, tidal volume increased until, above a given level, the cut-off phenomenon was again observed, in the phase they called Range 2.

Unanaesthetized animals

Gautier (1976) has induced hyperventilation by hypoxia and hypercapnia in the awake cat and compared the respiratory patterns with those of the anaesthetized animal. He showed that at the greater levels of ventilation of the cat when awake (approximately double the levels seen in anaesthesia at increased pCO_2), the pattern of respiration was different from that in anaesthesia and it varied with the nature of the respiratory stimulus. In particular, in normoxic hypercapnia T_I did not decrease at higher levels of ventilation, whereas in anaesthesia, T_I decreased as V_T increased, confirming the findings of Clark & von Euler (1972). Thus in the cat when awake and breathing carbon dioxide, the plot of V_T/T_I had a totally new contour; as V_T increased from initial levels of 30–40 ml to 75 ml, T_I increased from 1 sec to 1.3 sec; only as V_T increased further to 110 ml did T_I fall, to 0.9–1 sec. The relationship of V_T to the duration of expiration (T_E) was more consistent. It follows that with an early reduction of T_E there was an increase of T_I and only at higher levels of ventilation did the two show proportional decrements. In normocapnic hypoxia, although ventilatory levels increased in the awake animal, respiratory patterns were similar to those in the anaesthetized animal. Thus Gautier (1976) concludes that without the depressive influence of anaesthesia, carbon dioxide has a specific effect on respiratory patterns at lower levels of ventilation; it tends to increase the duration of inspiration despite the increases of lung inflation which would be expected to terminate inspiratory activity.

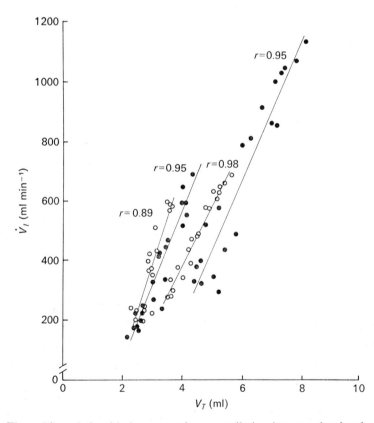

Fig. 1. The relationship between minute ventilation (measured as inspired volume, $\dot{V}_I$) and tidal volume (V_T) in four unanaesthetized rats (different symbols) in which ventilation increased in response to normoxic hypercapnia (unpublished data of R. L. Laughton).

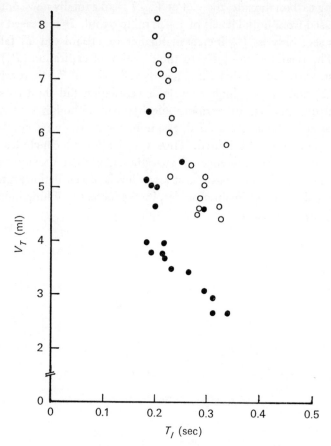

Fig. 2. The relationship between tidal volume (V_T) and inspiratory duration (T_I) in the unanaesthetized rat, ventilation increasing in response to normoxic hypercapnia. Results are from two experiments on one animal undertaken at an interval of approximately six weeks (unpublished data of R. L. Laughton). ' $\bigcirc$ ', first set of data; ' $\bullet$ ', second set of data obtained six weeks later.

In our laboratory, R. L. Laughton has undertaken similar studies on the rat. With only the minor restraint of a jacket, this animal will remain quiet in the body plethysmograph for 1–3 h. In our studies, the definitive respiratory measurements are made after a minimum period of 10 min in any particular gaseous environment.

In the control rat, typical respiratory frequency is 60–70 breaths min⁻¹ and tidal volume 2–3 ml. As carbon dioxide is added to the inspired mixture, frequency and tidal volume increase in a repeatable manner. The relationships of the inspired minute volume ($\dot{V}_I$) to V_T for four rats is shown in Fig. 1. In individual animals the correlation is close. When the

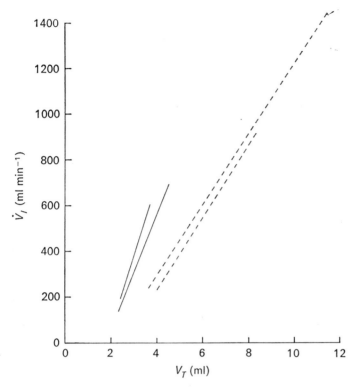

Fig. 3. The effect of combined pneumonectomy and contralateral vagotomy on respiratory patterns of the unanaesthetized rat. Results for two control animals (———) (as in Fig. 1) and two treated animals (- - - -). Ventilation ($\dot{V}_I$) is stimulated by normoxic hypercapnia. With loss of vagal input respiration is slower and deeper; increases are achieved with selective preference for tidal volume (V_T) (unpublished data of R. L. Laughton).

data are examined in terms of V_T/T_I, relationships are similar to those found by Clark & von Euler (1972), and to those found in hypoxia by Gautier (1976) (Fig. 2). There is no evidence that increases of carbon dioxide, at the lower levels, increase T_I. In the rat, T_I reaches a limiting value of approximately 0.2 sec.

However, the plot of T_I to T_E shows that T_E is reduced much more than T_I at lower levels of hyperventilation; this is consistent with the findings of Gautier, though quantitatively different. In shorter respiratory cycles, T_I and T_E diminish approximately equally in the rat as in other animals.

Effective vagotomy was achieved in rats by a combination of left pneumonectomy and right vagotomy (a technique of E. E. Nattie and D. Bartlett, personal communication). Control studies showed that pneumonectomy alone did not produce a significant change in breathing patterns.

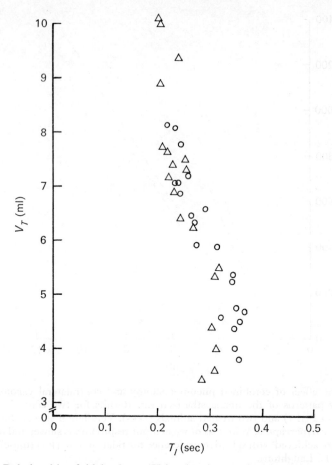

Fig. 4. Relationship of tidal volume (V_T) to inspiratory duration (T_I) in unanaesthetized rat after pneumonectomy with contralateral vagotomy. Results for two animals (different symbols). There is a phase of respiration (V_T 3.5–5.5 ml) where increased tidal volume does not shorten inspiratory duration (compare with Fig. 2).

After vagotomy, hypercapnia produced a different relationship of $\dot{V}_I$ to V_T, ventilation being achieved at greater tidal volumes with smaller increases of frequency. The difference in patterns is shown in Fig. 3. When the tidal volume:inspiratory time (V_T/T_I) relationship is considered (Fig. 4), it is found that for increases of tidal volume from 3.5 to 5.5 ml, T_I remains constant, as predicted from von Euler's model. At higher levels of inspired carbon dioxide, however, as tidal volume increases to 8–10 ml, T_I decreases in hyperbolic fashion to reach a new constant of approximately 0.2 sec.

From these experiments we conclude that in the unanaesthetized rat the pattern of hypercapnic respiration is influenced by unknown factors beyond those incorporated in the model of Clark & von Euler (1972); but the application of their theory can be seen in terms of the existence of an inspiratory characteristic and of volume-dependent cut-off.

Humans

A constant relationship between changes of frequency and tidal volume in human respiration was recognized by Hey et al. (1966). They considered individual subjects under the ventilatory stimulus of hypoxia, hypercapnia, exercise, metabolic acidaemia and drugs. Plotting the relationship of ventilation to tidal volume ($\dot{V}_E/V_T$, comparable with $\dot{V}_I/V_T$ of Figs. 1 and 3) they found the pattern to be the same, irrespective of the stimulus. An elevation of body temperature acted differently, increasing frequency more than volume. Otherwise, tidal volume increased linearly to a limiting value of approximately half the vital capacity.

Clark & von Euler (1972) studied human subjects during a rebreathing process which produced hyperoxic hypercapnia. As a preliminary the subjects hyperventilated so that initial tidal volumes were low. The pattern of V_T/T_I showed the characteristics seen in the cat under urethane anaesthesia: there was a phase of constant frequency while the tidal volume increased typically from 0.5 to 1.2 litres (Range 1); thereafter T_I diminished with increasing V_T (Range 2). Clark & von Euler concluded that the respiratory pattern is determined, in man, by the same basic mechanisms as in the cat but with quantitative differences. They were able to point out that Range 2 behaviour, where pulmonary inflation apparently cuts off inspiration, begins at the level of inflation where experimental vagal block in human subjects indicates that the Hering–Breuer reflex begins to operate (Guz, 1975). The data of Hey et al. (1966) do not show evidence of Range 1 behaviour, but this was explained by the information that the subjects breathed air with carbon dioxide in the control studies, apparently to stabilize their respiration; and so could already have been stimulated out of Range 1.

Newsom Davis & Stagg (1975) found only a reasonably close relationship between the tidal volume and inspiratory time of individual breaths of resting subjects (correlation of $V_T:T_I = 0.7$) and a poorer correlation of T_I to T_E ($r = 0.4$). They concluded that control was asserted on the rate of inspiratory airflow not on volume.

Specific evidence that frequency and tidal volume do not have a unique relationship in normal subjects has been obtained by a group in Tenney's laboratory at the Dartmouth Medical School (J. A. Daubenspeck, J. D. Sinclair, Stephen M. Tenney and S. M. Tenney, in preparation). They compared the patterns of breathing in hypoxic hypocapnia and hypoxic hypercapnia. Studies were made on young subjects at rest and at low levels of exercise; repeated studies were made so that the variation of oxygenation, carbon dioxide levels, and metabolism could be as small as possible; and the complex statistical problem of presentation of a valid mean for the results was met by use of a technique for directional analysis. The essential comparison was made by beginning with hypocapnic respiration; both at rest and exercise, at an alveolar oxygen tension (paO_2) of approximately 7 kPa, the subjects hyperventilated relative to their carbon dioxide production and lowered $paCO_2$ from normoxic levels. Controlled addition of carbon dioxide to the inspired gas was then used to bring their $paCO_2$ to the control level, allowing measurement of the increases of ventilation, frequency and tidal volume in the hypocapnic range. Further addition of carbon dioxide to inspired gas then allowed the same measurements in the hypercapnic range. In the hypocapnic range, increases of ventilation were largely achieved by increases of frequency; often enough, the tidal volume actually decreased as ventilation increased. The difference in patterns of respiration was statistically highly significant.

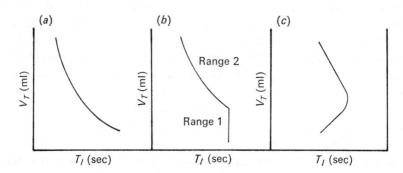

Fig. 5. A schematic representation of the relationship between tidal volume (V_T) and inspiratory duration (T_I) in different species and conditions. (a) V_T increases and T_I decreases at increased levels of ventilation. At later stages of the inspiratory time cycle, inspiratory effort is apparently cut off at lower lung volume by vagal input from pulmonary inflation receptors. Pattern seen in cat, anaesthetized with pentobarbitone, in hypercapnia (Clark & von Euler, 1972); cat, awake, in hypoxia (Gautier, 1976); rat, awake, in hypercapnia (unpublished data of R. L. Laughton). (b) At low levels of V_T, T_I remains constant; at higher levels of V_T, pattern as in (a). The input from inflation receptors is apparently inadequate, at lower tidal volume, to cut off inspiratory activity. Pattern seen in cat, urethane anaesthesia, hypercapnia; in man, awake, hypercapnia (Clark & von Euler, 1972) in rat with left pneumonectomy, right vagotomy (unpublished data of R. L. Laughton). (c) At low levels of V_T, T_I diminishes; at higher levels of V_T, pattern approximately as in (a). Apparently an additional factor affects the V_T/T_I pattern at low tidal volume. Cat, awake, hypercapnia (Gautier, 1976).

Looking at the range of studies of the neural determination of respiratory patterns, from those of anaesthetized animals to those of conscious human subjects, it is clear that there is much in common. The various patterns are shown schematically in Fig. 5. Experimental evidence of a drive, and of termination of inspiration by vagal input from pulmonary inflation receptors, though obtained in anaesthesia, can fairly be assumed to apply to the normal human subject. Clearly, however, this is not all. There may be significant species differences – for example between cat and rat. As well, there is repeated evidence of an unexplained influence of carbon dioxide. Its effects on mechanical properties of the lung, through diminution of airway resistance, or on carbon dioxide-sensitive inflation receptors (Guz, 1975), may contribute to the explanation; a direct effect of carbon dioxide on neural mechanisms controlling breathing remains a significant possibility.

CHEMICAL DRIVE TO RESPIRATION IN EXERCISE

Enhanced chemosensory drive to breathing in exercise has long represented an attractive hypothesis. Since arterial gas levels remain normal, and in the absence of evidence of a chemoreceptor contribution from muscle tissues, venous blood, or transferred metabolic gases, it is necessary to postulate a lowered threshold or increased sensitivity of the chemosensory mechanisms. Evidence for this remains as elusive as ever. Cunningham's group have considered that there was a lowering of threshold (Bannister, Cunningham & Douglas, 1954), that the threshold changed little but sensitivity doubled (Cunningham, Lloyd & Patrick, 1963) and that the threshold was lowered but sensitivity change was variable (Battacharyya et al., 1970). Other conflicting studies are reviewed by Guz (1975). In recent studies, Daubenspeck, Sinclair, Tenney & Tenney (1976) again found evidence of increased sensitivity without change of threshold.

Some disagreement on this fundamental question is easily explained because of the problems of end-tidal or arterial sampling, of obtaining stable respiratory patterns even in cooperative subjects, and of obtaining a valid statistical expression of results, particularly because of the effect of the 'dependent' variable, ventilation, on the 'independent' variable, $paCO_2$ (J. A. Daubenspeck, J. D. Sinclair, Stephen M. Tenney & S. M. Tenney, in preparation).

One factor which seemed likely to cause the difference in calculations of sensitivity was the way in which $paCO_2$ was determined from expired gases. It should be noted that arterial sampling is not feasible when experiments are repeated 8–12 times on each subject. Therefore $paCO_2$ is estimated, in the steady-state technique, either from end-tidal samples (corrected for predicted deviation of pCO_2 from arterial levels) or from expired pCO_2, using an assumed value for the physiological deadspace. In repeated studies on young subjects in our laboratory, I. H. Sarelius and M. Ward have looked at the effect of the value used for physiological deadspace. Ventilatory sensitivity to carbon dioxide at rest and exercise was estimated by use of end-tidal sampling and by use of a predicted deadspace volume, and values at the upper and lower levels of the 95 % confidence level of this prediction were estimated by using the new prediction data of Bradley, Harris *et al.* (1976). Each method produced evidence of a comparable increase of carbon dioxide sensitivity of ventilation during exercise. Thus neither the value used for deadspace volume, nor the use of end-tidal sampling, can explain the failure of some studies to demonstrate enhanced carbon dioxide sensitivity.

SPECIFIC NEURAL MECHANISMS IN EXERCISE

Accepting a basic control of respiratory patterns based on a bulbopontine pacemaker driven by a chemical input and switched according to the level of pulmonary inflation, it is necessary to add, in exercise, those modifications that are specific to the process. There is not yet an experimental basis, however, for such an elaboration of the simple model. There is not even reliable evidence concerning the significance of the accepted neural elements: a 'command' signal originating in the forebrain, an input from moving limbs and an input associated with rising body temperature.

Thus it is not yet possible even to put together a picture of human responses to the decision to exercise, let alone establish the significance of such ventilatory stimuli as come from joint movements (Barron & Coote, 1973), spindle excitation (Gautier *et al.*, 1969), or elevated body temperature (Vejby-Christensen & Strange Petersen, 1973). The jump from cellular mechanisms to integrated responses is far beyond the capacities of the current exercise physiologists.

Much of the experimental work from our own laboratory referred to in this article was carried out by Mrs Rosemary L. Laughton, whose contribution is gratefully acknowledged. The work was supported on a grant from the Medical Research Council of New Zealand.

REFERENCES

Andersen, P. & Sears, T. A. (1970). Medullary activation of intercostal fusimotor and alpha motoneurones. *J. Physiol., Lond.* **209**, 739–55.
Asmussen, E. (1973). Ventilation at transition from rest to exercise. *Acta physiol. scand.* **89**, 68–78.

Bannister, R. G., Cunningham, D. J. C. & Douglas, C. G. (1954). The carbon dioxide stimulus to breathing in severe exercise. *J. Physiol., Lond.* **125**, 90–117.

Barron, W. & Coote, J. H. (1973). The contribution of articular receptors to cardio-vascular reflexes elicited by passive limb movement. *J. Physiol., Lond.* **235**, 423–36.

Bhattacharyya, N. K., Cunningham, D. J. C., Goode, R. C., Howson, M. G. & Lloyd, B. B. (1970). Hypoxia, ventilation, P_{CO_2} and exercise. *Resp. Physiol.* **9**, 329–47.

Beaver, W. L. & Wasserman, K. (1970). Tidal volume and respiratory rate changes at start and end of exercise. *J. appl. Physiol.* **29**, 872–6.

Bertrand, A., Hugelin, A. & Vibert, J. F. (1974). A stereologic model of pneumo-taxic oscillator based on spatial and temporal distribution of neuronal bursts. *J. Neurophysiol.* **37**, 91–107.

Bradley, C. A., Harris, E. A., Seelye, E. R. & Whitlock, R. M. L. (1976). Gas exchange during exercise in healthy people. I: The physiological deadspace volume. *Clin. Sci. molec. Med.* **51**, 323–33.

Bradley, G. W., von Euler, C., Martilla, I. & Roos, B. (1975). A model of the central and reflex inhibition of inspiration in the cat. *Biol. Cybernetics* **19**, 105–16.

Clark, F. J. & von Euler, C. (1972). On the regulation of depth and rate of breath-ing. *J. Physiol., Lond.* **222**, 267–95.

Cunningham, D. J. C., Drysdale, D. B., Gardner, W. N., Jensen, J. L., Strange Peterson, E. & Whipp, B. J. (1977). Very small, very short-latency changes in human breathing induced by step changes of alveolar gas composition. *J. Physiol., Lond.* **266**, 411–21.

Cunningham, D. J. C., Lloyd, B. B. & Patrick, J. M. (1963). The relation between ventilation and end-tidal P_{CO_2} in man during moderate exercise with and without CO_2 inhalation. *J. Physiol., Lond.* **169**, 104–6P.

Daubenspeck, J. A., Sinclair, J. D., Tenney, Stephen, M. & Tenney, S. M. (1976). Ventilation in hypoxic hypocapnia. *J. Physiol., Lond.* **256**, 29–30P.

Dejours, P. (1964). Control of respiration in muscular exercise. *Handb. Physiol.* **1**(3), 631–48.

von Euler, C., Hayward, J. N., Martilla, I. & Wyman, R. J. (1973). Respiratory neurones of the ventrolateral nucleus of the solitary tract of cat: vagal input, spinal connections and morphological identification. *Brain Res.* **61**, 1–22.

Feldberg, W. (1976). The ventral surface of the brain stem: a scarcely explored region of pharmacological sensitivity. *Neuroscience* **1**, 427–41.

Feldberg, W. & Malcolm, J. L. (1959). Experiments on the site of action of tubo-curarine when applied via the cerebral ventricles. *J. Physiol., Lond.* **149**, 58–77.

Gautier, H. (1976). Pattern of breathing during hypoxia or hypercapnia of the awake or anesthetised cat. *Resp. Physiol.* **27**, 193–206.

Gautier, H., Lacaisse, A. & Dejours, P. (1969). Ventilatory response to muscle spindle stimulation by succinylcholine in cats. *Resp. Physiol.* **7**, 383–8.

Goodwin, G. M., McCloskey, D. I. & Mitchell, J. H. (1971). Circulatory and ventilatory responses to exercise. *J. Physiol., Lond.* **219**, 40–1P.

Guz, A. (1975). Regulation of respiration in Man. *A. Rev. Physiol.* **37**, 303–23.

Hanna, J. N., Hill, P. McN. & Sinclair, J. D. (1975). Human cardiorespiratory responses to acute cold exposure. *Clin. exp. Pharmac. Physiol.* **2**, 229–38.

Harris, E. A., Seelye, E. R. & Whitlock, R. M. L. (1976). Gas exchange during exercise in healthy people. II: Venous admixture. *Clin. Sci. molec. Med.* **51**, 335–44.

Hey, E. N., Lloyd, B. B., Cunningham, D. J. C., Jukes, M. G. M. & Bolton, D. P. G. (1966). Effects of various respiratory stimuli on the depth and frequency of breathing in Man. *Resp. Physiol.* **1**, 193–205.

Hodgson, H. J. F. & Matthews, B. P. C. (1968). The ineffectiveness of excitation of the primary endings of the muscle spindle by vibration as a respiratory stimulant in the decerebrate cat. *J. Physiol., Lond.* **194**, 555–63.

Kao, F. F., Lahiri, S., Wang, C. & Mei, S. S. (1967). Ventilation and cardiac output in exercise. Interaction of chemical and work stimuli. *Circulation Res.* **20** and **21**, Suppl. 1, 179–91.

Leusen, I. (1972). Regulation of cerebrospinal fluid composition with reference to breathing. *Physiol. Rev.* **52**, 1–56.

Marino, P. L. & Lamb, T. W. (1975). Effects of CO_2 and extracellular H^+ iontophoresis on single cell activity in the cat brainstem. *J. appl. Physiol.* **38**, 688–95.

Merrill, E. G. (1970). The lateral respiratory neurones of the medulla: their associations with nucleus ambiguus, nucleus retroambigualis, the spinal accessory nucleus and the spinal cord. *Brain Res.* **24**, 11–28.

Mitchell, R. A. & Berger, A. J. (1975). Neural regulation of respiration. *Am. Rev. resp. Dis.* **111**, 206–24.

Mitchell, R. A., Loeschcke, H. H., Massion, W. H. & Severinghaus, J. W. (1963). Respiratory responses mediated through superficial chemosensitive areas on the medulla. *J. appl. Physiol.* **18**, 523–33.

Newsom Davis, J. & Stagg, D. (1975). Inter-relationships of the volume and time components of individual breaths in resting man. *J. Physiol., Lond.* **245**, 481–98.

Pappenheimer, J. R., Fencl, V., Heisey, S. R. & Held, D. (1965). Role of cerebral fluids in control of respiration as studied in unanesthetised goats. *Am. J. Physiol.* **208**, 436–50.

Read, D. J. C. (1967). A clinical method for assessing the ventilatory response to carbon dioxide. *Aust. A. Med.* **16**, 20–32.

Saltin, B., Nazar, K., Costill, D. L., Stein, E., Jansson, E., Essen, B. & Gollnick, P. D. (1976). The nature of the training response; peripheral and central adaptations to one-legged exercise. *Acta physiol. scand.* **96**, 289–305.

Tenney, S. M. & Ou, L. C. (1977). Ventilatory response of decorticate and decerebrate cats to hypoxia and CO_2. *Resp. Physiol.* **29**, 81–92.

Trouth, C. O., Loeschcke, H. H. & Berndt, J. (1973). Histological structures in the chemosensitive regions on the ventral surface of the cat's medulla oblongata. *Pflügers Arch. ges. Physiol.* **339**, 171–83.

Vejby-Christensen, H. & Strange Petersen, E. (1973). Effect of body temperature on ventilatory transients at start and end of exercise in Man. *Resp. Physiol.* **17**, 315–24.

Wasserman, K. & Whipp, B. J. (1975). Exercise physiology in health and disease, *Am. Rev. resp. Dis.* **112**, 219–49.

Somatotopic organization of the cuneate nucleus in the cat, with special reference to forepaw cutaneous projections

M. AOKI

The cuneate nucleus of cat receives principally tactile information from the forelimb. Some myelinated primary afferent innervating specific types of cutaneous receptors on the hairy and glabrous skin (Jänig, Schmidt & Zimmermann, 1968; Aoki, 1977; Aoki & Yamamura, 1977) have been shown to ascend different parts of the dorsal funiculus (Uddenberg, 1968). However, the corresponding projection manner of such afferents to cuneate neurones within the nucleus still remains unclarified, although the general somatotopic representation of the body surface in the cuneate–gracile nuclei was previously described in the cat (Kruger, Siminoff & Witkovsky, 1961; Millar & Basbaum, 1975) and in other animals (Nord, 1967; Woudenberg, 1970; Hamilton & Johnson, 1973).

Therefore, an attempt was made to elucidate further the somatotopic organization of the cuneate nucleus. The present study was prompted by the recent anatomical finding that a cytoarchitectonic difference exists between the rostral and caudal part of the nucleus (Keller & Hand, 1970). For systematic explorations, first, the cuneate units encountered with micro-electrodes were identified by natural stimulation in terms of modality type. Second, their distributions in dorso-ventral and medio-lateral axes were explored at different rostro-caudal levels.

In this study, special attention was paid to the paw pad units, since they are considered to transmit particular sensory information, different from those of the hindlimb, to the higher levels of the central nervous system (Aoki & Yamamura, 1977).

METHODS

Subject preparation

Experiments were performed on 15 adult cats weighing 2.5–4.0 kg. Anaesthesia was induced with fluothane (halothane), and maintained by sodium pentobarbital administered intravenously in an initial dosage of 10–15 mg kg^{-1} body weight. Supplemental doses (5 mg) were administered as required. Routinely the animals were immobilized by gallamine trithio-dide (Teisan Co.) and artificially ventilated. Rectal temperature was maintained at about 37 °C by a heating pad and an infrared lamp. The animals were placed in a stereotaxic apparatus. An occipital craniotomy was made and the caudal part of cerebellum was aspirated with a glass pipette to expose the medulla. A warmed paraffin pool held by skin flaps covered the exposed medulla.

Recording and stimulation

The tungsten micro-electrodes, insulated (Insl-X) except for the tip, were introduced vertically into the surface of the cuneate nucleus under a dissecting microscope. Standard amplifying and recording systems were employed as previously described elsewhere (Aoki & Yamamura, 1977). Soma and axon spikes were differentiated by the criteria already established by Perl, Whitlock & Gentry (1962). Electrode penetrations were made at 250–500-μm intervals in medio-lateral and rostro-caudal axes. The read-ings of rostro-caudal and medio-lateral positions with respect to the obex were made by a micromanipulator attached to the stereotaxic frame.

The units encountered and isolated in each electrode penetration were identified in terms of tactile modalities, based on the response properties to adequate natural stimuli. The localizations and characteristics of the receptive fields (RFs) were marked on figurine drawings. Various forms of tactile stumuli were used; a cat vibrissa for bending a single hair, a glass probe with a finely polished tip of 0.5 mm in diameter to press 'touch' points, a blunt glass rod to press and move skin. When necessary, hairs were either clipped or depilated with a cosmetic depilatory.

On some occasions, response characteristics were further quantitatively examined using a vibrator (Ling Model 100) and the tuning curves for the pad units were constructed to discriminate them from each other. Move-ments of a small probe (0.2 mm in diameter) were monitored by a displace-ment transducer (7DCDT) mounted on the vibrator.

Histological reconstruction

At the end of each experiment, an electrolytic lesion (100 μA of cathodal current for 10 sec) was made in the final track and the electrode was left *in situ*. The animals were perfused with 10 % formalin and the paraffin oil covering the medulla was replaced by formalin. On the following day, the medulla was excised, removed from the animal and fixed in formalin for a week. Then the electrode was withdrawn and serial frozen sections of

30 μm thick were cut through the relevant region. The sections were stained
with cresyl violet. By these means it was possible to identify electrode
position with a clear electrode track and marking at its top.

RESULTS

In the present explorations, a total of 225 electrode penetrations were made
systematically into the cuneate (main) nucleus and the external nucleus
(ECU). The area where nuclear unit activities were recorded ranged
longitudinally approximately from 4 mm rostral to 4 mm caudal to the
obex. Fig. 1 illustrates a dorsal view of the medulla and location of the

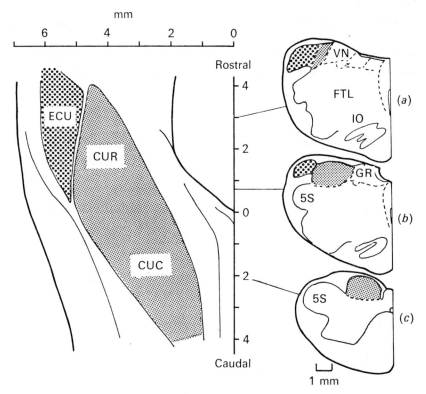

Fig. 1. A dorsal view of the medulla oblongata with the cerebellum removed. The
shaded area represents approximate unilateral locations of the main cuneate nucleus
and the external cuneate nucleus (ECU). The main cuneate nucleus is subdivided
into a rostral (CUR) and caudal (CUC) region. On the right-hand side are three
selected cross-sections at the different levels; (a) rostral, (b) intermediate and (c)
caudal. The rostro-caudal and medio-lateral scales are expressed in mm, using the
obex and the midline as the zero, respectively. Abbreviations: VN, vestibular
nucleus; FTL, lateral tegmental field; IO, inferior olive; GR, gracile nucleus;
5S, spinal trigeminal nucleus.

cuneate nucleus, reconstructed from the present electrophysiological and histological data. Since the cytoarchitecture of the main nucleus is not homogeneous in the rostral and caudal parts (Keller & Hand, 1970), they will be referred to separately as rostral (CUR) and caudal (CUC) parts in the following text.

Classification of units

According to the terminology used for the cutaneous receptors and the types of afferents projecting to the dorsal column (Burgess, Petit & Warren, 1968; Aoki & Yamamura, 1977; Iggo & Ogawa, 1977), cuneate units were classified into one of the following types. Hair units (G), activated by 'guard' hair bending, 'touch' units (T), excited by pressing sensitive spots, rapidly adapting (RA) and slowly adapting (SA) pad units, Pacinian corpuscle units (PC) in and around forepaw pads, and other unclassified units including probable 'field' type units. Deep or subcutaneous units represent those excited by muscle stretch or joint movements. The differentiation of these two units was not easy, but by careful exploration joint units which responded to a small angular displacement of a particular joint were sometimes identified.

G and T units were readily identified by airpuff or pressing the skin with a small probe, but identification of pad units needed vibratory stimulation.

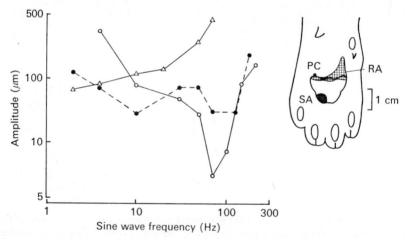

Fig. 2. An example of the response characteristics of three different cuneate units which had their peripheral receptive fields in and around the main pad. Three units; PC (Pacinian corpuscle, ○—○), RA (rapidly adapting, ●—●) and SA (slowly adapting, △—△) are discriminated by their tuning curves using vibratory stimuli. One-to-one following threshold (μm) is plotted against the sine wave frequency (Hz) on log–log coordinates. See text for details. Penetration was made at −1 mm caudal to the obex and 2 mm lateral to the midline.

As shown in Fig. 2, tuning curves for each of the pad units were different as described in the first-order afferents (Iggo & Ogawa, 1977). The best tuning frequency for PC units was found to be approximately 100 Hz, whereas RA units showed a relatively flat tuning curve. The best tuning frequency for SA units was 5 Hz or less.

Pattern of modality representation

In general agreement with the previous works (Kruger *et al.*, 1961; Gordon & Jukes, 1964; Millar & Basbaum, 1975), the somatotopic arrangement was organized in an orderly fashion, such that neighbouring peripheral body parts are represented in adjacent regions, though mixing of modalities in a single dorso-ventral penetration was commonly observed. Fig. 3 gives the dorso-ventral distribution of various tactile units responding to forepaw stimulation. As is illustrated, an electrode penetrated the CUC at 0.7 mm caudal to the obex and 3.7 mm lateral to the midline. There was a tendency to record the same types of units successively in clusters in adjacent regions. In some occasions, only G or T units were encountered in a single track. This result may indicate clustering of cells or 'cell nests' and suggests some degree of spatial segregation of modality subtypes within the nucleus. As an electrode was advanced deeper into the ventral region, peripheral RF locations made a progressive movement from the volar surface to the dorsal paw. It was of interest to note that the RF showed little overlap.

Medio-lateral distribution of the units

A series of electrode penetrations was made in a medio-lateral axis at 200–500-μm intervals, at different levels. Fig. 4 illustrates the results obtained from one of such penetrations in the CUC. In the row of electrode tracks 1, 2 and 3, the RF locations moved from the volar to the dorsal paw in dorso-ventral insertions. As the position of a track was placed to the more lateral side, the RF locations shifted gradually to the more radial side of the paw. In track 3, a unit recorded in the most ventral region had the RF in the neighbouring region of the starting RF, ending up just encircling the paw.

Another series of penetrations, 500 μm apart, was made 0.5 mm posterior to the obex in tracks 4 to 8. In these tracks, RFs were found predominantly on the forearm region. When the electrode tracks were placed at the more lateral side, RF locations shifted from the ulnar to the radial side. As the electrode was advanced deeper, the RFs moved to more proximal part of

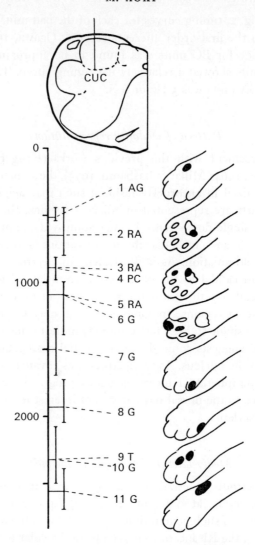

Fig. 3. A representative example of single dorso-ventral penetration through the caudal region of the cuneate nucleus (CUC). Vertical bars indicate the range in which the units were isolated from the noise level. Note that the units of similar modality subtypes are recorded in clusters. On the right-hand side, their peripheral receptive field locations are illustrated in black. Abbreviations: AG, axon spike of guard hair; RA, rapid adapting; PC, Pacinian corpuscle; G, guard hair; T, touch. Penetration was made at 0.7 mm caudal to the obex and 3.7 mm lateral to the midline. Vertical scale is expressed in μm.

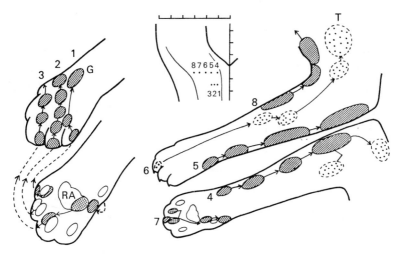

Fig. 4. Medio-lateral explorations of the units in the caudal region of the cuneate nucleus and the distribution patterns. Inset picture indicates the positions of electrode tracks which are numbered in sequence. Arrows connecting receptive fields (RFs) indicate that the RF locations make an orderly progression in each dorso-ventral penetration. The scale is 1 mm. Abbreviations: T, touch; G, guard hair; RA, rapidly adapting.

the forelimb. In tracks 6 and 8, only the same types of units were recorded in succession in a single track.

Fig. 5 illustrates another series of penetrations made at the more rostral part of the cuneate nucleus. In these tracks the units responding to pad stimulation were encountered much less frequently, but cutaneous units which had the RF on the fore- and upper-arm or deep units were predominantly recorded. Again the disto-proximal progression and ulnar-to-radial side shifting of the RFs were the general rule, except for track 4 in which the RF shifted to a distal part. The lateral tracks 5 and 6 were considered to have penetrated the ECU, because only deep units (some of them were identified as joint units) were recorded.

Modality representation in the external cuneate nucleus

Although a sufficient number of tracks was not made in the ECU to describe the topographical organization in detail, Fig. 6 demonstrates a typical track which was histologically verified as penetrating the ECU. In this track, only joint units were recorded. The position of the joint units moved from the distal digit to the more proximal shoulder joint as the electrode advanced deeper. This result leads to the suggestion that an

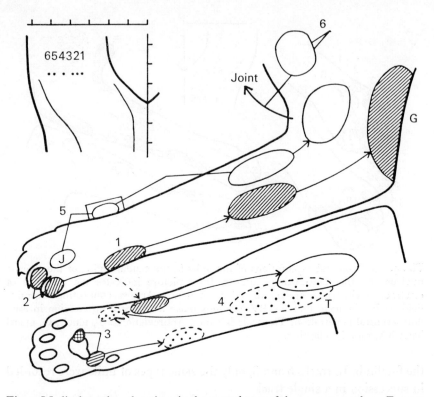

Fig. 5. Medio-lateral explorations in the rostral part of the cuneate nucleus. Format is the same as in Fig. 4. Note that units obtained in the most lateral penetrations (tracks 5 and 6) are deep units, activated either by joint movements or muscle stretch. Abbreviations: T, touch; G, guard hair; J, joint. The scales are in mm.

orderly topographic arrangement exists also in the ECU (O'Neal & Westrum, 1973). In other penetrations, deep units sensitive to muscle stretch were also encountered, as previously reported (Rosén & Sjolund, 1973).

Rostro-caudal distributions of the units

A series of penetrations was made rostro-caudally along the long axis of the nucleus. As illustrated in Fig. 7, 14 tracks were performed from the rostral to the caudal part of the nucleus at 500-μm intervals. The locations and the modalities of the RF are illustrated in the drawings along with the corresponding number of the electrode track(s). In tracks 5, 6, no distinct spikes were isolated and therefore they have been excluded from the figure. Track 1 apparently penetrated through the ECU, since only joint units were recorded. From tracks 2 to 7, electrodes penetrated the CUR and

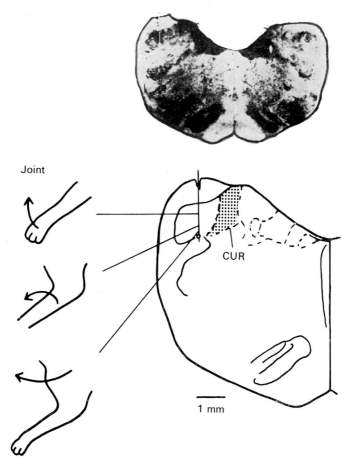

Fig. 6. An example of a lateral track penetrating the external cuneate nucleus. Three deep units which are excited by angular movements in the directions indicated by the arrows were obtained. CUR, rostral part of the cuneate nucleus. Subsequent histological examination identified the electrode track as being made at 2.2 mm rostral to the obex and 4.8 mm lateral to the midline.

from tracks 8 to 14, electrodes penetrated the CUC. As is shown in Fig. 7, the RF locations appeared in sequence in the immediate vicinity at all levels of the main nucleus as an electrode was placed to the more caudal positions. This result clearly demonstrated that the neurones receiving cutaneous inputs from the same regions on the body surface are arranged along the long axis of the nucleus.

DISCUSSION

For a full understanding of the mechanism of tactile information processing in such relay nuclei as the cuneate–gracile complex, it seems necessary to elucidate the following three points; first, responses to natural stimulation of the first-order units innervating specific types of cutaneous receptors and their projection manner (Brown, 1968; Petit & Burgess, 1968); second, the nature of linkages between the first- and second-order neurones, involving mode of convergence and divergence, inhibitory interactions (Armett, Gray, Hunsperger & Lal, 1962); third, the central destination of the relay cells to higher levels (Brown, Gordon & Kay, 1974) and other inputs. As regards the receptor types in the forelimb and their afferent projections, previous studies have demonstrated that forelimb cutaneous units are similar to those in the hindlimb (Aoki & Yamamura, 1977) and that the afferents occupy different depths in the dorsal funiculus (Uddenberg, 1968). However, corresponding topographic arrangement of the second-order units within the cuneate nucleus remained unclarified. The present explorations were concentrated on elucidating the above point.

Somatotopic representation

The present study has confirmed the gross somatotopy described earlier that, in the cuneate nucleus, the toes project to a larger region dorsally and to a smaller portion of the proximal forelimb in the ventral region (Kruger et al., 1961; Millar & Basbaum, 1975). The ulnar and radial sides of the forelimb project to the medial and lateral sides of cuneate nucleus, respectively and this concurs with the anatomical projection pattern of dorsal roots (Hekmatpanah, 1961). The present study, furthermore, revealed that the neurones responding to pad stimulation occupy a large part of the dorsal region of the CUC. Fig. 8 summarizes the somatotopic representation of the body surface, reconstructed by combining the present data and those of the previous works. Contrary to some previous views, rostro-caudal differentiation in terms of somatotopy was clearly elucidated.

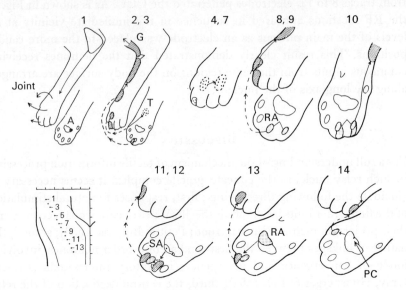

Fig. 7. A representative example of the rostro-caudal exploration along the long axis of the cuneate nucleus. Fourteen tracks at 500-μm intervals from the rostral to the caudal portion of the nucleus are shown in an inset diagram. The locations and characteristics of the receptive fields are illustrated in the pictures along with the corresponding number of the track(s). Abbreviations: A, axon spike; T, touch; RA, rapidly adapting; SA, slowly adapting; PC, Pacinian corpuscle. Scales are in mm.

Rostro-caudal differentiation

Since the RF size is related to peripheral location, it follows that cuneate units with smaller RFs on the forepaw are located in the CUC, whereas those with larger RFs on the proximal forelimb are located in the CUR. This may provide one possible explanation for the rostro-caudal differentiation of cutaneous units as to RF sizes in the gracile nucleus (Gordon & Jukes, 1964). This organization appears to be related to cytoarchitectonic differentiation in that large round cells are arranged in clusters in the CUC and smaller cells in the CUR (Kuypers & Tuerk, 1964; Keller & Hand, 1970).

Distribution of modality subtypes and its functional significance

Although no precise comparison of RF sizes was made, they were on average several times larger than those of the primary units (Welker & Johnson, 1965; Burgess et al., 1968; Aoki & Yamamura, 1977). Furthermore, each RF property was unimodal, suggesting convergence of the same modality units onto a single second-order neurone. However, in this study, there was also clear evidence of convergence of different modalities. Some RA units were not only activated by light pressure to the pad, but also by hair bending. Those units may correspond to 'pad-and-touch sensitive units' in the gracile nucleus (Gordon & Jukes, 1964). The tendency to record in succession the same modality in a single track suggests that clustering of units occurs within the nucleus. This finding is in accord with the differential distribution of afferents in the dorsal funiculus (Uddenberg, 1968) and in the peripheral nerve (Aoki & Yamamura, 1977).

From the standpoint of information processing, the following two interesting findings were obtained. First, the RFs of adjacent units were usually separated from each other as observed also in the peripheral nerve (Aoki & Yamamura, 1977). This arrangement may provide an efficient way of covering large areas with small groups of units. Second, PC units were frequently encountered in the CUC, intermingled with other pad units. Concerning the function of PC units, it was recently demonstrated that inhibition of other tactile units in the cuneate nucleus occurred following stimulation of Pacinian corpuscles (Bystrzycka, Nail & Rowe, 1977). This finding supports the notion that Pacinian corpuscles in and around the paw pads (Lynn, 1971) convey sensory information about changing aspects of tactile stimuli, temporarily inhibiting other channels (Andersen, Etholm & Gordon, 1970; Jabbur & Banna, 1970), through the cuneate nucleus to the cortex with great security of transmission (McIntyre, Holman & Veale, 1967).

To clarify further the functional organization of the nucleus, the central destination of other types of cuneate neurones comprising cuneo-thalamic relay cells and interneurones (Andersen, Eccles, Oshima & Schmidt, 1964; Andersen, Eccles, Schmidt & Yokota, 1964) must be extensively studied.

SUMMARY

This study was performed to elucidate further the topographic arrangement of the cuneate neurones excited by tactile stimulation to the forelimb skin of cats. Emphasis was laid on the projection from the volar surface of the fore paw to the nucleus. Under light pentobarbital anaesthesia, extracellular unit recordings were made by tungsten micro-electrodes. The cells were identified in terms of tactile modality subtypes based on response characteristics to various forms of light mechanical stimuli and their positions within the nucleus were histologically verified. (1) Mechanoreceptive neurones excited by stimuli to the forepaw pads were most frequently encountered in the caudal part and dorsally near the surface of the nucleus. (2) Receptive fields were arranged in an orderly fashion in the transverse

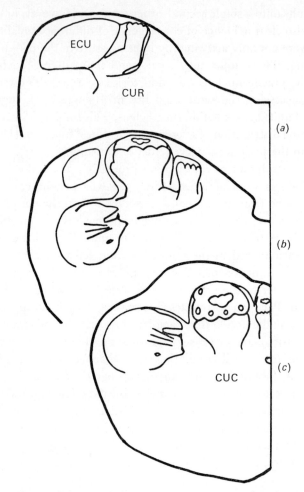

Fig. 8. A summary diagram of the somatotopic representation in the cuneate–gracile complex and spinal trigeminal nucleus at different levels (*a, b, c*). Somatotopia is visualized as a cat lying on its back, typically represented in intermediate level (*b*). The forepaw is represented mostly in the caudal level (*c*)). ECU, external cuneate nucleus; CUR, rostral part of the cuneate nucleus; CUC, caudal part of the cuneate nucleus.

direction across the nucleus, with the most lateral toe projecting to the medial side of the nucleus. They made, in dorso-ventral insertions, an orderly progression from the volar surface to the dorsal paw. (3) In single electrode tracks, there was a tendency to record similar types of tactile units in clusters. (4) In the more rostral part of the nucleus, the pad units were less frequently encountered but the whole part of forelimb was represented in the dorso-ventral plane. (5) In the most lateral tracks at the rostral levels, the units responding exclusively to either joint movement or muscle stretch were recorded. These tracks were histologically verified to be penetrating the ECU.

The author is grateful to Professor S. Mori for his encouragement during the experiment and critical reading of the manuscript. Valuable assistance from Dr T. Yamamura, Miss M. Moriguchi and Mr K. Miyakawa is also gratefully acknowledged.

REFERENCES

Andersen, P., Eccles, J. C., Oshima, T. & Schmidt, R. F. (1964). Mechanisms of synaptic transmission in the cuneate nucleus. *J. Neurophysiol.* **27**, 1096–116.

Andersen, P., Eccles, J. C., Schmidt, R. F. & Yokota, T. (1964). Identification of relay cells and interneurons in the cuneate nucleus. *J. Neurophysiol.* **27**, 1080–95.

Andersen, P., Etholm, B. & Gordon, G. (1970). Presynaptic and postsynaptic inhibition elicited in the cat's dorsal column nuclei by mechanical stimulation of skin. *J. Physiol., Lond.* **210**, 433–55.

Aoki, M. (1977). Distal slowing of conduction in forelimb and hindlimb myelinated cutaneous afferent fibers in cat. *Exp. Neurol.* **56**, 200–11.

Aoki, M. & Yamamura, T. (1977). Functional properties of peripheral sensory units in hairy skin of a cat's forelimb. *Jap. J. Physiol.* **27**, 1–11.

Armett, C. J., Gray, J. A. B., Hunsperger, R. W. & Lal, S. (1962). The transmission of information in primary receptor neurones and second-order neurones of a phasic system. *J. Physiol., Lond.* **164**, 395–421.

Brown, A. G. (1968). Cutaneous afferent fibre collaterals in the dorsal columns of the cat. *Exp. Brain Res.* **5**, 293–305.

Brown, A. G., Gordon, G. & Kay, R. H. (1974). A study of single axons in the cat's medial lemniscus. *J. Physiol., Lond.* **236**, 225–46.

Burgess, P. R., Petit, D. & Warren, R. M. (1968). Receptor types in cat hairy skin supplied by myelinated fibers. *J. Neurophysiol.* **31**, 833–48.

Bystrzycka, E., Nail, B. S. & Rowe, M. (1977). Inhibition of cuneate neurones: its afferent source and influence on dynamically sensitive 'tactile' neurones. *J. Physiol., Lond.* **268**, 251–70.

Gordon, G. & Jukes, M. G. M. (1964). Dual organization of the exteroceptive components of the cat's gracile nucleus. *J. Physiol., Lond.* **173**, 263–90.

Hamilton, T. C. & Johnson, J. I. (1973). Somatotopic organization related to nuclear morphology in the cuneate–gracile complex of opossums *didelphis marsupialis virginiana*. *Brain Res.* **51**, 125–40.

Hekmatpanah, J. (1961). Organization of tactile dermatomes, C_1 through L_4, in cat. *J. Neurophysiol.* **24**, 129–40.

Iggo, A. & Ogawa, H. (1977). Correlative physiological and morphological studies of rapidly adapting mechanoreceptors in cat's glabrous skin. *J. Physiol., Lond.* **266**, 275–96.

Jabbur, S. J. & Banna, N. R. (1970). Widespread cutaneous inhibition in dorsal column nuclei. *J. Neurophysiol.* **33**, 616–24.

Jänig, W., Schmidt, R. F. & Zimmermann, M. (1968). Single unit responses and the total afferent outflow from the cat's foot pad upon mechanical stimulation. *Exp. Brain Res.* **6**, 100–15.

Keller, J. H. & Hand, P. J. (1970). Dorsal root projections to nucleus cuneatus of the cat. *Brain Res.* **20**, 1–17.

Kruger, L., Siminoff, R. & Witkovsky, P. (1961). Single neuron analysis of dorsal column nuclei and spinal nucleus of trigeminal in cat. *J. Neurophysiol.* **24**, 333–49.

Kuypers, H. G. M. & Tuerk, J. D. (1964). The distribution of the cortical fibers within the nuclei cuneatus and gracilis in the cat. *J. Anat.* **98**, 143–62.

Lynn, B. (1971). The form and distribution of the receptive fields of Pacinian corpuscles found in and around the cat's large foot pad. *J. Physiol., Lond.* **217**, 755–71.

McIntyre, A. K., Holman, M. E. & Veale, J. L. (1967). Cortical response to impulses from single Pacinian corpuscles in the cat's hind limb. *Exp. Brain Res.* **4**, 243–55.

Millar, J. & Basbaum, A. I. (1975). Topography of the projection of the body surface of the cat to cuneate and gracile nuclei. *Exp. Neurol.* **49**, 281–90.

Nord, S. G. (1967). Somatotopic organization in the spinal trigeminal nucleus, the dorsal column nuclei and related structures in the rat. *J. comp. Neurol.* **130**, 343–55.

O'Neal, J. T. & Westrum, L. E. (1973). The fine structural synaptic organization of the cat lateral cuneate nucleus. A study of sequential alterations in degeneration. *Brain Res.* **51**, 97–124.

Perl, E. R., Whitlock, D. G. & Gentry, J. R. (1962). Cutaneous projection to second-order neurons of the dorsal column system. *J. Neurophysiol.* **25**, 337–58.

Petit, D. & Burgess, P. R. (1968). Dorsal column projection of receptors in cat hairy skin supplied by myelinated fibers. *J. Neurophysiol.* **31**, 849–55.

Pubols, B. H., Jr, Welker, W. I. & Johnson, J. I., Jr (1965). Somatic sensory representation of forelimb in dorsal root fibers of raccoon, coatimundi, and cat. *J. Neurophysiol.* **28**, 312–41.

Rosén, I. & Sjölund, B. (1973). Organization of group I activated cells in the main and external cuneate nuclei of the cat: convergence patterns demonstrated by natural stimulation. *Exp. Brain Res.* **16**, 238–46.

Uddenberg, N. (1968). Differential localization in dorsal funiculus of fibres originating from different receptors. *Exp. Brain Res.* **4**, 367–76.

Woudenberg, R. A. (1970). Projections of mechanoreceptive fields to cuneate–gracile and spinal trigeminal nuclear regions in sheep. *Brain Res.* **17**, 417–37.

Mechanoreceptors of the
glabrous skin and tactile acuity

M. ZIMMERMANN

The anatomical basis and physiological basis of the various qualities of cutaneous sensation – touch, pressure, vibration, tickle, cold, warm, itch, pain – have been the object of considerable speculation and controversy since the beginning of the nineteenth century. For example, histologists ascribed each of these qualities intuitively to one of the nerve and organs known at that time; these correlations, although wrong in most cases, are still contained in some textbooks of histology and of physiology.

Primarily within the past 20 years, considerable knowledge on the characteristics of skin receptors has been accumulated by recording from single afferent fibres, and by the use of quantitative types of natural receptor stimulation. In this paper a description is given of the types and characteristics of low-threshold mechanoreceptors present in the glabrous skin of the hand in primates and of the pad in cats, and of the possible sensory experience related to their excitation. No reference will be made to hairy skin.

TYPES AND CHARACTERISTICS OF MECHANORECEPTORS
IN THE GLABROUS SKIN

Three main types of sensitive mechanoreceptors with group II (Aβ) afferents, with conduction velocities of about 30 to 80 m sec^{-1}, have been identified in the glabrous skin of the cat, monkey and man, which are commonly designated as slowly adapting (SA), rapidly adapting (RA) and Pacinian corpuscle (PC) receptors. Fig. 1 gives a schematic synopsis of the discharge characteristics and the morphological correlations. In (*a*) responses to ramp indentations of the skin overlying the receptors have been chosen to depict their functional properties. It is seen that the SA receptors discharge at a frequency which rises in proportion to the indentation during the ramp, followed by a nearly constant response rate during the plateau of the stimulus. This discharge rate is a measure of the skin deformation (*S*) (see also Fig. 2).

Two types of SA receptors, SA-I and SA-II, can be discerned based on

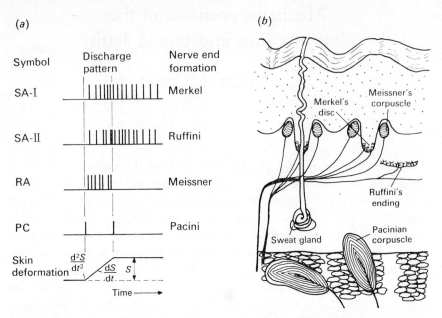

Fig. 1. Schematic synopsis of the mechanoreceptors of the glabrous skin. (*a*) Impulse discharges of four types of receptors to a ramp indentation of the skin surface, the time course being indicated in the last trace. (*b*) Location in the skin of the nervous end organs which have been attributed to either functional type of receptors shown in (*a*). SA = slowly adapting; RA = rapidly adapting; PC = Pacinian corpuscle; *S* = measure of skin deformation.

their physiological characteristics. SA-I responds mainly to deformations vertical to the skin surface, whereas the most effective stimulus for the SA-II receptor is horizontal stretching of the skin, and shearing forces to the fingernails in primates (Johansson, 1976). The discharge of the RA receptor is confined to the period of change of skin deformation, being a transient of constant velocity in this example. As will be shown later (Fig. 3), the RA receptor can be considered as a sensor for indentation velocity, $\mathrm{d}S/\mathrm{d}t$. The PC receptor produces impulses during neither the plateau, nor the constant velocity period, but at the times of change of velocity, i.e. when acceleration, $\mathrm{d}^2S/\mathrm{d}t^2$, occurs (see also context of Fig. 6). Thus, the three types of mechanoreceptors in the glabrous skin provide information on different parameters of skin stimuli: force of indentation, S, velocity, $\mathrm{d}S/\mathrm{d}t$, and acceleration, $\mathrm{d}^2S/\mathrm{d}t^2$. It is conceivable that the pattern of nervous outflow in all these afferents from a particular skin area is used by the central nervous system to synthesize the wealth of tactile sensations.

However, there are indications that the parameters S, $\mathrm{d}S/\mathrm{d}t$ and $\mathrm{d}^2S/\mathrm{d}t^2$ are not processed in the central nervous system by certain 'mathematical operations' commensurate with Newtonian mechanics, as would probably be suggested by an information scientist. For example, there is a rigid separation in the central somatosensory pathways of information from phasic and tonic skin receptors, yielding two separate projection areas of the primate hand in the postcentral gyrus SI area (Paul, Merzenich & Goodman, 1972), and probably an additional significant representation of PC afferents in SII (McIntyre, Holman & Veale, 1967). Specific inhibitory interactions occur in the central nervous system within neurone populations connected to either SA or RA receptors, but there is virtually no interaction between these systems (e.g. Gordon & Jukes, 1964; Jänig, Schmidt & Zimmermann, 1968b; Jänig, Schoultz & Spencer, 1977). Thus, the above classification of cutaneous mechanoreceptors is phenomenological rather than functional.

To understand better how the holistic internal image of tactile events is created, studies will be required on the spatio-temporal pattern of activity in a population of afferents, which includes all types of receptors. So far, however, only a few studies have been concerned with this difficult task (e.g. Fuller & Gray, 1966; Jänig et al., 1968a,b; Gardner & Spencer, 1972a,b; Johnson, 1974; Johansson & Vallbo, 1976).

Morphology

As shown in Fig. 1b, four types of nerve end organs have been established to correspond to the different functional classes of receptors. This correspondence has emerged mainly from studies in which a single afferent was first functionally classified by controlled mechanical stimulation, and then the skin explored histologically beneath the receptive field (e.g. Lynn, 1969; Jänig, 1971b; Iggo & Ogawa, 1977). In this way, the RA receptor has been identified as the Meissner corpuscle (or its analogue end formation in the cat), located in the dermal papillae. The clusters of Merkel cells located on the epithelial pegs or ridges are the SA-I receptors. Pacinian corpuscles, which are usually located in the subcutaneous fatty tissue or in association with joint capsules (Gray & Matthews, 1951) and interosseal membranes (Hunt & McIntyre, 1960), form the PC receptors. SA-II receptors have been associated with Ruffini-type endings so far only in hairy skin (Chambers, Andres, von Düring & Iggo, 1972); it is tempting to extrapolate this association to the glabrous skin. The morphological details of all types of endings have been recently reviewed (Andres & von Düring, 1973).

THE SA RECEPTORS AND SUBJECTIVE SCALINGS OF
INDENTATION INTENSITY

Both types of SA receptors, SA-I and SA-II, yield tonic responses to a maintained skin deformation. There is a continuous decline of the discharge frequency during a steady-state stimulus (Fig. 2a) due to adaptation: at sufficient intensity, however, the discharge may last for many minutes, or virtually infinitely.

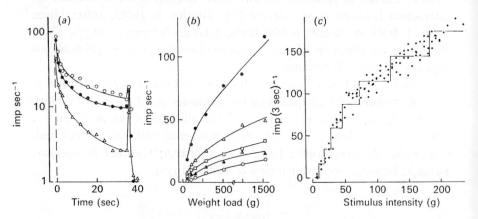

Fig. 2. Intensity coding in the slowly adapting receptor. (a) Time courses of discharges to different forces (g) loading the whole area (about 15 mm in diameter) of the large pad of the cat's foot, where the receptor was located. ' △ ', 155 g; ' ● ', 525 g; ' ○ ', 995 g. (b) Plot of the average frequencies (same receptor as in (a)) at various times of loading with the weights given on the abscissa. ' ● ', 1 sec; ' △ ', 2.5 sec; ' □ ', 5 sec; ' ▲ ' 10 sec; ' ○ ', 30 sec. Ten recordings were averaged in (a) and (b). (c) Plot of single measurements of discharge frequency in another SA receptor (imp (3 sec)$^{-1}$) versus stimulation intensity (force in g, stimulator probe 2 mm in diameter) which was varied at random; the inter-stimulus interval was 30 sec. The staircase function was drawn into the field of scatter of points. ((a), (b) from Jänig, Schmidt & Zimmermann, 1968a; (c) M. Zimmermann, unpublished.)

The frequency F at any time after stimulus onset bears a monotonically rising relationship to stimulus intensities (Fig. 2b) which cover the range of forces S acting upon the footpad of the cat in its daily life. The intensity plot, F versus S, may be fitted by a power function:

$$F = (S - S_0)^n \tag{1}$$

where S_0 is the threshold stimulus strength. Values found for the exponent n are usually in the range between 0.5 and 1.0.

However, other functions (log–log tan h) may be fitted adequately to the data (Knibestöl, 1975). Since neither of these mathematical functions can be derived from first principles of mechanoreceptor function, all of them are descriptive rather than logical. They are of some heuristic value insofar as inferences on the central nervous system operation might be deduced from comparison of such quantitative mathematical functions which relate either neurophysiological data, on the one hand, and subjective magnitude estimations, on the other hand, to the intensity of the stimulus (Mountcastle, 1974). Often it has been found that such subjective psychometric power functions have the same exponent as the power function relating receptor discharge frequency, or size of cortical-evoked potentials, to the stimulus intensity. However, there has been some debate about such correlations of neurophysiological and psychophysical measurements, and about the conclusions drawn from these on the central nervous system processing of stimuli to yield a conscious perception (Kruger & Kenton, 1973; Knibestöl & Vallbo, 1976).

Information transmission in SA receptors

The preceding considerations were mainly concerned with the type of stimulus that characteristically excites a particular receptor. In this section, the question about the reliability of stimulus encoding will be elucidated.

To yield quantitative data on information transmission in receptors, various methods have been applied. They all are based upon analysis of variability of the receptor discharge to precisely controlled stimuli. An example of such an analysis is given in Fig. 2c for an SA receptor. In a series of experiments, many stimuli have been applied at random intensity. It is clear from this graph that the discharge rate not only depends upon the stimulus intensity, but also exhibits fluctuations – biological noise – which is obviously inherent to the transduction process in the receptor. Therefore, the relationship between discharge rate and stimulus intensity is characterized by a band of uncertainty. The number of distinguishable states of stimulus intensities in this blurred relationship can be estimated by drawing a staircase function into the field of scatter of experimental points. This yields about eight states of stimulus intensity, information about which can be deduced from the receptor discharge. In terms of information theory, the measure of information (I) per stimulus contained in the afferent discharge of this SA receptor is calculated as

$$I = \log_2 8 = 3 \text{ bits per stimulus} \tag{2}$$

Apart from this simple graphic procedure, several mathematical approaches have been used to calculate the content of information of the receptor discharge (e.g. Werner & Mountcastle, 1965; Kruger & Kenton, 1973;

Dickhaus, 1976). They all yielded the same or similar results as the method used in Fig. 2c.

Thus, measurement of the inaccuracy in the encoding mechanism yields an estimate for the limitation of the receptor to transmit information on stimulus intensity. Corresponding approaches have been used, or should be used, to determine the performance of a receptor, or of a receptor population, in transmitting other parameters (e.g. spatial, temporal) of the wealth of stimuli impinging upon the skin surface.

Evaluations of this type have been of basic relevance when correlating these neuronal data with the corresponding psychophysical results, for example with the subjective scaling of intensity of stimuli, or the discrimination of differences in intensity between two successive stimuli (i.e. the difference limen, or just noticeable difference). It became evident from these comparisons that information contained in the population of primary afferents is employed to varying degrees by the central nervous system to produce a conscious sensory decision. For example, in psychophysical experiments, human subjects can discriminate approximately eight categories of intensities of skin indentation. This figure of 3 bits per stimulus is the same as that found in a single SA fibre (Fig. 2c). Therefore, it appears as if most of the many fibres excited even by a small surface area stimulation probe would either not contribute to the information of the perceived stimulus, or would compensate for information losses that occur, e.g. at the synaptic relays along the somatosensory pathways.

However, in extreme situations all the information available in a population of afferent fibres excited by a certain stimulus may be utilized to yield the maximum resolution obtained in a sensory discrimination task, as has been best demonstrated for the population of cold receptors in the primate hand (Darian-Smith & Johnson, 1977).

THE RA RECEPTOR AND ITS CAPACITY TO TRANSMIT INFORMATION ON VELOCITY OF MOVEMENT

The encoding of stimulus velocity dS/dt by the RA receptor is shown in Fig. 3a. The number of impulses per stimulus, or mean discharge frequency, is a monotonic function of stimulus velocity. The range of relevant velocities found in a systematic study was from 0.1 to beyond 100 m sec^{-1} (Dickhaus, 1976). A power function with $n = 0.7$ can be used to fit the data shown in Fig. 3a. However, other mathematical functions may yield better fits in other experiments (Knibestöl, 1973; see comments in the preceding paragraph on SA receptors).

Another way of establishing a stimulus–response relation is to plot the

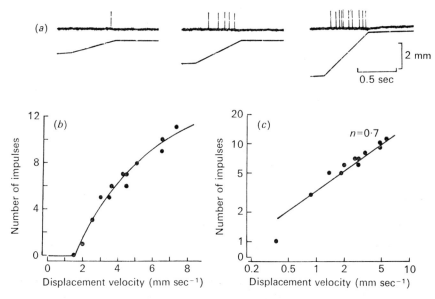

Fig. 3. Velocity coding in the rapidly adapting (RA) receptor. (a) Responses from an RA fibre (upper traces) of the cat's footpad to linearly rising skin indentations at various slopes (lower traces). (b) Number of impulses per stimulus at varying displacement velocities, the rise time of the stimulus being the same throughout (as in (a)). (c) Same data as in (a) and (b), plotted in log–log coordinates, after subtraction of the threshold velocity of 1.6 mm sec^{-1} from the actual indentation velocities. (M. Zimmermann, unpublished.)

interval Δt between successive spikes against indentation velocity. When using a Δt value averaged over the total time of discharge (Fig. 4a), the relationship is the inverse of that shown in Fig. 3. The same type of measurements from human RA receptors are plotted in log–log coordinates in Fig. 4b (upper graph). In this experiment, a power function could be fitted to the data, yielding an exponent of $n = -0.63$. Subjects were asked to make estimations of indentation velocity. These psychometric data could also be fitted by a power function, with $n = 0.61$. It was concluded from these results that the interspike interval Δt of the RA receptor discharge might be the neuronal parameter carrying the information on stimulus velocity in this experimental paradigm (Franzén & Lindblom, 1976).

Information transmission in RA receptors

The information content of RA afferent discharges has been studied recently (Dickhaus, Sassen & Zimmermann, 1976) by measuring the

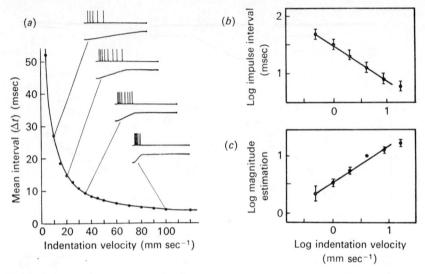

Fig. 4. Velocity coding in the rapidly adapting (RA) receptor and subjective velocity scaling. (*a*) The mean interval Δt between successive spikes of an RA receptor of the cat's pad is plotted against indentation velocity. At each given velocity, all the intervals obtained from 20 trials were averaged. (*b*) Mean interval from an RA receptor of the human hand plotted against velocity of skin indentation, log–log cordinates. (*c*) Subjective estimates of magnitude of indentation velocity to stimulation as in (*b*), log–log coordinates. (*a*) from Dickhaus, Sassen & Zimmermann, 1976; (*b*), (*c*) from Franzén & Lindblom, 1976.)

variability of the interval Δt between successive impulses of the discharge (see Fig. 4*a*). This analysis of variability revealed that, in particular, the interval Δt_1 between the first two impulses of the discharges had an extremely small standard deviation (s.d.), as is evident from Fig. 5*a* (curve labelled Δt_1). From the s.d. of interspike intervals, the information content in bits per stimulus was calculated and plotted in Fig. 5*b*. When only the first interval Δt_1 was used, this calculation yielded the highest values for the information, which were up to 4.7 bits per stimulus; that is, in the first interspike interval Δt_1 of a single RA afferent, up to 26 steps of indentation velocity can be discerned. However, when an increasing order n of interspike intervals of each discharge was included in the calculation, the information content markedly decreased (Fig. 5*b*).

The maximum rate of information transmission (in bits sec^{-1}) in RA receptors was evaluated by increasing the stimulus repetition rate to as much as 10 Hz in experiments such as those shown in Figs. 4 and 5. From analysis of the first interspike interval Δt_1, values of 30 bits sec^{-1} were obtained, which are extremely high when compared with the figure of

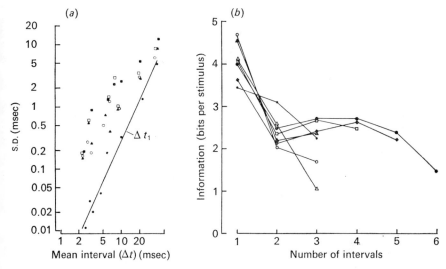

Fig. 5. Rapidly adapting (RA) receptors: discharge variability and information transmission. (*a*) The standard deviation (S.D.) (log–log scale) versus the mean duration of the interspike intervals Δt of the discharge of an RA receptor from the cat's pad. Data refer to first interval Δt_1 only (●), and to mean intervals including an increasing order of successive spikes of each discharge, respectively (○, 2 intervals; ▲, 3; □, 4; ■, 5). Each point has been determined from up to 12000 stimulations. (*b*) Information in bits per stimulus calculated for seven RA receptors (different symbols) from the variability of interspike intervals, in relation to the order n of successive intervals of each discharge that have been used for the calculation. (From Dickhaus, Sassen & Zimmermann, 1976.)

5–10 bits sec^{-1} estimated for the SA receptors (Werner & Mountcastle, 1965).

Thus, the RA receptor is well suited for providing information on a characteristic parameter of moving stimuli to the central nervous system.

THE PC RECEPTOR, A THRESHOLD SENSOR FOR ACCELERATION

This very rapidly adapting receptor has been identified as the Pacinian corpuscle by various authors (e.g. Gray & Matthews, 1951; Hunt & McIntyre, 1960; Lynn, 1969). It is most sensitive to mechanical transients, as is best demonstrated by using sinusoidal mechanical stimulation of the skin surface. In Fig. 6*e*, the minimum amplitude yielding one impulse per cycle of sinusoidal stimulation is plotted against the frequency of the sinusoid. A characteristic fall of this threshold of entrainment is visible with increasing frequency, being linear at a slope of -2 in the double logarithmic coordinate system used. This feature of the threshold fall lends

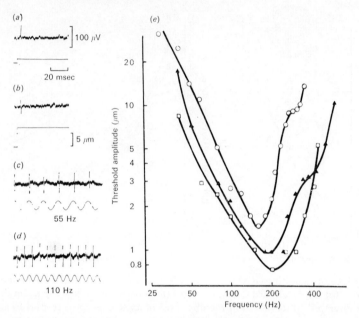

Fig. 6. Responses of Pacinian corpuscle (PC) receptors of the cat's foot. (a), (b) Impulse discharges (upper traces) during step indentations (lower traces) of the skin. (c), (d) Responses of the same unit to sinusoidal skin indentation at the given frequencies. (e) Plot of the minimum amplitude (ordinate) required to generate one spike per cycle of sinusoid stimulation, against the frequency of stimulation, for three PC units (different symbols). Log–log scale. ((a) to (d) M. Zimmermann, unpublished; (e) from Jänig, Schmidt & Zimmermann, 1968a.)

support to the assumption made above (Fig. 1) that the PC receptor might respond to acceleration; this will be shown by the following calculation.

The time course of skin displacement $S(t)$ during sinusoidal stimulation at frequency ω is given as

$$S(t) = A \times \sin\omega t \qquad (3)$$

where A is the amplitude of the sinusoidal oscillation. The acceleration b is the second time derivative

$$b = \mathrm{d}^2 S/\mathrm{d}t^2 = -A \times \omega^2 \sin\omega t \qquad (4)$$

Let us assume that a minimum threshold acceleration b_T is required to fire the receptor, whose threshold acceleration will be reached in each cycle when $\sin\omega t = -1$, i.e. at $\omega t = 270°$. This yields

$$b_T = \omega^2 \times A = \text{constant} \qquad (5)$$

That is, when the frequency ω is increased, the threshold acceleration b_T is reached at a smaller threshold amplitude A_0:

$$A_0 = b_T/\omega^2 \tag{6}$$

In a log–log coordinate system, this relationship between threshold amplitude A_0 and frequency $\omega = 2\pi f$ of sinus stimulation is linear with a slope of -2. This is verified by the data in Fig. 6e in the range up to 200 Hz.

The above considerations reveal that the threshold response and the cyclic entrainment of the PC receptor depend on a critical value of the acceleration of the skin displacement. However, values of suprathreshold accelerations are not encoded into the rate of discharge. This is a basic difference compared to the SA and RA receptors, which encode intensity and velocity values, respectively, over a considerable dynamic range of either variable (Figs. 2, 3, 4).

THRESHOLDS OF RECEPTORS AND OF PSYCHOPHYSICAL DETECTION

Indentation thresholds for all four types of glabrous skin mechanoreceptors are available from the cat, monkey and man. In Fig. 7a are shown the distributions of receptor populations in the human hand. No major differences exist among receptor samples from different locations, e.g. the finger tip and the palm of the hand (Johansson & Vallbo, 1976). The variances in threshold, evident from Fig. 7a, within the populations of each receptor type might in part be due to the different thicknesses of the epidermis/stratum corneum overlying the nerve end organs. This has been shown directly by ablating the stratum corneum of the cat's footpad (Jänig, 1971b), which resulted in a decrease in the thresholds of individual RA and SA units to between 10 % and 30 % of the initial values (Fig. 7b). Likewise, the lower thresholds of the PC units in the cat (median 1 μm, Jänig et al., 1968a) compared with man (10 μm, Fig. 7a) might be explained by the smaller dimensions of the cat's foot compared with the human hand: in the cat, both the average distance from the skin surface to this subcutaneously located receptor and the mass of tissue and bone to be accelerated by the stimulus are smaller. Both factors determine the smaller stimulus energy required at the skin surface to excite the PC units.

Two psychophysical paradigms have been applied to study the threshold of perception of mechanical skin stimuli in man and monkey: sinusoidal oscillation, which will be referred to in the subsequent paragraph, and single ramp stimuli. In these studies, direct or indirect comparisons were made between perception threshold and neuronal activity in the receptor afferents.

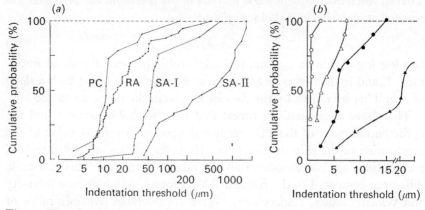

Fig. 7. Thresholds of cutaneous mechanoreceptors. (a) Cumulative probability distributions of indentation thresholds for samples of four kinds of receptors of the human hand, determined with stimuli having a slope of 4 mm sec^{-1}. PC = Pacinian corpuscle; RA = rapidly adapting; SA = slowly adapting. (b) Shifts of cumulative probabilities of indentation thresholds (rise time of stimulus 1 m sec) of RA and SA receptors by removing the stratum corneum of the cat's pad. ' ○ ', RA receptor before removal of the stratum corneum, ' ● ', after removal; ' △ ', SA receptor before removal of the stratum corneum, ' ▲ ', after removal. ((a) from Johannson & Vallbo, 1976; (b) from Jänig, 1971b.)

Ramp indentations of skin in humans: receptor thresholds and psychophysical stimulus detection

A direct comparison was possible in the approach by Johansson & Vallbo (1976), who recorded from single units in the nerve to the hand in human volunteers. They stimulated the site of minimum threshold of a skin receptor with mechanical transients at a velocity of 4 mm sec^{-1} (ramp stimuli, as in Fig. 3), and recorded simultaneously both the occurrence of an afferent spike and the probability of the subjective perceptual detection of the same near-threshold stimulus. In Fig. 8a, data from an RA receptor of the finger tip are plotted. Both the probability of occurrence of a spike and of the subjective yes response bore practically the same relationship to stimulus intensity. Such results were obtained with RA receptors at the lower end of the threshold distribution (Fig. 7a); it therefore seems unlikely that the point stimulus excited more than this particular fibre. That is, a single spike in a single RA fibre can produce a conscious perception of the mechanical event at the finger tip. Similar results were obtained with some PC receptors.

A different result emerged when investigating RA receptors located in the palm of the hand (Fig. 8b). Here, a much higher stimulus strength was required to produce perception than to fire the fibre. This suggests that

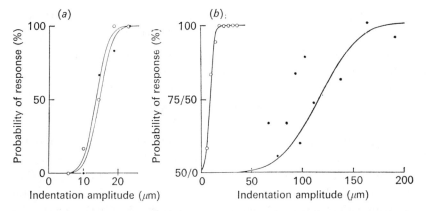

Fig. 8. Thresholds of rapidly adapting (RA) receptors and of stimulus detection. (*a*) The ordinate plots the probability of occurrence of a nerve impulse in a single RA afferent from the human finger tip (○) and of a yes response (●) of a detection task in the same subject, recorded simultaneously, when the stimulus intensity (abscissa) of successive stimuli was varied at random. (*b*) Similar to (*a*); however, an RA receptor of the palm of the hand was stimulated and recorded. The psychophysical detection paradigm was a 'two alternative forced choice' (2 AFC), the correct response probability (●) referring to the left-hand numbers of the ordinate scale ((*a*), (*b*) from Vallbo & Johansson, 1976).

many fibres have to be recruited by the stimulus to yield a consciously perceived neuronal input to the central nervous system.

When studying SA fibres with mechanical transients at a slope of 4 mm sec^{-1}, the thresholds for single impulse discharge were always higher than the perception thresholds (Johansson & Vallbo, 1976). This might be due to the generally higher SA receptor thresholds (Fig. 7*a*), the perception of the stimulus being determined by a nearby RA unit of lower threshold. Alternatively, the SA afferents might require spatial and/or temporal summation for detection, as do the RA units of the palm (Fig. 8*b*). That SA units really can be activated selectively to produce a conscious perception has been concluded from psychophysical measurements in which the velocity of the mechanical stimulus was varied (Lindblom, 1974): when the velocity of the stimulus was decreased below 0.1 mm sec^{-1}, which is the lower limit of critical slopes for RA and PC units, the threshold of perception rose steeply from about 6 μm to 80 μm. It is, however, not clear from this work whether activation of a single SA receptor was sufficient to produce conscious sensation.

NEURONAL AND PSYCHOPHYSICAL RESPONSES TO SINUSOIDAL
STIMULI; FREQUENCY DISCRIMINATION

Sinusoidal skin stimulation in the frequency range up to 400 Hz has been used by Mountcastle and his colleagues to study the detection threshold, quality of sensation and discrimination of intensity and of frequency (pitch) in human and monkey subjects, and to relate these psychophysical results to the neuronal parameters in SA, RA and PC afferents.

The perception threshold of sinusoidal stimuli to the human and monkey hand in relation to sinusoid frequency is shown by the curve in Fig. 9*a*. That monkey and man have identical threshold curves could be established by psychophysical (man) and behavioural (monkey) experiments using the same stimulation set-up (Mountcastle, LaMotte & Carli, 1972). The shaded regions depict the threshold response areas of many individual RA

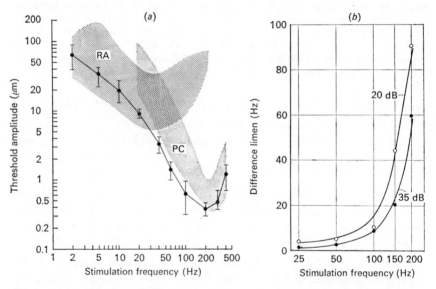

Fig. 9. Sinusoidal skin stimulation: receptor and perception thresholds, frequency discrimination. (*a*) The shaded areas indicate the relationship between thresholds and the frequency of mechanical skin stimulation, of rapidly adapting (RA) and Pacinian corpuscle (PC) receptors, respectively, of the monkey. The related neurones in SI cortex have their threshold curves also in these areas. The line is the averaged threshold for detection of sinusoidal stimulation in the monkey and man. Log–log scale. (*b*) The curves plot the difference in frequency (ordinate) of two successive trains of sinusoidal stimulations that just are recognized as different, by human subjects in relation to base frequency (abscissa, log scale). The curves have been obtained at different levels of intensity, as indicated in dB. ((*a*) according to data from Mountcastle, Talbot, Sakata & Hyvärinen, 1969 and Mountcastle, LaMotte & Carli, 1972; (*b*) according to Goff, 1959.)

and PC afferents from the monkey's hand. Two types of threshold could be established: the absolute threshold, i.e. when an occasional spike appeared during a train of sinusoid cycles, and a somewhat higher tuning threshold, when one nerve impulse was evoked per sinusoid cycle, locked in phase with the stimulus (see also Fig. 6).

Absolute and tuning thresholds characteristically decrease and then increase with increasing frequency, as indicated for RA and PC units, respectively, by the corresponding shaded region in Fig. 9a. It is apparent that the perception threshold parallels the RA threshold curves in the frequency range up to about 50 Hz, and the PC curves above 50 Hz. PC thresholds are generally higher than RA thresholds below 50 Hz, but the reverse is true above 50 Hz. Thus, near-threshold, sinusoidal skin stimulation is a means to activate selectively receptors of either population, and to study the sensations evoked by them.

It is now clear that in the monkey and man the absolute threshold of the most sensitive RA and PC fibres of the fingers is correlated with perception of stimuli (Mountcastle *et al.*, 1972), in good agreement with the results of Vallbo and his collaborators reviewed in the preceding paragraph. The cyclic entrainment of the RA fibres above the tuning threshold is a prerequisite for frequency discrimination, as will be referred to below.

Quality of perception of sinusoidal skin stimulation

Apart from attributing the psychophysical threshold to excitation of RA and PC receptors in either frequency range below and above 50 Hz it is now well established that the quality of sensation is dichotic: subjects denote their experiences as either flutter or as vibration, below and above 50 Hz, respectively. This duality persists also at stimulus intensities far above threshold, i.e. when both types of receptors are activated.

Inhibitory interactions of sinusoidal skin stimuli

The threshold of detection of a mechanical skin stimulus by human subjects can be influenced by concurrent ('conditioning') mechanical stimulation at a nearby skin site. In a systematic investigation (Ferrington, Nail & Rowe, 1977) it was shown that a conditioning low-amplitude vibration at 300 Hz was particularly effective in elevating the detection threshold for any kind of mechanical skin displacement (i.e. tonic, low- or high-frequency sinusoidal). In neurophysiological investigations using the same paradigms of skin stimulation in cats it could be shown that the PC afferents particularly exerted powerful inhibition in the cuneate nucleus (Bystrzycka,

Nail & Rowe, 1977): neurones of all classes according to their excitatory drive from the various types of slowly and rapidly adapting cutaneous mechanoreceptors were under this inhibitory control by the PC afferents. It can be concluded from these parallel investigations in man and the cat that an important function of the PC receptors might be to gain control of afferent inflow in all types of cutaneous mechanoreceptors.

Discrimination of frequency of sinusoidal skin stimulation

The capability of discrimination between slightly different frequencies of two successive trains of sinusoidal stimulation has been investigated in man and the monkey. In Fig. 9b the difference limen (i.e. the just noticeable difference) of human subjects is plotted against frequency of stimulation, at intensity levels 20 dB and 35 dB above threshold. Below 50 Hz, the difference limen is small, e.g. 2 and 4 Hz at 25 and 50 Hz, respectively; that is, changes in frequency of less than 10 % are recognized. Above 50 Hz there is a steep increase, the difference limen reaching 30 % at 200 Hz. This change of performance at about 50 Hz has also been suggested to be a basic difference in the central representation of RA and PC afferents: SI cortical neurones connected to PC afferents fail to show any cyclic entrainment which occurs in the afferent fibres, whereas SI neurones related to RA afferents reproduce the cyclic responses of the primary afferent fibres (Mountcastle, Talbot, Sakata & Hyvärinen, 1969).

The frequency discrimination in the flutter range (determined by excitation of RA receptors) has been extensively studied by LaMotte & Mountcastle (1975). At a base frequency of 30 Hz the difference limen was 2.7 Hz in the monkey and 1.8 Hz in man, in good agreement with previous findings (Goff, 1959; Fig. 9b). Comparison in this study with the neuronal responses in afferent fibres of monkeys revealed that frequency discrimination required the stimulus intensity to exceed the tuning threshold of RA afferents. Therefore, this psychophysical capability was attributed to the time information encoded in the dominant interval between nerve impulses, which is the same as the cycle time of the sinusoidal stimulus.

RECEPTIVE FIELD SIZE, INNERVATION DENSITY AND TWO-POINT DISCRIMINATION

The discrimination of spatial details of objects we touch with our hands is a major capacity mediated by the population of glabrous skin mechanoreceptors and their central representation. Either the size of the receptive fields or the innervation density might be a factor in determining two-point

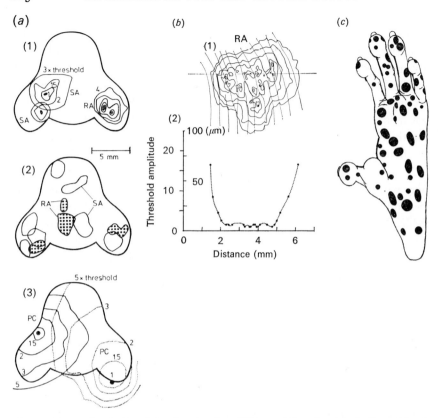

Fig. 10. Receptive fields of slowly adapting (SA), rapidly adapting (RA) and Pacinian corpuscle (PC) receptors. (a1) Receptive fields of SA and RA receptors in the cat's footpad established by systematic exploration with a small probe driven by a piezoelectric stimulator. The contours are the iso-threshold lines, the stimulus intensities are indicated in multiples of the minimum threshold. (a2) Receptive fields of seven SA and four RA receptors established as the iso-threshold line at twice minimum threshold. (a3) Receptive field contours of two PC units, measured as in (a1). (b1) The closed lines are the iso-threshold contours of an RA receptor of the human finger tip, each indicating a stimulus strength between 4 μm and 66 μm. Thin lines indicate the papillary ridges. Same length calibration as in (b2). (b2) Threshold profile across the receptive field shown in (b2), taken along the thin horizontal line in (b1). Stimulus intensities are given in multiples of minimum threshold (left-hand orindate scale) and in μm. (c) Composite diagram of 56 RA and SA receptive fields determined in the monkey's foot. The actual length of the foot is proximately 120 mm. ((a) from Jänig, Schmidt & Zimmermann, 1968a; (b) from Johansson, 1976; (c) from Lindblom, 1965.)

threshold, a simple measure used to characterize spatial discrimination and its variation on the skin surface (Weber, 1835; Weinstein, 1968). Both parameters will be considered here.

Examples of receptive fields of single afferent fibres in the cat, monkey and man are shown in Fig. 10. It is evident that the size of a receptive field as determined by controlled stimulation of the skin depends on stimulus intensity. For comparison it is convenient to give the receptive field size at a stimulus strength that is twice the minimum. This value agrees fairly well with field measurements performed by use of von Frey hairs under microscopic magnification. No significant differences in field dimensions exist between RA and SA-I receptors (Jänig et al., 1968a; Talbot, Darian-Smith, Kornhuber & Mountcastle, 1968; Knibestöl, 1973, 1975), whereas the fields of the PC units are considerably larger. The larger fields of the PCs probably are due to their deeper locations, and their common association with joint ligaments and bones which might act as collectors and transmitters of mechanical energy. Usually the SA-II receptors also have fields of responsiveness which are much larger than those of RA and SA-I, perhaps due to their deeper location and sensitivity to lateral skin stretching. Since not much information is available on the SA-II receptors of the glabrous skin and their significance in subjective sensation, they will not be considered further here.

It is seen from Fig. 10a and b that the receptive fields of RA and SA receptors contain several spots where the threshold is at a minimum. This reflects the anatomical observation that each afferent fibre has several end organs. In the cat's foot, the number of receptor endings per afferent nerve fibre has been estimated as 6 to 10 (Jänig, 1971a), whereas in man the range is 4 to 7 for SA and 12 to 17 for RA units (Johansson, 1976). On the other hand, direct anatomical observations (Andres & von Düring, 1973) and estimations based on innervation density (Jänig, 1971a) revealed that some, if not most, of the SA and RA receptive sites are innervated by more than one nerve fibre. This interdigitation of fibres in the skin probably increases the sensitivity of fibres for small skin displacements; information on the site of stimulation, however, might be blurred by the increase and overlap of fields compared with the case of a mosaic-like distribution of single nerve endings.

Available data on average receptive field sizes are pooled in Table 1. It is evident that the mean field size generally increases with species dimensions. The exceptionally large fields of responsiveness in man of the PC receptors should be noted.

The innervation density of the different types of fibres in the human hand has been estimated (Johansson & Vallbo, 1976) from single fibre experiments and some supplementary data (Buchthal & Rosenfalck, 1966). The results are pooled in Fig. 11b for three regions of the hand, as is outlined in Fig. 11a. An outstandingly high density of RA and SA-I receptors in the finger tip is evident from this representation. In Fig. 11a, the spatial acuity of these three skin areas is plotted; the acuity values are simply the inverse

Table 1. *Survey of receptive field sizes of single fibres from RA and SA receptors, and from PC receptors, in glabrous skin regions of different species*

Species	Region	Field sizes (mm²) of		Authority
		RA, SA	PC	
Rat	Hind foot	1–2	?	Sanders and Zimmermann, unpublished
Cat	Hind foot	3.5	15	Jänig et al., 1968a and Zimmermann, unpublished
Monkey	Finger tip	3.8	?	Talbot et al., 1968
	Palm of hand	11.1	?	
Monkey	Foot	9.9	?	Lindblom, 1965
Man	Finger tip	27	2400	Knibestöl, 1973, 1975
	Palm of hand	82		

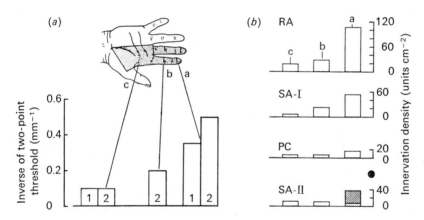

Fig. 11. Two-point discrimination and innervation density of the human hand. (a) The columns plot tactile acuity, which has been defined here as the inverse value of two-point threshold measured by Weber (columns 2) and by Weinstein (columns 1). Values are referred to the corresponding areas a,b,c of the hand. (b) Innervation densities in the three regions a,b,c of the hand (as indicated in (a)) of four types of receptors, estimated from the probability of occurrence of either type of unit in a large sample of recordings in human subjects. The SA-II units indicated by the hatched area were related to the bed of the fingernail rather than to the glabrous skin. ((a) according to Weber, 1835; Weinstein, 1968; (b) from Johansson & Vallbo, 1976.) RA = rapidly adapting; SA = slowly adapting; PC = Pacinian corpuscle.

of the two-point thresholds of Weber (1835) and Weinstein (1968). From comparison of Figs. 11a and 11b it becomes clear that the gradient of spatial acuity from the palm to the finger tip parallels the innervation density of RA and SA-I afferents. Thus, it is conceivable that the innervation density of these types of receptors is an important variable in determining the spatial discriminative capacity of the human hand.

The diameters of receptive fields have sometimes been suggested to be a primary factor in spatial resolution. However, the average field diameters of neither the SA-I nor the RA afferents are consonant with this hypothesis, as is shown in Table 2.

Table 2. *Comparison of two-point thresholds (mean of Weber, 1835 and Weinstein, 1968) and the average diameters of RA and SA receptive fields of the human hand. The field diameters were calculated from field areas given by Knibestöl (1973, 1975) on the assumption that areas were circular*

| Region | Two-point threshold (mm) | Field diameter (mm) of | |
		SA	RA
Palm of hand	11	10.6	9.8
Finger tip	2.4	5.1	7.1
Palm:finger ratio	4.6	2.1	1.4

CONCLUSIONS

The glabrous skin as a sensory surface appears to have evolved from the more primitive hairy skin, by the development of a peripheral and central neuronal apparatus subserving the outstanding discriminative capabilities for spatial and temporal details. These skin areas mediate the newborn's earliest impressions from the outside world. They are employed in active exploration of the environment during development by the child, by which, in cooperation with the visual system, the image of three-dimensional space is created in the brain.

What are the major differences underlying this enormous increase in tactile performance of glabrous compared with hairy skin? Apart from the large area and high specialization of the SI cortex related to the glabrous skin regions, it is mainly the presence here of large numbers of Meissner corpuscles, RA receptors. I would like to emphasize the important role of the RA receptors.

(1) They occur in a higher density than all other receptors in the primate hand, the RA population probably providing the major contribution to spatial acuity.

(2) Compared with other mechanoreceptors, the precision and the rate of information transmission on the velocity of moving stimuli are greatest for RA receptors.

(3) A single impulse from a finger tip RA receptor yields a conscious perception, i.e. there is an enormous synaptic security and no contamination with noise for these afferents along the ascending pathways.

(4) The high performance in the time domain, measured as frequency discrimination, is mediated only by the RA receptors.

It is clear that the other types of receptors certainly contribute to the complete internal representation of tactile events; however, what makes the human hand so superior is the population of RA receptors. One factor has not been considered in this article: the effect of active movement of the hand on the information transmitted by the cutaneous mechanoreceptors, particularly by the RA receptors. After the basic characteristics of the receptors have been established, it is most desirable to include active movement in neurophysiological and psychophysical studies of the skin senses. This would possibly elucidate the physiological basis of the classic statement, made half a century ago, by the psychophysiologist David Katz (1925): '*Die Bewegung ist ein elementarer gestaltender Faktor der Tast-phänomene*' – movement is a basic factor giving shape to the phenomena of tactile perception.

The author wishes to thank Mrs Almuth Manisali for the graphics, Mrs Ursula Nothoff for the bibliography and typing of the manuscript and Dr Earl Carstens for improving the English. The author's work contained in this article has been supported by the Deutsche Forschungsgemeinschaft.

REFERENCES

Andres, K. H. & von Düring, M. (1973). Morphology of cutaneous receptors. In *Handbook of sensory physiology*, vol. II, ed. by A. Iggo, pp. 3–28. Berlin: Springer-Verlag.

Buchthal, F. & Rosenfalck, A. (1966). Evoked action potentials and conduction velocity in human sensory nerves. *Brain Res.* **3**, 1–122.

Bystrzycka, E., Nail, B. S. & Rowe, M. (1977). Inhibition of cuneate neurones: its afferent source and influence on dynamically sensitive 'tactile' neurones. *J. Physiol., Lond.* **268**, 251–70.

Chambers, M. R., Andres, K. H., von Düring, M. & Iggo, A. (1972). The structure and function of the slowly adapting type II mechanoreceptor in hairy skin. *Q. Jl exp. Physiol.* **57**, 417–45.

Darian-Smith, J. & Johnson, K. O. (1977). Temperature sense in the primate. *Br. med. Bull.* **33**, 143–8.

Dickhaus, H. (1976). Neurophysiologische Untersuchungen sowie system- und informationstheoretische Analyse der Eigenschaften von rasch adaptierenden Mechanoreceptoren der unbehaarten Haut der Katzenfußsohle. Dissertation Technische Hochschule Karlsruhe.

Dickhaus, H., Sassen, M. & Zimmermann, M. (1976). Rapidly adapting cutaneous mechanoreceptors (RA): coding variability and information transmission. In *Sensory functions of the skin in primates*, ed. Y. Zotterman, pp. 45–54. Oxford: Pergamon Press.

Ferrington, D. G., Nail, B. S. & Rowe, M. (1977). Human tactile detection thresholds: modification by inputs from specific tactile receptor classes. *J. Physiol., Lond.* **272**, 415–33.

Franzén, O. & Lindblom, U. (1976). Coding of velocity of skin indentation in man and monkey. A perceptual–neurophysiological correlation. In *Sensory functions of the skin in primates*, ed. Y. Zotterman, pp. 55–65. Oxford: Pergamon Press.

Fuller, D. R. G. & Gray, J. A. B. (1966). The relation between mechanical displacements applied to a cat's pad and the resultant impulsive patterns. *J. Physiol., Lond.* **182**, 465–83.

Gardner, E. P. & Spencer, W. A. (1972a). Sensory funneling. I. Psychophysical observations of human subjects and responses of cutaneous mechanoreceptive afferents in the cat to patterned skin stimuli. *J. Neurophysiol.* **35**, 925–53.

Gardner, E. P. & Spencer, W. A. (1972b). Sensory funneling. II. Cortical neuronal representation of patterned cutaneous stimuli. *J. Neurophysiol.* **35**, 954–77.

Goff, G. D. (1959). Differential discrimination of frequency of cutaneous mechanical vibration. Ph.D. thesis, Univ. of Virginia, Charlottesville.

Gordon, G. & Jukes, M. G. M. (1964). Dual organization of the exteroceptive components of the cat's gracile nucleus. *J. Physiol., Lond.* **173**, 263–90.

Gray, J. A. B. & Matthews, P. B. C. (1951). Response of Pacinian corpuscles in the cat's toe. *J. Physiol., Lond.* **113**, 475–82.

Hunt, C. C. & McIntyre, A. K. (1960). Characteristics of responses from receptors from the flexor longus digitorum muscle and the adjoining interosseus region of the cat. *J. Physiol., Lond.* **153**, 74–87.

Iggo, A. & Ogawa, H. (1977). Correlative physiological studies of rapidly adapting mechanoreceptors in cat's glabrous skin. *J. Physiol., Lond.* **266**, 275–96.

Jänig, W. (1971a). The afferent innervation of the central pad of the cat's hind foot. *Brain Res.* **28**, 203–16.

Jänig, W. (1971b). Morphology of rapidly and slowly adapting mechanoreceptors in the hairless skin of the cat's hind foot. *Brain Res.* **28**, 217–31.

Jänig, W., Schmidt, R. F. & Zimmermann, M. (1968a). Single unit responses and the total afferent outflow from the cat's footpad upon mechanical stimulation. *Exp. Brain Res.* **6**, 100–15.

Jänig, W., Schmidt, R. F. & Zimmermann, M. (1968b). Two specific pathways to the central afferent terminals of phasic and tonic mechanoreceptors. *Exp. Brain Res.* **6**, 116–129.

Jänig, W., Schoultz, T. & Spencer, W. A. (1977). Temporal and spatial parameters of excitation and afferent inhibition in cuneothalamic relay neurons. *J. Neurophysiol.* **40**, 822–35.

Johansson, R. (1976). Receptive field sensitivity profile of mechanosensitive units innervating the glabrous skin of the human hand. *Brain Res.* **104**, 330–4.

Johansson, R. & Vallbo, A. B. (1976). Skin mechanoreceptors in the human hand: neural and psychophysical thresholds. In *Sensory functions of the skin in primates*, ed. Y. Zotterman, pp. 171–84. Oxford: Pergamon Press.

Johnson, K. A. (1974). Reconstruction of population response to a vibratory stimulus in quickly adapting mechanoreceptive afferent fiber population innervating glabrous skin of the monkey. *J. Neurophysiol.* **37**, 48–72.

Katz, D. (1925). *Der Aufbau der Tastwelt.* Leipzig: Joh. Ambrosius Barth.

Knibestöl, M. (1973). Stimulus–response functions of rapidly adapting mechanoreceptors in the human glabrous skin area. *J. Physiol., Lond.* **232**, 427–52.

Knibestöl, M. (1975). Stimulus–response functions of slowly adapting mechano-receptors in the human glabrous skin area. *J. Physiol., Lond.* **245**, 63–80.

Knibestöl, M. & Vallbo, A. B. (1976). Stimulus–response functions of primary afferents and psychophysical intensity estimation on mechanical skin stimulation in the human hand. In *Sensory functions of the skin in primates*, ed. Y. Zotterman, pp. 201–13. Oxford: Pergamon Press.

Kruger, L. & Kenton, B. (1973). Quantitative neural and psychophysical data for cutaneous mechanoreceptor function. *Brain Res.* **49**, 1–24.

LaMotte, R. H. & Mountcastle, V. B. (1975). Capacities of humans and monkeys to discriminate between vibratory stimuli of different frequency and amplitude: a correlation between neural events and psychophysical measurements. *J. Neurophysiol.* **38**, 539–59.

Lindblom, U. (1965). Properties of touch receptors in distal glabrous skin of the monkey. *J. Neurophysiol.* **28**, 966–85.

Lindblom, U. (1974). Touch perception threshold in human glabrous skin in terms of displacement amplitude on stimulation with single mechanical pulses. *Brain Res.* **82**, 205–10.

Lynn, B. (1969). The nature and location of certain phasic mechanoreceptors in the cat's foot. *J. Physiol., Lond.* **201**, 765–73.

McIntyre, A. K., Holman, M. E. & Veale, J. L. (1967). Cortical responses to impulses from single Pacinian corpuscles in the cat's hind limb. *Exp. Brain Res.* **4**, 243–55.

Mountcastle, V. B. ed. (1974). *Medical physiology*, vol. I. Saint Louis: C. V. Mosby.

Mountcastle, V. B., LaMotte, R. H. & Carli, G. (1972). Detection thresholds for stimuli in humans and monkeys: comparison with threshold events in receptive afferent nerve fibers innervating the monkey hand. *J. Neurophysiol.* **35**, 122–36.

Mountcastle, V. B., Talbot, W. H., Sakata, H. & Hyvärinen, J. (1969). Cortical neuronal mechanisms in flutter-vibration studied in unanesthetized monkeys. Neuronal periodicity and frequency discrimination. *J. Neurophysiol.* **32**, 452–84.

Paul, R. L., Merzenich, M. & Goodman, H. (1972). Representation of slowly and rapidly adapting cutaneous mechanoreceptors of the hand in Brodman's areas 3 and 1 of *Macaca mulatta*. *Brain Res.* **36**, 229–49.

Talbot, W. H., Darian-Smith, I., Kornhuber, H. H. & Mountcastle, V. B. (1968). The sense of flutter-vibration: comparison of the human capacity with response patterns of mechanoreceptive afferents from the monkey hand. *J. Neurophysiol.* **31**, 301–34.

Vallbo, A. B. & Johansson, R. (1976). Skin mechanoreceptors in the human hand: neural and psychophysical thresholds. In *Sensory functions of the skin in primates*, ed. Y. Zotterman, pp. 185–99. Oxford: Pergamon Press.

Weber, E. H. (1835). Über den Tastsinn. *Arch. Anat. Physiol. wiss. Med.* 152–9.

Weinstein, S. (1968). Intensive and extensive aspects of tactile sensitivity as a function of body part, sex, and laterality. In *The skin senses*, ed. D. R. Kenshalo, pp. 195–222. Springfield, Illinois: Charles C. Thomas.

Werner, G. & Mountcastle, V. B. (1965). Neural activity in mechanoreceptive cutaneous afferents: stimulus–response relations, Weber functions, and information transmission. *J. Neurophysiol.* **28**, 359–97.

Mechanisms of nociception*

A. IGGO

This short paper reviews my involvement in the study of pain mechanisms through laboratory investigations using experimental animals, principally directed at peripheral sensory receptor mechanisms in the skin and viscera and more recently at the neuronal mechanisms in the spinal cord. This interest in neurophysiology was fostered in its early stages by Archie McIntyre. My starting point was a study of visceral reflexes and the sensory innervation of the abdominal viscera. Since most of the sensory nerve fibres innervating the viscera are very thin, it was essential to develop refinements of the recording techniques available in 1950 for the electrophysiological study of small axons. These developments, including both micro-dissecting techniques and electrophysiological techniques, have made it possible since 1954 (Iggo, 1954, 1958) to examine as single sensory units, first the visceral and later the cutaneous and muscle non-myelinated sensory axons (Fig. 1*a*). These very small nerve fibres had, since the pioneer work of Adrian and Zotterman in the 1930s, been regarded as 'pain fibres'. The opportunity to make an exact single unit analysis was very exciting and not all of them turned out to be nociceptors.

Continuing advances in technology have yielded new information in many laboratories and have, among other things, led to the direct recording of sensory nerve fibre activity from the peripheral nerves of conscious human subjects (Hagbarth, 1976). The results of this new approach have, in general, confirmed earlier and contemporary studies in laboratory animals, of the kind reported.

SPECIFICITY OF RECEPTORS

Johannes Müller has a famous name associated especially with his law of Specific Nerve Energies, which laid the foundation for von Frey's (1895) codification of a specific correlation between the structure of cutaneous

---

* This article is based on an address to the Deutsche Arbeitskreis für Neurovegeta-tive Therapie at Bad Nauheim on 11 June 1977.

receptors and the sensations of touch, etc. This idea was hotly debated and a contrary view, that there were no specific nerve endings (receptors), was postulated by Weddell and his collaborators. They proposed a temporo-spatial pattern theory, which was soon challenged by physiological evidence for functional specificity. This was followed up by combined morpho-logical and electrophysiological studies, and it is now possible to assert with great confidence that the cutaneous sensory receptors have a high degree of specificity of function correlated with distinctive morphological structures (Iggo, 1974). I have chosen as an example a mechano- (or touch) receptor in skin originally described in the frog by Merkel (1880), and later in human and mammalian skin by Pinkus (1905). It was not until 1960 that, partly by the work of Hunt & McIntyre (1960) and then by the combined use of electrophysiological and morphological techniques, its function as a tactile receptor was actually established. There is still no direct and unequivocal evidence that in man it functions as a tactile sense organ, although there is strong presumptive evidence that this is so. This cutaneous receptor, now generally known as the type I slowly adapting mechanoreceptor (Iggo & Muir, 1969), typically responds to steady touch with a continuous discharge of nerve impulses, and fires at its maximum rate at innocuous (that is non-damaging) intensities of stimulation. This and other evidence clearly rejects any suggestion that this kind of receptor contributes directly to the sensa-tion of pain, and makes it necessary to seek elsewhere for 'pain' receptors.

Detailed and exacting studies of the responses of sensory fibres in peri-pheral nerves, using quantitatively controlled mechanical and thermal stimuli, have established that there are three main classes of sensory recep-tors in the skin – (a) mechanoreceptors, (b) thermoreceptors and (c) noci-ceptors. The detailed evidence has been reviewed elsewhere (Burgess & Perl, 1973; Iggo, 1974), and I will not repeat it here; instead I will con-centrate on the nociceptors. These can be defined as receptors that are excited by injurious, or potentially injurious, physical or chemical stimula-tion of the body tissues in which they lie; and that are insensitive to non-injurious stimuli.

NOCICEPTORS

Using the argument that what is painful for man also excites nociceptors in animals, it was first established by physiological studies that there are sensory receptors in the skin and deeper tissues which are excited by injurious intensities of stimulation. Two major kinds are known: (a) mechanical nociceptors and (b) thermal nociceptors. Both kinds are innervated by thin, myelinated or non-myelinated nerve fibres. The *mechanical nociceptors*, as their name indicates, are excited by injurious mechanical stimuli, such as pin-pricks or squeezing the skin or deeper tissues with toothed dissecting forceps. The response may continue indefinitely while the stimulus is applied, and the receptors can, therefore, provide a long-term signal for pain. Each receptor has a small receptive field (a few square millimetres).

The mechanical nociceptors are unresponsive to thermal stimuli below 50 °C, although they may be excited briefly by higher temperatures. Such very high temperatures are not normally reached in conscious subjects, and are certainly very injurious to the tissues, leading perhaps to pathological

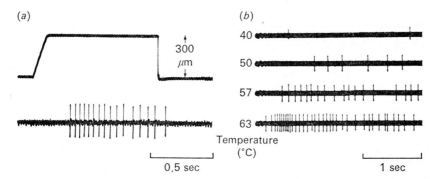

Fig. 1. (a) Response of a C mechanoreceptor (lower trace) to indentation of the skin (upper trace) recorded from a single non-myelinated sensory nerve fibre. (b) Response of a C thermal nociceptor during application to the skin of a metal rod at the temperatures indicated (from Iggo, 1977).

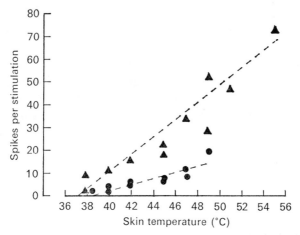

Fig. 2. Sensitization of the response of a C thermal nociceptor to heating the skin at random intensities indicated on the abscissa (● initial responses) and after 30 min of testing at 3-min intervals (▲) (from Beck, Handwerker & Zimmermann, 1974).

changes. In conscious subjects the normal protection of the tissues from these overtly damaging temperatures is due to another set of 'pain' receptors, the *thermal nociceptors*, which are excited at lower temperatures. Some examples are illustrated in Fig. 1. The temperatures shown in Fig. 1b refer to the stimulator, not to the skin. More exact skin temperature measurements from the work of Beck, Handwerker & Zimmermann (1974) are given in Fig. 2 and show that the thresholds range from 40 to 50 °C and that within the range of 40 to 50 °C a single sensory receptor responds quantitatively to the skin temperature. Many of these receptors can also be excited by severe mechanical stimuli. Application of the Hagbarth and Vallbo micro-neurographic recording technique in man has provided confirmation of the existence of these thermal nociceptors, with thresholds corresponding to those at which pain can be evoked in the conscious human subjects.

CHEMALGIA

A very characteristic feature of 'peripheral pain' is that the intensity of feeling is not constant – the same stimulus may at different times be 'painless' and 'painful'. There may be many reasons for this; one typical example is the enhanced sensitivity associated with inflammation of the tissues. Many investigators have searched for chemicals that are produced in the inflamed tissues and which will cause pain when applied to the skin. The list includes histamine, acetylcholine, 5-hydroxytryptamine, bradykinin and the prostaglandins.

There is no doubt that the excitability of the nociceptors can be altered by tissue damage. The thermal nociceptors may be made more sensitive (hyperpathia) or less sensitive, according to the experimental conditions chosen. Attempts have, therefore, been made to test the responses of the nociceptors to suspected chemalgogenic agents. Early results in my laboratory indicated that chemicals such as histamine, 5-hydroxytryptamine and bradykinin could excite the nociceptors, but that the action was non-specific (Fjällbrant & Iggo, 1961). More recently, several laboratories have explored this problem (Mense & Schmidt, 1974, in Kiel; Beck *et al.*, 1974, in Heidelberg; Perl, Kumazawa, Lynn & Kenins, 1976, in Chapel Hill, USA) and have established that bradykinin is a very potent excitant for nociceptors, thus confirming the earlier behavioural and reflex studies of Lim *et al.* (1967) in the USA and Besson, Conseiller, Hamann & Maillard (1972) in Paris.

The tissue reactions are no doubt very complex, but it is probable that the inflammatory processes in the tissues cause the release of chemicals that either directly, or indirectly, excite the nociceptors. For example, the prostaglandins do not directly excite the nociceptors, but may increase the potency of other chemicals such as bradykinin and 5-hydroxytryptamine (Chahl & Iggo, 1977).

Pathological conditions may significantly alter the peripheral 'pain' mechanisms. Traumatic injury can cause analgesia by complete or partial denervation of the tissues, and recovery from injury may be associated with aberrations of the normal mechanisms. These include (*a*) increased sensitivity of the regenerating nerve tip, which may convert a non-painful into a painful stimulus (Brown & Iggo, 1963; Wall & Gutnick, 1974); (*b*) initiation of impulses in dorsal root ganglia after distal transection of the peripheral nerve (Kirk, 1974) and (*c*) the almost terrifying problem of *causalgia*, in which nerve injury can lead to the initiation of severe pain from unstimulated tissue.

These various disorders are, or may be, of peripheral origin, but it would be naive to assume that the central nervous system does not very significantly influence the response of an individual to painful stimuli. Indeed, it is one of the paradoxes of 'pain' that the same physical stimulus may, or may not, actually cause pain. It is necessary to search for an answer in the central nervous system, and in the last decade my own experimental studies have taken up this search. I have felt confident that the problem was ready for attack because of the exact quantitative knowledge that we now have about the peripheral sensory receptor mechanisms, some of which I have just briefly reviewed.

SPINAL CORD NOCICEPTIVE MECHANISMS

The high degree of peripheral specificity which is such a feature of the cutaneous receptors might lead to the expectation that a similar kind of neuronal specificity existed in the central nervous system. To some degree this is the case, especially for the mechanoreceptor (tactile) pathways which reach the cerebral cortex along fairly well-defined and topographically organized paths (the dorsal column–lemnisco–thalamo–cortical tract). The afferent fibres from nociceptors, however, enter the dorsal horn of the spinal cord and there is now strong evidence to show that they project indirectly to the higher levels of the nervous system through the spinothalamic tract. They do not project directly into the dorsal columns.

The possible complexity of these spinal cord mechanisms was dramatized by the 'gate theory' of Melzack & Wall (1965), which sought to explain pain by invoking a mechanism that filtered the nervous activity entering the dorsal horn. I do not propose to consider this theory further in any detail, since it has been adequately discounted in a recent critical review by Nathan (1976).

Instead, I will turn immediately to a consideration of recent experimental evidence about the actual, rather than the theoretical, neuronal mechanisms in the spinal cord. Once again animal models are chosen, since they provide the opportunity for an exacting quantitative analysis. Electrophysiological techniques have been used in collaboration with Zimmerman and Handwerker in Heidelberg and with Drs Ramsey, Ogawa, Cervero, Molony and Handwerker in Edinburgh.

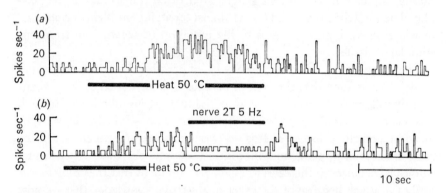

Fig. 3. Segmental inhibition. (a) Response of a class 2 dorsal horn neurone to heating the skin to 50 °C. (b) The suppressing effect on this response of electrical stimulation of the large mechanoreceptor afferent fibres in the plantar nerves at an intensity twice the nerve threshold (2T) and at a frequency of 50 per second (50 Hz) (from Handwerker, Iggo & Zimmermann, 1975).

First a brief review of the main kinds of neurone that were analysed. There are three main categories – *class* 1, which are excited only by an input from the sensitive mechanoreceptors, *class* 2, which are excited, in addition, by the nociceptors, and *class* 3, which are excited only by the nociceptors. From the point of nociception it might be thought that attention should be concentrated on class 3, and perhaps also on class 2, since these are excited by nociceptors. However, there are cogent reasons why this may be a mistake, and some results illustrated in Fig. 3 show what they are.

SEGMENTAL INHIBITORY MECHANISMS

The response to a peripheral noxious stimulus in a class 2 neurone may be a vigorous discharge of impulses, but as Fig. 3 shows, this can be almost completely abolished by the addition of an input from the mechanoreceptor afferent fibres, even though the latter on their own may be excitatory to the same neurone. Although this is evidence for complex neuronal networks and interactions, the effect is clear. A neurone that is excited by nociceptor stimulation can be switched off by simultaneous innocuous inputs from other cutaneous receptors. This kind of mechanism may contribute to the efficacy of counter-irritant therapy, and may even perhaps help to account for some of the analgesic effects of acupuncture.

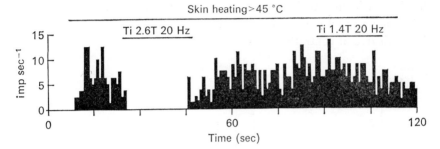

Fig. 4. Segmental inhibition. Response of a class 3 (nociceptor specific) dorsal horn neurone to heating the skin above 45 °C; the inhibition of this response by stimulation of large cutaneous afferent fibres in the tibial nerve at an intensity 2.6 times the nerve threshold and a frequency of 20 sec^{-1} (Ti 2.6T 20 Hz) and the lack of effect of stimulating the large muscle afferent fibres in the same nerve and at the same frequency by using a lower stimulus intensity of 1.4 times the nerve threshold (Ti 1.4T 20 Hz) (from Cervero, Iggo & Ogawa, 1976).

The class 3 (nociceptive) neurones show a similar response, although they are not directly excited by an input from the sensitive mechanoreceptors. A typical response is shown in Fig. 4. The dorsal horn neurone was excited by heating the skin, and while this effective stimulus was maintained, the cutaneous mechanoreceptor afferent fibres in the peripheral nerve were stimulated. They were able, by an action in the spinal cord, to block completely the response of the class 3 neurone to the nociceptor input. On withdrawal of the mechanoreceptor input, the noxious input was once again effective.

These results establish that there are effective mechanisms at the spinal level for modulating the onward transmission of information from the peripheral nociceptors. The exact mechanisms are still under investigation, but the recent successful recording of unitary neuronal activity in substantia gelatinosa neurones (Cervero, Molony & Iggo, 1977) is an indication of the direction that future progress may take.

DESCENDING INHIBITORY MECHANISMS

The segmental inhibitory mechanisms are only part of the story. The dorsal horn neurones are also under powerful descending inhibitory control, although to different degrees. There is a selective control, shown in Fig. 5. Class 2 (mechano- and nociceptor-driven) dorsal horn neurones may have their excitability by nociceptors almost completely suppressed, while leaving unchanged their responses to an input from the mechanoreceptors. The class 3 (nociceptor-driven) neurones, on the other hand, are under a lesser degree of central control.

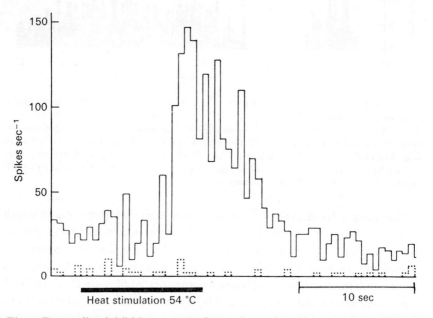

Fig. 5. Descending inhibition expressed on a class 2 dorsal horn neurone. With the spinal cord blocked rostral to the neurone there was a vigorous response of the cell to heating the skin to 54 °C, whereas when the cord was intact there was no response from the cell (from Handwerker *et al.*, 1975). Dashed record, spinal cord intact; solid record, spinal cord blocked.

These new results provide an exciting challenge to our understanding of the mechanisms of pain, as well as indicating the possible manipulation of the mechanisms, with the object of alleviating suffering.

CENTRAL PATHWAYS

It would be reasonable to postulate that the 'pain' would be experienced as a result of activity in the peripheral and spinal neurones that I have described only if some of the dorsal horn neurones projected to higher levels of the nervous system. Recent work in several laboratories has established that at least some of the class 2 and 3 neurones do project via the spinothalamic tract and so there is presumptive evidence that they are involved in pain. Furthermore, the central projection of the class 3 (nociceptor-driven) neurones provides a basis for a fairly clean-cut and specific pain pathway, leaving the class 2 (mixed-input) neurones in a supportive role.

REFERENCES

Beck, P. W., Handwerker, H. O. & Zimmermann, M. (1974). Nervous outflow from the cat's foot during noxious radiant heat stimulation. *Brain Res.* **67**, 373–86.

Besson, J. M., Conseiller, C., Hamann, K.-F. & Maillard, Marie-Claude. (1972). Modifications of dorsal horn cell activities in the spinal cord, after intra-arterial injection of bradykinin. *J. Physiol., Lond.* **221**, 189–205.

Brown, A. G. & Iggo, A. (1963). The structure and function of cutaneous 'touch corpuscles' after nerve crush. *J. Physiol., Lond.* **165**, 28–9P.

Burgess, P. R. & Perl, E. (1973). Cutaneous mechanoreceptors and nociceptors. In *Handbook of sensory physiology*, vol. II, ed. A. Iggo, pp. 29–78. Berlin: Springer-Verlag.

Cervero, F., Iggo, A. & Ogawa, H. (1976). Nociceptor-driven dorsal horn neurones in the lumbar spinal cord of the cat. *Pain* **2**, 5–24.

Cervero, F., Molony, V. & Iggo, A. (1977). Extracellular and intracellular recordings from neurones in the substantia gelatinosa Rolandi. *Brain Res.* **136**, 565–9.

Chahl, L. A. & Iggo, A. (1977). The effect of bradykinin and prostaglandin E_1 on rat cutaneous afferent nerve activity. *Br. J. Pharmac.* **59**, 343–7.

Fjällbrant, N. & Iggo, A. (1961). The effect of histamine, 5-hydroxytryptamine and acetylcholine on cutaneous afferent fibres. *J. Physiol., Lond.* **156**, 578–90.

von Frey, M. (1895). Beiträge zur sinnesphysiologie der haut. III. *Ber. Verh. K. Sächs. Ges. Wiss.* **47**, 166–84.

Hagbarth, K. E. (1976). Microneurography in man. In *Sensory functions of the skin in primates*, ed. Y. Zotterman, pp. 129–36. Oxford: Pergamon Press.

Handwerker, H. O., Iggo, A. & Zimmermann, M. (1975). Segmental and supraspinal actions on dorsal horn neurons responding to noxious and non-noxious skin stimuli. *Pain* **1**, 147–65.

Hunt, C. C. & McIntyre, A. K. (1960). An analysis of fibre diameter and receptor characteristics of myelinated cutaneous afferent fibres in cat. *J. Physiol., Lond.* **153**, 99–112.

Iggo, A. (1954). Receptors in the stomach and bladder. *J. Physiol., Lond.* **126**, 29–30P.

Iggo, A. (1958). The electrophysiological identification of single nerve fibres, with particular reference to the slowest conducting vagal afferent fibres in the cat. *J. Physiol., Lond.* **142**, 110–26.

Iggo, A. (1974). Cutaneous receptors. In *The peripheral nervous system*, ed. J. I. Hubbard, pp. 347–404. New York & London: Plenum.

Iggo, A. (1976). Peripheral and spinal 'pain' mechanisms and their modulation. In *Advances in pain research and therapy*, ed. J. J. Bonica & D. Albe-Fessard, vol. 1, pp. 381–94. New York: Raven Press.

Iggo, A. (1977). Cutaneous and subcutaneous sense organs. *Br. med. Bull.* **33**, 97–102.

Iggo, A. & Muir, A. R. (1969). The structure and function of a slowly-adapting touch curpuscle in hairy skin. *J. Physiol., Lond.* **200**, 763–96.

Kirk, E. J. (1974). Impulses in dorsal spinal nerve roots in cats and rabbits arising from dorsal root ganglia isolated from the periphery. *J. comp. Neurol.* **155**, 165–75.

Lim, R. K. S., Miller, D. G., Guzman, F., Rodgers, D. W., Rogers, R. W., Wang, S. K., Chao, P. Y. & Shih, T. Y. (1967). Pain and analgesia evaluted by the intraperitoneal bradykinin-evoked pain method in man. *Clin. Pharmac. Ther.* **8**, 521–42.

Melzack, R. & Wall, P. D. (1965). Pain mechanisms: a new theory. *Science* **150**, 971–9.

Mense, S. & Schmidt, R. F. (1974). Activation of group IV afferent units from muscle by algesic agents. *Brain Res.* **72**, 305–10.

Merkel, F. (1880). *Über die Endigungen der sensiblen Nerven in der Haut der Wirbeltiere*. Rostock: H. Schmidt.

Nathan, P. W. (1976). The gate-control theory of pain (a critical review). *Brain* **99**, 123–58.

Perl, E. R., Kumazawa, T., Lynn, B. & Kenins, P. (1976). Sensitization of high threshold receptors with unmyelinated (C) afferent fibres. In *Somatosensory and visceral receptor mechanisms*, ed. A. Iggo & O. B. Ilyinsky. *Prog. Brain Res.* **43**, 263–77.

Pinkus, F. (1905). Über Hautsinnesorgane neben dem menschlichen Haar (Haarscheiben) und ihre vergleichend-anatomische Bedeutung. *Arch. mikrosk. Anat. EntwMech.* **65**, 121–79.

Wall, P. R. & Gutnick, M. (1974). Ongoing activity in peripheral nerves: the physiology and pharmacology of impulses originating from a neuroma. *Exp. Neurol.* **43**, 580–93.

The developmental view of memory

R. F. MARK

Memory is the most difficult topic for neurophysiological research. In contrast with our knowledge of the way in which the moment to moment function of the nervous system controls behaviour, which is adequate in principle although crying out for details, there is no universally accepted theory as to how information is stored in the nervous system or how behaviour separated from environmental stimuli by more than a second or so is organized. And yet the way in which all mobile creatures, from bacteria to man, profit from experience in generating behaviour patterns is one of the most striking and immediate observations in biology. This ability is obviously of such evolutionary importance and so ubiquitous that one cannot help thinking that the fundamental mechanism for the modifiability of individual behaviour might be quite simple and indeed common to all creatures that can do it. The problem of the carriage of inheritance between generations was solved in evolution once and for all by the nucleic acids. The problem of rapid communication between cells for the organization of movement and the reception of environmental stimuli was solved in a similar manner by nerve cells with excitable membranes and specific connections between them that are basically identical from jelly fish to man. Why should not the long-term or permanent storage of information have a similar parsimonious mechanism at its base.

To embark on the search for such a key is an unashamedly reductionist approach to nature, a tactic that is becoming increasingly unfashionable nowadays. After all, what use is the molecular structure of DNA to an embryologist and what use is the knowledge of the biophysics of the nerve impulse to a psychologist? Would the knowledge of a fragment of biochemistry be any help in the understanding of the form and content of our own memories? And if not, which is the case, would it not be more beneficial to the human condition to enquire directly into the anatomy and physiology of the brain of man or his close relatives and search for a structure of memory where we know it has the most direct influence on mankind?

There is much research on memory which attempts to do this but the gains after about 50 years of collective effort have been very small. The reasons for what, on the whole, one must call failure combine the technical and the logical. One problem is that, even now, our methods for probing the structure and function of the brain have turned out to be impossibly crude when applied to a question as delicate as the nature of memory. The main conclusion to be drawn, which dates from the work of Lashley (1950) in the earlier part of this century, is that memory does not reside in any particular part of the brain. There is no all-purpose memory bank, filing system or tape recorder which the real-time nervous system can access; the idea that there should be one has no stronger foundation than straight introspection. We speak of searching our memory as we speak of searching a library. One has to go to the library, therefore might not our inner consciousness go to the memory. If there was such a structure accessible to current technology the combined approaches of brain lesions, electrical and chemical stimulation and recording in learning or remembering animals would certainly have shown it up by now.

There are two ways out of this impasse. Either one can suppose that the memory functions scattered through the different parts of the nervous system are too subtle to be revealed by present methods, or one can suppose that the memory structure does not exist. That this latter conclusion is by no means outrageous has been strongly argued over a lifetime's work by Young (1964). Learning, from the zoological rather than the psychological viewpoint, is not a matter of accumulation of new information but of acquiring profitable patterns of behaviour. The patterns of behaviour that can be acquired in any species have been rigorously selected in evolution to include only those that are of proven survival value. Thus a solitary wasp has an excellent spatial memory for the location of its nest but probably not very much else. A honeybee has an excellent and readily erasable memory for the colour of the flower upon which it is foraging. A ewe has an accurate memory for its own particular lamb. The examples, so familiar from natural history, are endless and in fact form most of the subject of ethology. Such constraints on learning are readily studied in the laboratory. Rats easily learn to avoid a food of novel taste if they are made ill within a few hours of its ingestion. Chickens, on the other hand, do not form taste aversions but readily associate food of a new colour or texture with illness (Martin & Bellingham, 1977). Rats feed often in the dark, relying on non-visual cues to find food; chickens feed during the day by visual search. How reasonable that the learning proclivities of a species are so firmly attached to species-specific behaviour.

There has never been the need in evolution to provide an animal with an uncommitted computer to govern its behaviour. The inheritance of behaviour ensures that the behavioural techniques for survival are perfectly matched to the morphological specializations fitting a species to a way of life. Similarly, there has never been a need to equip an animal with an uncommitted memory bank in which to store its experiences. Most passing events are meant to be forgotten. Only those which interact with an inherited behaviour pattern are relevant and the adaptability of behaviour is confined to such patterns.

This argument is simple but has profound implications. The current political controversy about inherited behaviour in man apart, there is no doubt that for most of the animal kingdom the substrate for behaviour is inherited. The material in nine tenths of Wilson's (1975) book is an impressive compilation of this. The way of dealing with those environmental situations which lock in with survival of a population are too important to be left to individual decision. They develop in the main through programmes intrinsic to the nervous system, comfortably in advance of their usefulness. This is not to say that the environment of an individual is unimportant but to emphasize that the programme is canalized in the embryological sense. Experience can modify but not reverse the direction of flow or the goal towards which the developmental processes are directed.

Set in the context of the richness and diversity of animal behaviour, learning is no longer so impressive a phenomenon. It consists of comparatively minute adjustments of individual behaviour sealed imperviously within the rigid limits to behaviour of the species. Inexorably the next argument follows. Learning is the life-long modification of development. And if so, the machinery of learning is the same as that of development, the only difference being that the environment can modify its pathway to a limited extent. Therefore, for clues as to the process one must look at the parts of the nervous system that are so essential that their development is worth an expensive investment of inherited instruction. Understanding of how these instructions became incorporated in the individual nervous system has more chance of revealing how memory works than looking for the biological equivalent of the magnetized module.

The development of the nervous system is, in the end, the patterning of synaptic connections between neurones so that a predictable output follows a standard input. From this point of view the easiest mechanism for long-term adaptable behaviour or learning is for there to have been a change in the distribution or efficiency of synaptic connections on a particular behavioural pathway. It is not the only possible mechanism. Modification of discharge frequencies due to a permanent change in the membrane properties of neurones of a dynamic network could also produce a reliable and stable change in input–output relations. Nevertheless the importance of exactly specified interneuronal connections in the function of those parts of the nervous system that have been studied in most detail suggests that the synapse should be the most easily modifiable unit. The question most relevant to a mechanism for memory, therefore, is by what mechanism do synaptic connections form and by what mechanism could the signalling function of the nervous system modify the patterns of connectivity.

The most comprehensive theory of neuronal connectivity is Sperry's (1963) theory of chemospecificity. In its most extreme form it postulates that the connections which are made during embryological development do so by matching chemical signals on the pre- and postsynaptic membranes. Each cell in a set of presynaptic neurones, for example those from the retina projecting to the brain are presumed to have a specific chemical tag which differentiates them from their neighbours in the same array. Each postsynaptic cell has a similar marker and synapses form between only those pairs that have matching affinities, the recognition being made by a chemical mechanism in the membranes which may be similar in kind to that by which lymphocytes recognize antigens in the immune response. These ideas had great historical importance because they could account for maladaptive connections that can be forced to form by surgical interference with the developing or regenerating brain. They emphasized the importance of intrinsic or native factors in embryogenesis of the brain and formed a very effective counter to preceding theories which postulated a primary formative role for experience in the sculpting of the nervous system. It is a difficult theory to extend to the functional modifiability of connections because this would require a change in the nature of chemical markers so that synapses could be shed from or attracted to new postsynaptic membranes

More recently, however, this theory has had to be radically modified by the discovery of a second process in the formation of topographically ordered sets of interneuronal connections. This is the process of competition (Prestige & Willshaw, 1975). It is still given that certain embryological forces, presumed chemical, do favour the formation of roughly appropriate connections; the optic nerve projects to one region of the brain and the auditory to another. Within each array of cells, however, exact point-to-point specifications are not accurately coded. What is, is the rank order of connections. The topography of the presynaptic cells is reinstated in the spatial order in which their connections form over the sheet of post-synaptic receiving cells. If half of the presynaptic cells are removed surgically, for example, the remainder form connections in the correct order but expand to innervate the whole population of postsynaptic cells.

Similarly if half the postsynaptic elements are removed the presynaptic fibres rearrange themselves to form an orderly but compressed map. In fact it is very easy in embryogenesis to force spatially incorrect connections to form by removing one source of the normal competing innervation. Therefore during normal development, within a class of neurones endeavouring to form connections, incorrect trial connections are not prohibited by chemical mismatch but are in fact bound to form. Sorting of connections by competition amongst synapses more or less well fitted to a given postsynaptic cell constitutes the mechanism whereby the last fine details of patterns of connectivity are established and maintained. What constitutes the organizing gradient and what are the mechanisms of competition are active topics of neuroembryological research at the moment, but the details do not concern us now. The important lesson comes from the general principle that the fine details in the topographical order of the nervous system come from competition between elements of a like nature for space on a corresponding array of cells.

Let us now take this idea from topographically ordered pathways, where the facts can be studied by anatomical and physiological methods, to the parts of the brain which organize whole behavioural sequences, where as yet most methods of study are depressingly inadequate. The same principles would suggest that the nerve cells involved should form their connections between corresponding sets by the same competitive mechanism. Here, however, the sets encode the classes of appropriate sensory stimuli for a particular act or the converse, the classes of behavioural response that may be triggered off by a given sensory stimulus. In contrast with the fixed wiring patterns of the more peripheral sensory or motor parts of the nervous system, these connections are multiple, each representing a possible mode of response, but the modes are not of infinite variety. They represent a selection of the comparatively restricted behavioural patterns that form the inherited behavioural repertoire of the animal. The problem in learning then is to select amongst this restricted range of behaviours those that are most effective.

Let us now suppose that the competitive principle still applies but that in these reaches of the nervous system it is not the mere presence of competing pathways which eliminates others but something related to their signal-carrying functions. A pathway that is effective in that it is often used will competitively displace connections on little used pathways, thus progressively enhancing its probability of use and progressively rendering the use of competing responses less likely. Learning has occurred, not by the acquisition of new patterns of behaviour but by eliminating unwanted patterns. This selective view of the learning process accords precisely with that of Young (1966) and is familiar enough to psychologists and to some physiologists (McIntyre, 1973). It has the following important features.

Learning is by selection from pre-existing patterns of behaviour built in by developmental mechanisms and inherited in a way to make them responsive to evolutionary pressures. The converse of this is rather frightening. Nothing can be learned unless the circuitry for it already exists in the brain. That the limits to animal adaptability are built into their brains is acceptable to most people, that the limits to human knowledge are similarly constrained may be harder to accept. Nevertheless the importance of inherited and intrinsic factors in human behaviour is being heard of more and more in psychology, particularly with respect to perception and the early psychological development of children. The limitation may in fact have no significance. Perhaps the human brain is already so complicated that the rate at which its interneuronal connections can be explored to produce all the available patterns of human behaviour may be slower than the rate of evolution of the species.

Secondly, the memory store does not exist apart from the neural circuitry for behavioural acts. The actual record consists of small modifications of intercellular connectivity of parts of the brain that must be used for a particular kind of behaviour. A memory bank for deposit, storage, search and retrieval has no physical reality in the brain.

Thirdly, the cellular mechanisms for changing connections are the same as for their sorting in development, at least as far as the control of growth and the selection and maturation of synapses are concerned. There is already abundant evidence that these processes can be influenced at certain stages in development by environmental stimuli, the most striking example being in the effect of binocular competition in the development of functional connections in the mammalian visual cortex (Wiesel & Hubel, 1963). The theory only requires that in the higher reaches of the nervous system development never stops, and that the impulse activity of nerve cells continues to influence the selection or suppression of synapses (Mark, 1974).

What are the advantages and disadvantages of this framework for research into memory? One pressing advantage is that whereas direct memory research, at least in those aspects concerned with brain mechanisms, is confused, contradictory and pervaded with a feeling of hopelessness (Gibbs & Mark, 1973), neuroembryological research is flourishing. Many new preparations and techniques are being developed and principles are beginning to emerge. This is the field which can supply ideas about mechanisms controlling the formation and function of synapses and biochemical or other tools which may be adapted to research on memory. The main disadvantage was mentioned at the beginning of this essay. When we finally know the way in which the biochemical switch is thrown to alter neuronal circuits we still shall not know much about memory. We shall not know what is encoded or how it is selected from the barrage of information surrounding an animal, just as knowledge of the DNA molecule cannot explain morphogenesis. On the other hand, such reductionist knowledge is better than none at all. It can provide tools for the real study of memory just as inhibitors of genetic transcription provide tools for the study of development and it can also provide a key to the kind of coding processes that could be used in the brain. Lastly, it could provide a means of controlling memory, the alleviation of failing memories or the elimination of undesirable ones, a piece of biological engineering so fraught with possibilities for good and evil as to make the concern about recombinant DNA seem trivial by comparison.

REFERENCES

Gibbs, M. E. & Mark, R. F. (1973). *Inhibition of memory formation.* New York: Plenum.

Lashley, K. S. (1950). In search of the engram. In *Physiological mechanisms in animal behaviour, Soc. exp. Biol. Symp.* No. 4, pp. 454–82. New York: Academic Press.

Mark, R. F. (1974). *Memory and nerve cell connections.* Oxford: Clarendon Press.

Martin, G. M. & Bellingham, W. P. (1977). Paper presented at the Fourth Experimental Psychology Conference, Newcastle, Australia.

McIntyre, A. K. (1973). Introductory remarks. In *Rep. Aust. Acad. Sci.* No. 16, pp. 5–7. Australia: Griffin Press.

Prestige, M. C. & Willshaw, D. J. (1975). On a role for competition in the formation of patterned neural connexions. *Proc. R. Soc., B* **190**, 77–98.

Sperry, R. W. (1963). Chemoaffinity in the orderly growth of nerve fibre patterns and connections. *Proc. natn. Acad. Sci.* **50**, 703–10.

Wiesel, T. N. & Hubel, D. H. (1963). Single cell responses in striate cortex of kittens deprived of vision in one eye. *J. Neurophysiol.* **26**, 1003–17.

Wilson, E. O. (1975). *Sociobiology, the new synthesis.* Cambridge, Massachusetts: The Belknap Press of Harvard Univ. Press.

Young, J. Z. (1964). *A model of the brain.* Oxford: Clarendon Press.

Young, J. Z. (1966). *The memory system of the brain.* Oxford: Oxford Univ. Press.

Mechanisms of plasticity in mammalian nerve–muscle interactions

R. A. WESTERMAN

It was Archie McIntyre's own concern for certain aspects of nervous system plasticity (Eccles & McIntyre, 1951, 1953) which first fostered my own interest in this field. This present contribution examines some largely unreported nerve–muscle interactions which occur if mammalian skeletal muscle is partly deprived of its motor nerve supply (Chan, Westerman & Ziccone, 1977). Various mechanisms are known whereby the partly denervated muscle recovers some or all of its contractile force. These are illustrated diagrammatically in Fig. 1, and will be defined briefly in introduction.

In this study I wished to examine the extent and relative contributions of these different compensatory mechanisms during the recovery from partial denervation.

Collateral sprouting (Fig. 1a)

Recovery of skeletal muscular contraction after partial transection of a motor nerve may be observed prior to the regeneration of the severed axons, and was first suggested by Exner (1885) to result from collateral growth of intact fibres to reinnervate the muscle. Subsequent workers (Weiss & Edds, 1946; Edds, 1950, Hoffman, 1950; Morris, 1953; Cöers & Woolf, 1959) conclusively demonstrated the occurrence of collateral sprouting from remaining undamaged axons which make new functional connections with denervated muscle fibres. This reinnervation occurs preferentially at the site of existing endplates rather than from new endplates (Bennett & Pettigrew, 1975) and has a rapid onset of one to two weeks not only at the peripheral nerve–muscle junction (Van Harreveld, 1945; Guth & Brown, 1965a, b) but also at the partially denervated sympathetic ganglion (Murray & Thompson, 1957) but not for spinal axons (Liu & Chambers, 1958).

Compensatory hypertrophy (Fig. 1b)

Acute hypertrophy of muscle fibres is a characteristic response of skeletal muscle to an increased work load (Denny-Brown, 1960; Gordon, Kasimierz & Fritts, 1967; Ianuzzo, Gollnick & Armstrong, 1976). It results in an increase in diameter of muscle fibres of different histochemical types and is thus not restricted to only certain classes of motor units (Gutmann, Schiaffino & Hanzlikova, 1971). Although such compensatory hypertrophy occurs relatively rapidly, there is some delay between the formation of new muscle mass and its ability to generate tension (Lesch et al., 1968). In partly denervated muscle, compensatory hypertrophy of innervated muscle fibres and atrophy of the denervated fibres is described (Denny-Brown, 1960; Brown, Jansen & Van Essen, 1976), suggesting that innervated fibres in a partly denervated muscle are subjected to an increased work load, so this mechanism might be expected to contribute especially to the early recovery of tension occurring in the first few weeks after partial denervation, before sprouting and regeneration are complete.

Regeneration of severed axons (Fig. 1c)

In addition to the rapid reinnervation by sprouting of intact axons, some of the damaged motor nerve fibres to a partly denervated muscle may regenerate more slowly. On reaching the muscle, regenerating axons may reinnervate any remaining denervated muscle fibres, or form new endplates at other loci on already innervated muscle fibres (Guth, 1962; Bennett & Pettigrew, 1974a) or, as in Fig. 1d, complete with existing innervation at original endplates (Bennett & Pettigrew, 1974b; Hoh, 1975; Bennett, McLachlan & Taylor, 1973). Many questions relating to such competitive interactions between axons in multiply innervated muscle of neonatal rat have recently been elegantly answered by Brown et al. (1976) and are discussed in a review of the specificity of nerve–muscle interactions by Fambrough (1976).

Selective reinnervation (Fig. 1d)

As a further complication, most muscles are composed of several different fibre types which are distinguished on morphological, histochemical and physiological grounds (Denny-Brown, 1929; Burke et al., 1971; Ariano, Armstrong & Edgerton, 1973; Burke, Levine, Tsairis & Zajac, 1973). The question of whether selectivity of mammalian muscle fibre types by reinnervating axons occurs during regeneration or sprouting is still considered open by Fambrough (1976), but in the toad, Hoh (1971) has provided evidence of genuine selective reinnervation of fast and slow muscles by corresponding nerve fibres, and this is further discussed by Mark (1974). By contrast, in the rat such a degree of selectivity does not occur (Miledi & Stephani, 1969; Hoh, 1975), and even some data on the goldfish (Mark, 1974) have been more cautiously reinterpreted (Scott, 1977). Other mechanisms which may occur during the recovery from partial denervation include various modifications of endplates and motor unit properties which are considered in the discussion.

AIMS

In view of the considerable plasticity of the nerve–muscle linkage (Brown et al., 1976; Fambrough, 1976) and the demonstrable changes in the remaining motoneurones supplying a partly denervated muscle (Kuno, Miyata & Muñoz-Martinez, 1974a, b; Kuno, 1975; Huizar, Kuno, Kudo & Miyata, 1977), we chose the paradigm of partial denervation to study the question of selective reinnervation in the heterogeneous fast-twitch muscle flexor digitorum longus (FDL) of the cat. Our particular aim was to examine whether the normal distribution of the various motor unit types in the FDL would persist without alteration following either short or long postoperative recovery from partial denervation. We expected that during both the shorter and longer recovery period intact skeletomotor axons should exhibit sprouting at their peripheral terminals. In addition, with longer recovery, regeneration of and reinnervation by previously damaged axons should occur with possible resulting plasticity of connections, so evidence of selective reinnervation was sought.

STRATEGY AND METHODS

In the cat the motor nerve supply to the FDL muscle is provided from the ventral roots of lumbar segments 6 and 7. Our experiments (shown diagrammatically in Fig. 1) utilize partial de-efferentation and deafferentation of the hindlimb muscles by complete extradural transection of either L6 or L7 spinal nerve roots at initial aseptic operations under halothane anaesthesia in young cats weighing between 0.5–1.6 kg. The extradural section of both dorsal (DR) and ventral (VR) spinal roots of one segment was performed instead of extradural VR section alone because it was technically easier, the certainty of total VR section was greater and de-afferentation does not appear to affect the extent of sprouting (Barker & Ip, 1966) or the action potentials of deafferented motoneurones (Kuno et al., 1974b; more importantly, Al-Amood (1973) detected no changes in

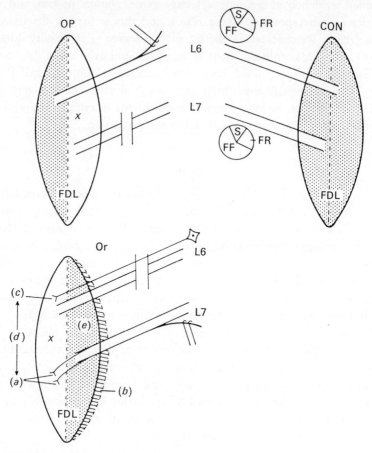

Fig. 1. Partial denervation (at left) by L6 or L7 dorsal and ventral root (VR) transection may result in adaptations of all components of the motor unit. Intact axons to innervated muscle (stippled areas) may sprout (*a*) to reinnervate denervated muscle fibres *x*; and compensatory hypertrophy of innervated fibres (*b*) may also occur during the shorter postoperative survivals. During the long postoperative intervals regeneration of previously severed axons (*c*) may occur, with reinnervation of any remaining denervated muscle fibres or possibly competition (*d*) with sprout-innervated muscle fibres, resulting in selective reinnervation of certain types of motor units. The properties of original motor units or their distribution apparently may undergo transformation (*e*). Possible modifications of endplates are not illustrated diagrammatically. Sectorgrams at top right depict symmetry of the distribution of motor unit types in both L6 and L7 VR: see Table 1*b* and text. Motor units in partly denervated muscle are isolated by splitting the intact VR (both left) during terminal experiment. OP = partly denervated muscle; CON = control muscle; FDL = flexor digitorum longus.

isometric or isotonic contraction characteristics of FDL muscle up to 8 weeks after deafferentation.

In order to distinguish better the nature and extent of the various plastic changes which may occur during recovery from such partial denervation we allowed postoperative survivals between 3 and 20 weeks (short term) or in excess of 46 weeks (long term).

Although adaptations of all components of the motor unit can occur (Burke & Edgerton, 1975), transformation of original motor unit properties (Fig. 1e) was not expected in our experimental model, but evidence suggesting that this may occur will be presented and discussed (see also Chan et al., 1977).

At the final acute experiment both hindlimbs were totally denervated except for the nerve to the lateral head of the flexor digitorum longus (LFDL). The extent of partial denervation which had been produced by the original transection of one VR was estimated at the start of the acute experiment by comparing each fractional contribution of L6 and L7 to the LFDL peripheral nerve tetanic tension (P_0) of whole muscle in the control hindlimb at optimum length. This assumes symmetry in the proportion of innervation to LFDL by corresponding root segments on each side. Although between animals there is considerable variation in the proportion of the muscle innervated by each of the two roots, in a single animal there is remarkable similarity between the pattern of innervation on the left- and right-hand side (Buller & Pope, 1977). We also tested this in five control animals and proved symmetry better than 95 % in all instances. The amount of recovery of tension by various mechanisms within the partly denervated muscle can be readily assessed if the maximum tetanic tension (P_0) produced by the intact root to the partly denervated muscle (OP) is expressed as a percentage of control muscle maximum tetanic tension (CON) produced by stimulating the whole LFDL muscle nerve on the control side (i.e. $\%P_0$ OP/CON). Single motor units were isolated by dissection of VR filaments and identified as unitary by the all-or-none behaviour of the antidromic action potential, isometric twitch and LFDL surface electromyograph, using the techniques described by Bagust, Lewis & Westerman (1973). Axonal conduction velocities were calculated from the latency of response to muscle nerve stimulation and the conduction distance. Motor unit tension measurements were made with the muscle held at whole muscle twitch optimum (Burke et al., 1973) and also at the length for the motor unit tetanus optimum (Sriratana, 1977). Test procedures for each unit included recording the unfused tetanic response for 'sag' and the ability to maintain tension during prolonged tetanic stimulation (fatigue index). Units were classified physiologically into four types on

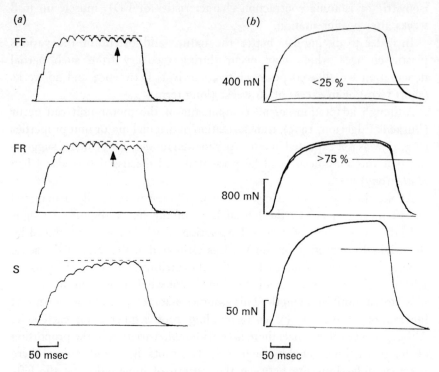

Fig. 2. Records of innervated fast fatigue (FF), fatigue-resistant (FR) and slow (S) motor units classified according to the test criteria of Burke, Levine, Tsairis & Zajac (1973) for (a) 'sag' (arrowed) during an unfused tetanus and (b) the fall in tension during prolonged tetanic stimulation ('fatigue'). 'Sag' was tested by stimulating with a pulse train of 500-msec duration and an inter-pulse interval of 125 % of the unit's contraction time. Resistance to fatigue was measured using 40 sec^{-1} tetani of 330-msec duration repeated 1 sec^{-1} for 120 sec and the photographically superimposed mechanical responses to the first and last tetanus in the repetitive sequence are shown at right. For the FF unit, final tetanic tension is less than 25 % of initial tension; for the FR and S units final tension does not fall below 75 % of initial tension in this fatigue test.

the basis of the 'sag' and 'fatigue' criteria used for the heterogeneous muscle medial gastrocnemius by Burke *et al.* (1973): FF (fast fatigue), FR (fatigue-resistant) both show sag, and S (slow) is fatigue resistant and non-sagging. Those units exhibiting sag and a degree of fatigue intermediate between FR and FF were designated FI, but they were few and have been included with FF for most statistical comparisons (Table 2; Figs. 5, 6). In Fig. 2 examples of such sag (at left) and fatigue behaviour (at right) of each of three types of motor units are shown: from above FF, FR, S.

RESULTS

Altered distribution of motor unit types in the partly denervated LFDL is
shown in Table 1. In total, 162 normal motor units in the LFDL were
isolated from either L7 ($n = 120$) or L6 ($n = 45$) VR of 13 control animals,
and on physiological classification these showed (in Table 1*b*) a distribution
of motor unit types similar to that which Ariano *et al.* (1973) obtained
histochemically for the same LFDL muscle (Table 1*a*). When compared to

Table 1. *Motor unit distribution*

LFDL	*n* MUs/cats*	FF	FI	FR	S
	(a) Histochemical				
(Ariano, Armstrong & Edgerton, 1973)	279/3	63 %		23 %	14 %
	(b) Physiological				
(Sriratana, 1977) L7	120/9	57	14	26	23
(Chan *et al.*, 1977) L6	45/5	24	7	10	4
Normal MU	165/14	79	21	36	26
		49 %	13 %	22 %	16 %
		62 %			
	(c)				
Reinnervated MUs	85/3	50	14	14	7
10–21 weeks postoperative		59 %	17 %	17 %	8 %
		76 %			
	(d)				
Reinnervated MUs	121/6	74	19	16	12
46–51 weeks postoperative		61 %	16 %	13 %	10 %
		77 %			

* Column *n* indicates both the number of motor units (MUs) and cats, separated
by an oblique slash. In subsections 1*b*, 1*c*, 1*d* the number of MUs in each class
is given together with their percentage of the total population.
LFDL = lateral head of the flexor digitorum longus; FF = fast fatigue;
FI = fatigue intermediate between FF and FR; FR = fatigue-resistant;
S = slow.

the distribution of normal motor unit types, reinnervated motor unit
populations for both the short-term (Table 1*c*) and long-term (Table 1*d*)
recovery of partly denervated LFDL were significantly different at
$P < 0.05$. There was an increase in the relative proportion of faster, more
fatiguable FF and FI motor units in both short- and long-term reinnervated
populations of approximately the same amount, viz. from 62 to 76 or 77 %.
This increase of FF and FI numbers was accompanied by corresponding

reductions in the proportions of FR and S motor units (Table 1*b* cf. *c, d*). Because motor units from partly denervated muscle are isolated randomly by splitting from the intact VR, the distribution of muscle motor unit types depends on the axon types present in the remaining untransected VR and should not be influenced by the degree of terminal sprouting from those axons. Such sprouting only alters the size of motor units, not their number. We therefore tested the assumed symmetry of the distribution of motor unit types between L6 and L7 VRs in control animals (Table 1*b*) and found it good except for the smaller sample of S units in L6 VR. With the longest postoperative intervals some functional regeneration of severed axons to the LFDL had occurred from transected VRs and the observed change in motor unit distribution might reflect selective reinnervation by FF axons preferentially. But because the same type and amount of modification of the motor unit spectrum is observed with shorter postoperative intervals in the complete absence of any detectable regeneration of axons in the transected VR (Table 1*c*) we believe that it cannot reflect selective reinnervation and may be related to a transformation of motor unit properties consequent upon an increased activation of remaining intact units (Pette, Smith, Staudte & Vrbova, 1973; Burke, Rudomin & Zajac, 1976). Possible mechanisms which would result in such actual or apparent transformation of motor units will be discussed.

Increased size of motor units in partly denervated LFDL is depicted in Table 2 and Figs. 5 and 6. Motor units size (i.e. tension development) reflects the diameter and number of muscle fibres within the unit, and is readily assessed if unit tetanic tension is expressed as a percentage of whole muscle tetanic tension. In Table 2 the normal mean motor unit sizes in descending order are seen to be FF and FI, FR and S and by contrast, the motor unit sizes in reinnervated LFDL show that FR units are on the average larger than FF units (Table 2). This suggests that the FR motor units, which are more specialized for tonic activity, increase their territory by more successful and extensive axon sprouting or by exhibiting relatively more compensatory muscle fibre hypertrophy. Either of these adaptations alone or in combination would lead to an increased size of motor unit. It is not possible to distinguish between sprouting or hypertrophy by simply measuring unit maximum tetanic tension, but the results of preliminary histochemical measurement of one pair of operated-side and control-side fibre diameters in LFDL muscles from one short-term experiment of 3 weeks postoperative survival after an estimated 80 % denervation are shown in Figs. 3 and 4, and suggest that sprouting predominates. The control LFDL is represented in Figs. 3*a* and 4*a* and the partly denervated muscle below shows a greater variation in the density of glycogen staining of

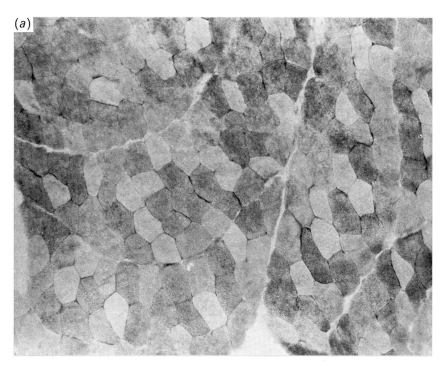

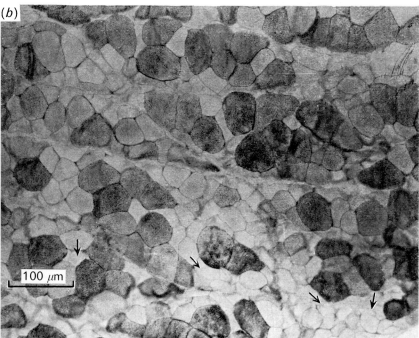

Fig. 3. Frozen sections 12-μm thick and stained by the periodic acid–Schiff method (Burke *et al.*, 1973). (*a*) Control lateral head of the flexor digitorum longus (LFDL) muscle; (*b*) 80 % denervated LFDL muscle after 3 weeks postoperative recovery. In both specimens density of glycogen staining indicated by the darkness of the individual myofibre profiles is variable, as are the transverse diameters of fibres. The obviously smaller, paler fibres such as those arrowed in (*b*) are histologically consistent with recently denervated fibres.

Table 2. *Motor unit size*

Motor unit type	Normal MU % P_0		All reinnervated MU % P_0 CON
FF (+FI)	0.08–2.24	Range	0.07–9.00
	0.80 ± 0.11	Mean ± s.e.	1.54 ± 0.25
	$n = 31$		$n = 162$
FR	0.04–1.34	Range	0.04–7.63
	0.62 ± 0.13	Mean ± s.e.	1.76 ± 0.38
	$n = 10$		$n = 30$
S	0.03–0.64	Range	0.09–2.42
	0.29 ± 0.13	Mean ± s.e.	0.92 ± 0.18
	$n = 4$		$n = 20$

Motor unit (MU) tetanic tension for both normal and all reinnervated motor units are expressed as a percentage of whole control muscle tetanic tension (P_0 CON). The range of MU size as well as mean ± standard error (s.e.) and number (n) of motor units in each MU class is given.
FF = fast fatigue; FI = fatigue intermediate between FF and FR; FR = fatigue-resistant; S = slow; P_0 = nerve tetanic tension.

individual myofibre profiles as well as transverse diameter. The obviously smaller, extremely pale fibres (arrowed) in Fig. 3*b* are histologically characteristic of muscle fibres denervated longer than a week. There is an irregular distribution of denervated fibres alone and in small groups among innervated fibres of normal appearance and fewer than half of the fibres appear denervated. The histograms in Fig. 4 were obtained using a Zeiss TGZ3 particle size analyser as described by Westerman & Wilson (1968), and show a bimodal distribution of fibres in which the smallest diameter group (dotted line) morphologically resembles denervated fibres, and this group constituted 43 % of the population measured. The histograms of Fig. 4 show that innervated fibres in partly denervated muscle (crosshatched) have a mean fibre diameter of 41 μm slightly below the control LFDL muscle fibre mean of 43 μm. Thus in the 3-week postoperative recovery after 80 % denervation, although there was a notable degree of sprouting which resulted in a threefold increase of tension from the remaining intact L6 VR to produce 64 % of the whole muscle tetanic tension on the control side, there is histologically no evidence of a significant amount of hypertrophy at this early stage of recovery from a large partial denervation. It is suggested that extensive sprouting and reinnervation provided the majority of recovery of tension in this instance, and the degree of unrecovered tension (36 %) agrees well with the proportion of still denervated fibres determined histologically (43 %).

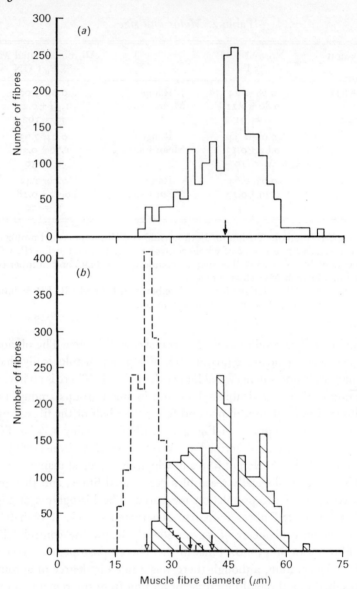

Fig. 4. Histograms of muscle fibre diameters from the control and partly denervated lateral head of the flexor digitorum longus (LFDL) muscles illustrated in Fig. 3. Representative regions of both LFDL muscles were photographed, printed and measured (Westerman & Wilson, 1968). The distribution of 2010 control muscle fibres (a) and 3700 fibres in the partly denervated LFDL (b) is given with the mean diameter 43 μm and 34 μm, respectively, shown by solid arrows. The broken portion of the lower histogram depicts diameters of 1590 fibres which histologically appear still denervated (arrowed muscle fibres in Fig. 3b; data from 2110 apparently innervated fibres in partly denervated LFDL are crosshatched; the means of these two sub-populations, 21 μm and 41 μm, respectively, are indicated by open arrows.

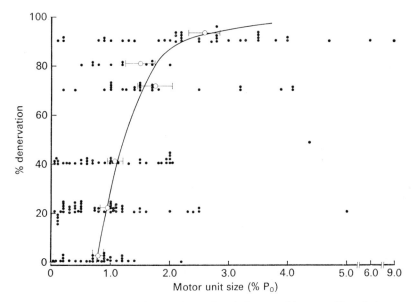

Fig. 5. Scattergram combining all 140 fast fatigue and intermediate fatigue motor units from short- and long-term reinnervations (●). The % denervation (estimated from terminal experiment) is plotted against the individual motor unit size, expressed as the % tetanic tension (P_0) of the motor unit/whole control lateral flexor digitorum longus tetanic tension. Means (open circles) and standard error bars are shown for each of the different extents of partial denervation for which motor units were obtained.

Increased mean motor unit size is proportional to the extent of partial denervation. This is most obvious in the Fig. 5 scattergram data combining all 140 FF and FI units from both long- and short-term reinnervations. The estimated percentage of denervation is plotted against the individual motor unit (MU) size P_0–expressed as %P_0 MU/CON LFDL. Fig. 5 shows that the mean motor unit size does not increase very much for small percentages of denervation but, with only a solitary exception, the largest motor units were obtained after recovery from very extensive denervations. The largest unit P_0 was 9% of the entire control muscle tetanic tension in an animal with 21 weeks recovery from a 94% denervation from which 24 units were isolated *in toto*. This raises the question of how many α motor neurones supply the normal LFDL in the cat? Boyd & Davey's (1966) data indicate 205, which would give a mean size for normal LFDL motor units of 0.5% P_0 whole muscle. Compare Table 2 – the overall mean unit size for normals is 0.71%, which may reflect some bias in VR splitting sampling techniques.

Increased motor unit size is less marked with longer postoperative recovery.

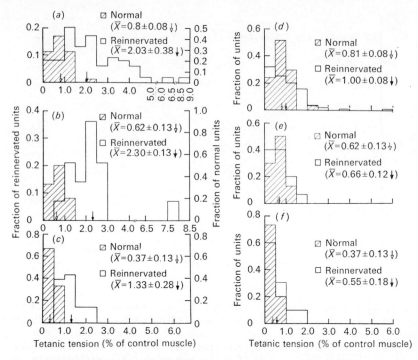

Fig. 6. (*a*), (*b*), (*c*) Histograms of the distribution of motor unit size (% tetanic tension of the motor unit/whole control lateral flexor digitorum longus tetanic tension as in Fig. 5) and of the 85 short-term reinnervated motor units of each type. From above, (*a*) Fast fatigue (FF) and intermediate fatigue (FI) types, (*b*) fatigue-resistant (FR) and (*c*) slow (S) motor units, with solid arrows indicating the means. Crosshatching depicts the distribution of the normal motor units of each type with mean (open arrow) and standard error of the mean for each distribution. Histograms (*d*), (*e*), (*f*) are for corresponding classes of motor units (FF and FI, FR and S, respectively) from long-term reinnervation. Symbols as for Fig. 6*a*, *b*, *c*.

Fig. 6*a*,*b*,*c* shows histograms of the distribution of size (unit tetanic tension expressed as a percentage of control muscle) for all 85 short-term reinnervated motor units of each type. The largest proportional increase of motor unit size occurs in the FR group, followed by the S type, with FF units showing the smallest increase relative to their large initial size in control muscles. This suggests that FR units may exhibit a greater sprouting capability than S or FF units. This is in contrast with the increased number of FF units seen in both groups of reinnervated motor units in Table 1*c*,*d*.

To indicate further the degree of plasticity in this neuromuscular interaction, Fig. 6*d*,*e*,*f* presents histograms of the motor unit sizes for the 120 long-term reinnervated units. Among this group of long-term reinnervated units there are some which are obviously larger than the normal motor units of corresponding physiological type, but the mean sizes of these reinnervated unit populations are only a little larger than those of the normals. We would interpret this in the light of the demonstration of some degree of reinnervation by axons which had regenerated through the originally severed root. Such regeneration contributed amounts of tension of up to 30 % of whole control LFDL tension, and seemed loosely related to the extent of the initial denervation in this group of six animals. We also sought evidence of polyneuronal innervation of the partly denervated muscle in each of these animals, using the method of Brown & Matthews (1960) where the sum of tetanic tensions of two almost equal divisions of the VRs was greater than the tetanic tension elicited by stimulating both divisions simultaneously.

The degree of polyneuronal innervation estimated in this way was small in all six animals belonging to the long-term recovery group. The greatest amount was 7.6 % and in three of the six it was less than 2 % – which corresponds to that observed in the adult medial head of the flexor digitorum longus (MFDL) muscle (Bagust et al., 1973). We thus suggest that after 11–12 months of post-denervation recovery the reduction in mean motor unit size may reflect the arrival of regenerating axons in the muscle which have successfully reinnervated some remaining denervated fibres. This alone would be unlikely to explain the extent to which the distribution of motor unit size in long-term reinnervated motor units has moved towards the normal. However, if regenerating axons converge onto sprout-reinnervated muscle fibres and compete successfully by functionally eliminating or suppressing some of these previously active sprout synapses (Bagust et al., 1973; Mark, 1974; Brown et al., 1976) then both the reduction in mean motor unit size with more prolonged postoperative recovery and the presence of a small amount of convergent polyneuronal innervation in half of the animals would be expected in this group. It is likely that the duration of any such competition is brief (Bagust et al., 1973; Tate & Westerman, 1973; Bennet & Pettigrew, 1974b), and the functional elimination of previous sprout-innervated neuromuscular synapses is largely complete after the 11–12 months of recovery allowed in the long-term reinnervated group.

Muscle fibre hypertrophy occurs with longer recovery and very extensive denervation. In another animal, whose LFDL had been estimated as 94 % denervated, with 21 weeks postoperative recovery allowed, similar histology and fibre diameter counts have been performed and the results displayed in Fig. 7, with the total fibre counts normalized to facilitate comparison with Fig. 4b.

The histograms shown in Fig. 7 were obtained from the muscle fibre

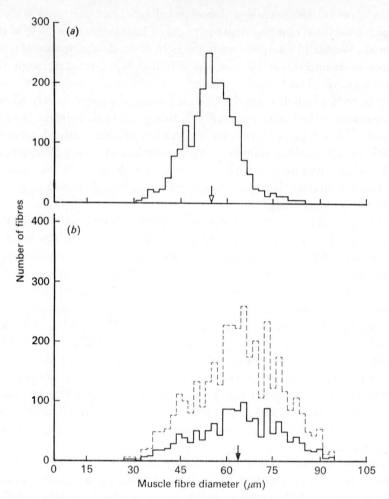

Fig. 7. Data obtained from a different animal to that illustrated in Figs. 3 and 4, 21 weeks after a 94 % denervation of the lateral head of the flexor digitorum longus (LFDL). Histograms of 2148 muscle fibre diameters from control (*a*) and 1400 fibres from partly denervated LFDL (*b*) show distribution and arrowed mean (55 and 63 μm, respectively) for each population. The broken histogram in (*b*) indicates the distribution of muscle fibre diameters for the partly denervated LFDL normalized to the same total fibre count (3700) as Fig. 4*b* (80 % denervation, 3 weeks recovery), to facilitate comparison between the two animals.

diameter counts made on sections of control muscle (*a*) and partly dener-vated muscle (*b*) stained for glycogen (PAS) and succinic dehydrogenase. There is some degree of hypertrophy evident when the distribution of fibre diameters in partly denervated LFDL are compared with the diameters of normal fibres. The mean diameters are 55 μm for normals and 63 μm for reinnervated. Although these data only represent one animal, they suggest that hypertrophy of sprout-innervated muscle fibres does occur with a longer recovery period of 20 weeks, following an extensive denervation.

Contractile performance of whole LFDL muscle alters after partial denervation. Table 3 compares some whole muscle parameters for 13 LFDL muscles partly denervated to varying extents (> or < 50%) with postoperative recovery periods between 3 and 21 weeks, with contralateral control muscle data. The muscle wet weights were not significantly different for the control and the two denervated groups nor were the time for half relaxation and twitch/tetanus ratio. By contrast, the contraction time of > 50% denervated LFDL muscle was significantly reduced ($P < 0.05$). This slightly faster contraction time is in contradistinction to the lengthening of contraction time seen for completely denervated fast-twitch muscle (Lewis, 1973) and the absence of any significant difference in the < 50% denervated group. Motoneurones in fast-twitch muscle FDL are characterized by a shorter after-hyperpolarization (Eccles, Eccles & Lundberg, 1958; Devanandan, Eccles & Westerman, 1965) compared with that for the slow-twitch soleus. Some motoneuronal changes have been demonstrated in intact soleus motoneurones functionally connected to innervated muscle fibres in partly denervated soleus muscle (Huizar *et al.*, 1977) about 3 weeks after the partial denervation. They observed a significant reduction in the duration of the after-hyperpolarization and of the twitch contraction times of such units, but there is no indication for how long these changes might persist, or whether they occur in the motoneurones supplying intact motor units in partly denervated fast-twitch muscle.

However, the present observations would be compatible with such a change in intact motoneurones and it should be pointed out that in all our partly denervated LFDL the contraction time was shorter, even if only marginally. Although both axonal conduction velocity and duration of after-hyperpolarization are both closely related to contraction times of the innervated muscle fibres (Devanandan *et al.*, 1965) the data of Huizar *et al.* (1977) make it evident that it is the duration of the after-hyperpolarization that is more closely related to the contraction properties of muscle fibres. This illuminates the observations of Salmons & Vrbova (1969) that the contractile characteristics of mammalian fast and slow muscle are influenced by the impulse pattern arriving at the muscle and the degrees of activity of the muscle.

Table 3. *Whole muscle data*

Whole muscle Denervation	% OP/CON		Ratio OP/CON Uncut VR P₀	Twitch duration		Ratio Twitch/tetanic tension
	Muscle wet weight	P_0		Time to peak (msec)	Half-relaxation time (msec)	
Normal $n = 10$ *MFDL (Lewis et al., 1973)	100.00	100.0	—	25.2	24.7	0.22
Complete Denervation (4/52) $n = 10$	57.0	30.0	—	44.9	24.7	0.64
Normal $n = 10$ LFDL (Chan et al., 1977)	100.0	99.0	—	31.4	27.5	0.24
†Reinnervated > 50 % ($\bar{X} = 83$ %) LFDL $n = 9$	99.9	79.9	5.49	27.5	27.2	0.25
< 50 % ($\bar{X} = 14$ %) $n = 11$	97.6	94.5	1.14	30.3	27.9	0.26

* MFDL is the medial head of the flexor digitorum longus muscle referred to as the flexor hallucis longus (FHL) by Lewis (1973).
† All reinnervated lateral flexor digitorum longus (LFDL) muscles from both short- and long-term groups have been pooled here and subdivided only by the extent of original partial denervation (means in parentheses). P_0 = maximum tetanic tension at optimum length; VR = ventral root.

DISCUSSION

Although partial deafferentation accompanied each denervation in this study there is evidence that neither the muscle isometric or isotonic contractile properties (Al-Amood, 1973), the intact motoneuione properties or action potentials (Kuno et al., 1974a,b), are affected by the former procedure, nor is the extent of sprouting (Barker & Ip, 1966). We consider that the remarkable symmetry reported by Buller & Pope (1977) and the measured symmetry of better than 95 % between LFDL tetanic tension developed by each pair of corresponding ventral roots in each of our five normal animals permits some semi-quantitative conclusions to be drawn from the data we obtained in partly denervated fast-twitch muscle. The increase of motor unit mean size after recovery from partial denervation is greatest for shorter durations of postoperative recovery and more extensive denervation. It is clear from comparing the muscle fibre diameters in Figs. 3 and 4 with those in Fig. 7 that hypertrophy of muscle fibres can provide a variable compensatory mechanism by which contractile force of a muscle can recover after partial denervation. Both animals had extensive denervation – 80 % and 94 %, respectively – but the population of fibres which morphologically appeared denervated (Fig. 4b, dotted line) after the short recovery of 3 weeks is not evident in the second animal with 21 weeks postoperative recovery (Fig. 7b). Indeed the mean fibre diameter for the partly denervated LFDL in this animal was 63 μm compared with that for the contralateral control muscle, 55 μm. This degree of hypertrophy corresponds to a mean increase of 34 % in the cross-sectional area of muscle fibres, with a resulting proportional increase of tension production by the muscle. It is clear that sprouting in this animal must have provided the major contribution to recovery of force development in this severely denervated muscle because its L6 VR innervation developed a maximum tetanic tension of 17.3 N, compared with a force production of 1.63 N from L6 VR to the normal muscle and 27.7 N from the entire innervation to the normal LFDL of the contralateral hindlimb. These data indicate more than a tenfold increase in the maximum tension produced by tetanic stimulation of L6 VR, and a recovery of tension to 62 % of control (17.3/27.7).

In this animal with 94 % denervation of one LFDL, one may gauge the full extent of the neuromuscular adaptations where only 6 % of the normal complement of axons have participated in axon sprouting and muscle fibre hypertrophy, so that few morphologically recognizable denervated fibres remained after 21 weeks (cf. Figs. 7b, 4b). We isolated a total of 24 motor units from the unsevered L6 VR and the sum (17.5 N) of their tetanic tensions corresponded closely to the 17.3 N obtained from previously stimulating that L6 VR before splitting. This suggests that probably all the functional units had been isolated from L6 VR, but this number of motor units (24) corresponds to 12 % of the 205 α axons counted by Boyd & Davey (1966) and considerably exceeds the estimate based on the extent of denervation (94 %). This could simply be ascribed to bilateral asymmetry and inter-animal variation, but an alternative possibility suggests that several larger γ motor axons may have participated in the sprouting to innervate some of the extrafusal fibres. If one includes the 129 γ axons counted by Boyd & Davey (1966) in the total pool of motor axons available for sprouting (129 + 205) we see that in the L6 VR there could be 6 % of 334, i.e. 20 α and γ motor axons in total. This is much closer to the 24 actually found, among which were six fibres with conduction velocities below 55 m sec^{-1}, longer twitch contraction times and the overall mean conduction velocities were 5 m sec^{-1} below the average for our population of normal units. The normal occurrence of a small number of β axons, which exhibit both fusimotor and skeletomotor connections (Bessou, Emonet-Dénand & Laporte, 1965; Barker, Stacey & Adal, 1970), adds credibility to the suggestion of fusimotor axon sprouting. Although provocative, this suggestion can be tested experimentally, and would be best examined in extensive partial denervations.

Another finding deserving comment in the long-term recovery group is the elimination of much of the marked increase of motor unit size which is so evident in the short-term group (cf. Figs. 5 and 6). Demonstrably functional regenerated axons in the transected root and the presence in three of the six animals of small amounts of convergent innervation from the two VRs suggest that reinnervation by regenerating axons and subsequent competitive interaction with sprout-innervated fibres may provide one mechanism by which such a change in motor unit size is brought about. This conforms with the remarkable regenerative propensity of the nervous system to form new synapses (Cotman & Lynch, 1976).

This study does not provide evidence as to what is the signal for sprouting, but the occurrence of similar terminal sprouting from motor axons is described in botulinum-toxin treated muscle where endplates are functionally ineffective (Duchen & Strich, 1968) and the recent reports of Brown & Ironton (1977) and Brown, Goodwin & Ironton (1977) using botulinum toxin suggest that the presence of denervated muscle fibres is an adequate stimulus for axon terminal sprouting. Interestingly, in five animals of our study the presence of < 10 % partial denervations did not provoke either recovery of the small loss of tension or muscle wet weight, both deficits remaining over 6 weeks, and the isometric twitch contraction time was not significantly shorter in these muscles. This is in sharp contrast with the extensive recovery of muscle tension, wet weight and the rapid onset of sprouting accompanied by reduced twitch contraction time as early as 3 weeks after larger denervations. Thus, there may be a 'threshold-concentration' of denervated fibres necessary to evoke collateral sprouting and the other changes seen during recovery from more extensive partial denervations.

The finding most difficult to explain is the effect of partial denervation on the apparent distribution of three different physiological types of muscle fibres, FF, FR, S. After both short- and long-term reinnervation there is an increased proportion of the faster, more fatiguable fibres (Table 1c,d). This changed distribution of fibre types is unlikely to reflect asymmetries of fibre types between the two VRs innervating the LFDL muscle and is not explicable purely in terms of different extents of sprouting by the three fibre types.

The possibility of neuromuscular failure at fine axon terminals after sprouting is more likely in the earliest stages of reinnervation and would result in the apparent rapid failure of the unfused tetanic response during the fatigue test. The use of paired shocks at graded intervals and the recording of the electromyogram during the tetanization of motor units suggested that after 6 postoperative weeks the neuromuscular connections were very secure. In two animals, tested at 3 and 4 weeks postoperatively, intermittent transmission failure was evident in several units (not included in the population) and was partially length-dependent. Similar instability of immature neuromuscular linkage has been observed in the neonatal kitten (Bagust et al., 1973) for both fast- and slow-twitch muscle. In one long-term reinnervated animal the sudden onset of neuromuscular failure occurred when muscle ischaemia resulted from arterial obstruction. Therefore, until histochemical confirmation of the altered motor unit distribution is provided, one cannot completely exclude neuromuscular failure as being in part responsible for the apparently increased proportion of faster, more fatiguable motor units.

Although less likely, structural or functional modification of endplates may in some way contribute to the apparent preponderance of FF and FI units. First, there are structural alterations – even in healthy muscle where the motor supply is wholly intact some sprouting occurs and Tuffery (1971) has demonstrated that rather than a continual turnover of neuro-muscular junctions during life, as proposed by Barker & Ip (1966), the individual synapses become more complex during adult life due to a process of sprouting, usually from the last node of Ranvier which forms connections with and thus elaborates the original endplate. In addition to structural complexity, more complex functional interactions are possible at the neuromuscular synapse and depend upon differing degrees of depression and facilitation of transmitter release at original neuromuscular junctions by comparison with newly reinnervated synapses (Bennett *et al.*, 1973). It is difficult to predict whether this would be likely to produce the observed shift in the population distribution of motor unit types but it would rely on different degrees of facilitation and depression at new synapses of the three different motor unit types and no information is available on the duration or extent of altered synaptic transmission at sprout-innervated junctions (E. M. McLachlan, personal communication). Although some early muscle wasting occurs it is unlikely that purely mechanical effects of partial denervation could so alter the contractile properties of FR and S units that they behave as FF units in sag and fatigue tests. Certainly, tetanic stimulation at other than the unit optimum length can enhance or obscure the 'sag' behaviour of units (Sriratana, 1977), but is unlikely to explain the increased fatiguability seen in the unit population of partly denervated LFDL.

The most attractive explanation of the altered motor unit distribution relates to the activity of remaining motoneurones in a partly denervated muscle. If one class of muscle fibre is reinnervated by axon sprouts of a different type, it is possible that the periphery alters the axon properties (Price, 1974; Lewis, Bagust, Webb, Westerman and Finol, 1977) although the reverse interaction is more generally conceded to occur (Salmons & Vrbova, 1969; Pette, Staudte & Vrbova, 1972). Particularly in view of our whole muscle contraction time data (Table 3) we consider that much if not all of the apparent transformation of motor unit types to a faster, more fatiguable unit population is consistent with the findings of Kuno *et al.* (1974*a,b*) and Huizar *et al.* (1977). They studied the discharge behaviour of intact motoneurones to partly denervated soleus muscle and found shortening of the after-hyperpolarization of intact soleus motoneurones still innervating denervated muscle as early as 3 weeks postoperatively and accompanied by a shortened twitch contraction time to peak. These changes are occurring in intact motoneurones which are involved in sprouting. Of equal importance are the findings and views of Vrbova and her colleagues (Salmons & Vrbova, 1969; Pette *et al.*, 1972, 1973) that the long-term activation history of the unit may determine its properties. Buller & Pope (1977) have extended Vrbova's chronic stimulation experiments to the cat FDL, and provide incontrovertible evidence that the pattern of nerve impulses reaching the muscle profoundly influences its contractile characteristics. It does so by determining the nature of the protein synthesis within the individual postsynaptic muscle fibres. Certainly these observations provide a very plausible explanation for the observed changes in motor unit distribution being due to a transformation of muscle fibre type. This is also consistent with the transformation of fibre type distribution observed histochemically in a variety of training and overload situations (Burke & Edgerton, 1975). The apparent change in distribution of fibre types is quite accessible to histochemical confirmation, and a detailed histochemical evaluation of our partly denervated LFDL muscles is in progress.

It is unwise to make projections too far removed from one's data. Nevertheless there are significant changes observable in the intact motor units innervating the partly denervated LFDL, and these occur with differing time courses. Much of the earliest recovery of tension following extensive lesions appears to result from axon sprouting and is accompanied by a large increase in average motor unit size. Some hypertrophy of innervated muscle fibres may occur but this is more evident after long recovery than at 3 weeks. The degree to which motor size increases is proportional to the extent of denervation, but becomes less marked with increasing postoperative recovery period up to one year. In those animals with longest recoveries there was evidence of regeneration and reinnervation by previously severed axons, and the possibility of some degree of neuromuscular convergence and competition cannot be excluded. The possibility of fusimotor sprouting to reinnervate extrafusal muscle fibres is highly speculative, as is the question of a 'threshold concentration' of denervated muscle fibres being necessary to stimulate rapid axon sprouting, but both suggestions are testable. The apparently altered distribution of motor unit types in partly denervated LFDL so that FF and FI types predominate remains to be confirmed histologically.

In conclusion it is apparent that a variety of adaptive processes may contribute to the changing nerve–muscle interactions during recovery from partial denervation. These mechanisms operate rather flexibly with differing speeds and to varying degrees, dependent largely upon the extent of the imposed neural deficit.

The work described in this paper has been carried out in collaboration with the following colleagues and students: K. A. Tate, S. P. Ziccone, H. S. Chan. I am grateful for the histochemical assistance of Dr Xenia Dennett and Miss Jandri Hoggins, who are continuing this study. The financial support of the Australian Research Grants Committee Project No. D1-75/15084 is most gratefully acknowledged. I am also indebted to Drs Elspeth McLachlan, L. Austin, B. G. Cragg and D. J. Tracey for their criticisms and helpful discussion.

REFERENCES

Al-Amood, W. S. (1973). Study of the effects of deafferentation on the isometric and isotonic contraction properties of fast and slow-twitch muscles of the cat pelvic limb. Ph.D. thesis, University of Bristol.

Ariano, M. A., Armstrong, R. B. & Edgerton, V. R. (1973). Hindlimb muscle fiber populations of five mammals. *J. Histochem. Cytochem.* **21**, 51–5.

Bagust, J., Lewis, D. M. & Westerman, R. A. (1973). Polyneuronal innervation of kitten skeletal muscle. *J. Physiol., Lond.* **229**, 241–55.

Barker, D. & Ip, M. C. (1966). Sprouting and degeneration of mammalian motor axons in normal and deafferented skeletal muscle. *Proc. R. Soc., B* **163**, 538–54.

Barker, D., Stacey, M. J. & Adal, M. N. (1970). Fusimotor innervation of the cat. *Phil. Trans. R. Soc., B* **258**, 315–46.

Bennett, M. R., McLachlan, E. M. & Taylor, R. S. (1973). The formation of synapses in reinnervated mammalian striated muscle. *J. Physiol., Lond.* **233**, 481–500.

Bennett, M. R. & Pettigrew, A. G. (1974a). The formation of synapses in striated muscle during development. *J. Physiol., Lond.* **241**, 515–45.

Bennett, M. R. & Pettigrew, A. G. (1974b). The formation of synapses in reinnervated and cross-innervated striated muscle during development. *J. Physiol., Lond.* **241**, 547–73.

Bennett, M. R. & Pettigrew, A. G. (1975). The formation of neuromuscular synapses. *Cold Spring Harb Symp. quant. Biol.* **40**, 409–24.

Bessou, P., Emonet-Dénand, F. & Laporte, Y. (1965). Motor fibres innervating extrafusal and intrafusal muscle fibres in the cat. *J. Physiol., Lond.* **180**, 649–72.

Boyd, I. A. & Davey, M. R. (1966). The composition of peripheral nerves. In *Control and innervation of skeletal muscle*, ed. B. L. Andrew, pp. 35–52. Dundee: Univ. St Andrews Press.

Brown, M. C., Goodwin, G. M. & Ironton, R. (1977). Prevention of motor nerve sprouting in botulinum toxin poisoned mouse soleus muscles by direct stimulation of the muscle. *J. Physiol., Lond.* **267**, 42–3P.

Brown, M. C. & Ironton, R. (1977). Suppression of motor nerve terminal sprouting in partially denervated mouse muscles. *J. Physiol., Lond.* **272**(1), 70P.

Brown, M. C., Jansen, J. K. S. & Van Essen, D. (1976). Polyneuronal innervation of skeletal muscle in newborn rats and its elimination during maturation. *J. Physiol., Lond.* **261**, 387–422.

Brown, M. C. & Matthews, P. B. C. (1960). An investigation into the possible existence of polyneuronal innervation of individual skeletal muscle fibres in certain hind limb muscles of the cat. *J. Physiol., Lond.* **151**, 436–57.

Buller, A. J. & Pope, R. (1977). Plasticity in mammalian skeletal muscle. *Phil. Trans. Soc., B* **278**, 295–305.

Burke, R. E. & Edgerton, V. R. (1975). Motor unit properties and selective involvement in movement. In *Exercise and sports science reviews*, ed. J. H. Wilmore & J. Keogh, vol. 3, pp. 38–81. New York: Academic Press.

Burke, R. E., Levine, D. N., Tsairis, P. & Zajac, F. E. (1973). Physiological types and histochemical profiles in motor units of the cat medial gastrocnemius. *J. Physiol., Lond.* **234**, 723–48.

Burke, R. E., Levine, D. N., Zajac, F. E., Tsairis, P. & Engle, W. K. (1971). Mammalian motor units: physiological–histochemical correlation in three types in cat gastrocnemius. *Science* **174**, 709–12.

Burke, R. E., Rudomin, P. & Zajac, F. E. (1976). The effect of activation history on tension production by individual motor units. *Brain Res.* **109**, 515–29.

Chan, H. S., Westerman, R. A. & Ziccone, S. P. (1977). Plasticity of motor units in the partly denervated fast-twitch muscle flexor digitorum longus. *Proc. Aust. physiol. pharmac. Soc.* **8**, 183*P*.

Cöers, C. & Woolf, A. L. (1959). *The innervation of muscle.* Oxford: Blackwell.

Cotman, C. W. & Lynch, G. S. (1976). Reactive synaptogenesis in the adult nervous system. In *Neuronal recognition*, ed. S. H. Barondes, pp. 69–108. New York: Plenum.

Denny-Brown, D. (1929). The histological features of striped muscle in relation to its functional activity. *Proc. R. Soc.*, *B* **104**, 371–411.

Denny-Brown, D. (1960). Experimental studies pertaining to hypertrophy, regeneration and degeneration. In *Neuromuscular disorders*, Proceedings of Association for Research in Nervous and Mental Disease, ed. R. D. Adams, L. M. Eaten & A. M. Sky, pp. 147–96. Baltimore: Williams & Wilkins.

Devanandan, M. S., Eccles, R. M. & Westerman, R. A. (1965). Single motor units of mammalian muscle. *J. Physiol., Lond.* **178**, 359–67.

Duchen, L. W. & Strich, S. J. (1968). The effect of botulinum toxin on the pattern of innervation of skeletal muscle in the mouse. *Q. Jl exp. Physiol.* **53**, 84–9.

Eccles, J. C., Eccles, R. M. & Lundberg, A. (1958). The action potentials of the alpha motoneurones supplying fast and slow muscles. *J. Physiol., Lond.* **142**, 275–91.

Eccles, J. C. & McIntyre, A. K. (1951). Plasticity of mammalian monosynaptic reflexes. *Nature, Lond.* **167**, 466–72.

Eccles, J. C. & McIntyre, A. K. (1953). The effects of disuse and of activity on mammalian spinal reflexes. *J. Physiol., Lond.* **121**, 492–516.

Edds, M. V. (1950). Collateral regeneration of residual motor axons in partially denervated muscles. *J. exp. Zool.* **113**, 517–51.

Exner, S. (1885). Notiz zu der Frage von der Faservertheilung mehrerer Nerven in einem Muskel. *Pflügers Arch. ges. Physiol.* **36**, 572–6.

Fambrough, D. M. (1976). Specificity of nerve–muscle interactions. In *Neuronal recognition*, ed. S. H. Barondes, pp. 25–67. New York: Plenum.

Gordon, E. E., Kasimierz, K. & Fritts, M. (1967). Adaptations of muscle to various exercises: studies in rats. *J. Am. med. Assoc.* **199**, 103–8.

Guth, L. (1962). Neuromuscular function after regeneration of interrupted nerve fibers into partly denervated muscle. *Exp. Neurol.* **6**, 129–41.

Guth, L. & Brown, W. C. (1965*a*). The sequence of changes in cholinesterase activity during reinnervation of muscle. *Exp. Neurol.* **12**, 329–36.

Guth, L. & Brown, W. C. (1965*b*). Changes in cholinesterase activity following partial denervation, collateral reinnervation and hyperneurotization of muscle. *Exp. Neurol.* **13**, 198–205.

Gutmann, E., Schiaffino, S. & Hanzlikova, V. (1971). Mechanism of compensatory hypertrophy in skeletal muscle of the rat. *Exp. Neurol.* **31**, 451–64.

Hoffman, H. (1950). Local reinnervation in partially denervated muscle: a histophysiological study. *Aust. J. exp. Biol. med. Sci.* **28**, 383–97.

Hoh, J. F. Y. (1971). Selective reinnervation of fast-twitch and slow-graded muscle fibres in the toad. *Exp. Neurol.* **30**, 263–76.

Hoh, J. F. Y. (1975). Selective and non-selective reinnervation of fast-twitch and slow-twitch rat skeletal muscle. *J. Physiol., Lond.* **251**, 791–801.

Huizar, P., Kuno, M., Kudo, N. & Miyata, Y. (1977). Reaction of intact spinal motoneurones to partial denervation of the muscle. *J. Physiol., Lond.* **265**, 175–91.

Ianuzzo, C. D., Gollnick, P. D. & Armstrong, R. B. (1976). Compensatory adaptations of skeletal muscle fiber types to a long-term functional overload. *Life Sci.* **19**, 1517–24.

Kuno, M. (1975). Responses of spinal motor neurones to section and restoration of peripheral motor connections. *Cold Spring Harb. Symp. quant. Biol.* **40**, 457–63.

Kuno, M., Miyata, Y. & Muñoz-Martinez, E. J. (1974a). Differential reaction of fast and slow α-motoneurones to axotomy. *J. Physiol., Lond.* **240**, 725–39.

Kuno, M., Miyata, Y. & Muñoz-Martinez, E. J. (1974b). Properties of fast and slow alpha motoneurones following motor reinnervation. *J. Physiol., Lond.* **242**, 273–88.

Lesch, M., Parmley, W. W., Hamosh, M., Kaufman, S. & Sonnenblick, E. H. (1968). Effects of acute hypertrophy on the contractile properties of skeletal muscle. *Am. J. Physiol.* **214**, 685–90.

Lewis, D. M. (1973). The effect of denervation on the differentiation of twitch muscles in the kitten hind limb. *Nature New Biol.* **241**, 285–6.

Lewis, D. M., Bagust, J., Webb, Sandra M., Westerman, R. A. & Finol, H. J. (1977). Axon conduction velocity modified by reinnervation of mammalian muscle. *Nature, Lond.* **270**, 745–6.

Liu, C. N. & Chambers, W. W. (1958). Intraspinal sprouting of dorsal root axons: development of new collaterals and preterminals following partial denervation of the spinal cord in the cat. *Archs Neurol. Psychiat., Chicago* **79**, 46–61.

Mark, R. F. (1974). Selective reinnervation of muscle. *Br. med. Bull.* **30**, 122–6.

Miledi, R. & Stephani, E. (1969). Non-selective reinnervation of slow and fast muscle fibres in the rat. *Nature, Lond.* **222**, 569–71.

Morris, D. D. B. (1953). Recovery in partly paralysed muscles. *J. Bone Jt Surg.* **35** B(4), 650–60.

Murray, J. G. & Thompson, J. W. (1957). The occurrence and function of collateral sprouting in the sympathetic nervous system of the cat. *J. Physiol., Lond.* **135**, 133–162.

Pette, D., Smith, M. E., Staudte, H. W. & Vrbova, G. (1973). Effects of long-term electrical stimulation on some contractile and metabolic characteristics of fast rabbit muscles. *Pflügers Arch. ges. Physiol.* **338**, 257–72.

Pette, D., Staudte, H. W. & Vrbova, G. (1972). Physiological and biochemical changes induced by long-term stimulation of fast muscle. *Naturwissenschaften*, **59**, 469–70.

Price, D. L. (1974). The influence of the periphery on spinal motor neurons. *Ann. N.Y. Acad. Sci.* **228**, 355–63.

Salmons, S. & Vrbova, G. (1969). The influence of activity on some characteristics of mammalian fast and slow muscles. *J. Physiol., Lond.* **201**, 535–47.

Scott, S. A. (1977). Maintained function of foreign and appropriate junctions of reinnervated goldfish extraocular muscles. *J. Physiol., Lond.* **268**, 87–110.

Sriratana, D. (1977). A comparison of the response characteristics of motor units in the cat's medial and lateral flexor digitorum longus muscle. *Proc. Aust. physiol. pharmac. Soc.* **8**, 180P.

Tate, K. A. & Westerman, R. A. (1973). Polyneuronal self-reinnervation of a slow-twitch muscle (soleus) in the cat. *Proc. Aust. physiol. pharmac. Soc.* **4**, 174–5.

Tuffery, A. R. (1971). Growth and degeneration of motor endplates in normal cat hind limb muscles. *J. Anat.* **110**, 221–47.

Van Harreveld, A. (1945). Reinnervation of denervated muscle fibres by adjacent functioning motor units. *Am. J. Physiol.* **144**, 477–93.

Weiss, P. A. & Edds, M. V. (1946). Spontaneous recovery of muscle following partial denervation. *Am. J. Physiol.* **145**, 587–607.

Westerman, R. A. & Wilson, J. A. F. (1968). The fine structure of the olfactory tract in the teleost *Carassius carassius* L. *Z. Zellforsch.* **91**, 186–200.

The sites and significance of brain angiotensin receptors

J. I. HUBBARD

THE PERIPHERAL AND CENTRAL ACTIONS OF ANGIOTENSIN II

Angiotensin II (AII) is an octapeptide, formed in the peripheral blood from its decapeptide precursor angiotensin I (AI) by a converting enzyme, as it passes through the lung. AI is in turn formed from a liver-synthesised precursor (angiotensinogen) by the kidney-liberated enzyme, renin (E.C 3.4.4.15). In the peripheral vasculature, AII is the most potent pressor agent known, having direct excitatory effects on arteriolar smooth muscle and indirectly stimulating the release of aldosterone, thus promoting sodium retention.

Some 17 years ago cross-circulation experiments indicated that AII also has effects in the central nervous system (Laverty, 1960, 1963; Bickerton & Buckley, 1961). These effects are now known to include a pressor action, a drinking response and an increase in the release of both antidiuretic hormone (ADH) and adrenocorticotropic hormone (ACTH). The central actions of AII have been much reviewed, especially in recent years, both specifically (Ferrario, Gildenberg & McCubbin, 1972; Severs & Daniels-Severs, 1973; Barker, 1976; Ganong, 1976) and as a part of general reviews of AII biochemistry and actions (Page & Bumpus, 1974; Regoli, Park & Rioux, 1974; Ganten et al., 1976; Peach, 1977), the regulation of blood pressure by the central nervous system (Onesti, Fernandes & Kim, 1976; Buckley, 1977; Buckley & Jandhyala, 1977), or the regulation of drinking (Fitzsimons, 1972; Peters, Fitzsimons & Peters-Haefli, 1975).

At present the time there is considerable controversy over the link, if any, between the peripheral and central actions of AII. If AII crosses the blood–brain barrier at all it does so at a very slow rate (Ganten et al., 1971b; Richardson & Beaulnes, 1971; Volicer & Loew, 1971; Ramsay & Reid, 1975). Parts of the brain lie outside the blood–brain barrier and are affected by AII (see pp. 335, 338). Other sites, within the blood–brain barrier, are also excited by AII (pp. 336–337). It seems probable that such sites are excited by AII which is formed in the brain. The strongest evidence that

AII is produced in the brain comes from the finding that renin and AI are as active centrally in producing the effects of injected AII as AII itself. Moreover, their actions are inhibited by a specific antagonist of AII, saralasin acetate (Epstein, Fitzsimons & Rolls, 1970; Fitzsimons, 1971; Cooling & Day, 1974; Reid & Ramsay, 1975).

A particularly thorough study of this point has been made by Gagnon and his colleagues, using the isolated rat pituitary. They have demonstrated (Gagnon, Cousineau & Boucher, 1973) that AII stimulates ADH release and that AI had a similar effect most probably by conversion to AII since the converting enzyme is known to be present in the rat pituitary (Yang & Neff, 1973), and AI is ineffective in the presence of an inhibitor of the converting enzyme (Sirois & Gagnon, 1975), which does not itself influence the pituitary AII receptors (Gagnon & Sirois, 1976).

Enzymes similar to kidney renin (iso-renin) and lung converting enzyme have been found in the brain (Roth, Weitzman & Piquilloud, 1969; Nahmod, Goldstein & Finkielman, 1971; Ganten *et al.*, 1971*b*, 1976; Yang & Neff, 1972; Daul, Heath & Garey, 1975). Recently, however, it has been found that the renin activity in the dog brain was possibly due to the lysosomal enzyme cathepsin D (E.C 3.4.4.23), suggesting that the apparent iso-renin effect was non-physiological (Day & Reid, 1976).

Pressor activity, identified as AI and AII by radioimmuno-assay and bioassay, has been reported in brain and cerebrospinal fluid (Fischer-Ferraro *et al.*, 1971; Ganten *et al.*, 1971*b*). Again it seems likely that these reports were premature. The pressor material detected by bioassay of brain extracts is probably not AII, for its effects were not blocked by saralasin acetate (Horvath, Baxter, Furby & Tiller, 1976). Results with the radioimmunoassay, which in any case shows much smaller amounts of AII than the bioassay (Horvath *et al.*, 1976), are flawed by the finding that angiotensinases present in brain extracts (Goldstein *et al.*, 1972) gave false positive readings. When this error was eliminated by inactivating the angiotensinases, very little, if any, AII activity could be detected (Reid, Day, Moffat & Hughes, 1977; J. Horvath, personal communication). The material detected in cerebrospinal fluid is probably not AII but an inert heptapeptide fragment (Hutchinson & Csicmann, 1977).

A physiological role for AII in the brain is suggested by the finding of angiotensin binding activity, which shows the characteristics expected of receptor binding in the rat and calf brain (McLean, Sirett, Bray & Hubbard, 1975; Bennett & Snyder, 1976; Sirett, McLean, Bray & Hubbard, 1977*b*; Sirett, Bray & Hubbard, 1977*a*). The independence of the brain iso-renin–angiotensin system from the peripheral system is shown by the persistence of the components of the brain system in normal concentration in rats with experimental pituitary tumors and very low peripheral renin levels (Ganten *et al.*, 1975*a*). The link between the peripheral and central limbs of the renin–angiotensin system appears to be the sympathetic nervous system (Zanchetti & Stella, 1976; Ganong, 1976). How the central system is activated is still obscure.

THE REGIONAL SITES OF ACTION OF ANGIOTENSIN II

Pressor effects

Intra-arterial injections. The first central effects of AII to be reported were found in cross-perfused animals. Laverty (1960, 1963), for instance, found that a nervously mediated vasconstriction occurred in a rat hindlimb, isolated from its body except for the nerves and perfused via its femoral artery with blood from a donor, when AII was injected into the host circulation. These experiments were not conclusive evidence for a brain action of AII, for AII also excites sympathetic ganglia (Kaneko, McCubbin & Irvine, 1961; Lewis & Reit, 1965). However, about the same time Bickerton & Buckley (1961) cross-perfused the head of a dog, isolated from its circulation and connected to its body only by its spinal cord, with blood from a donor animal. When AII was injected into the donor's circulation the systemic arterial pressure of the recipient as well as the donor rose. This result clearly implicated the central nervous system in the effect of AII.

Injection of AII into the carotid and vertebral arteries showed that only injections into the vertebral artery had a pressor effect (Yu & Dickinson, 1965) and Gildenberg (1969), by selectively ligating the basilar, vertebral and carotid arteries, showed that the area which had to be perfused for the central AII effect was the caudal pons and medulla. In this region the blood–brain barrier is only absent for the area postrema (AP) which lies in the lower medulla at the junction of the fourth ventricle with the spinal canal. It consists of bilateral, small, well-vascularised regions lying superficially in the floor of the ventricle. This consideration led Joy & Lowe (1970) to thermocoagulate the AP of greyhounds. This procedure abolished the pressor response to intravertebral infusion of AII. Confirmatory evidence for the role of the AP was provided by the demonstration that cooling the AP reversibly abolished the pressor response to AII injected into the intravertebral artery (Gildenberg, Ferrario, Alfidi & McCubbin, 1971), while microinjections of AII into the AP caused a pressor response (Ueda, Katayama & Kato, 1972) whereas injection of AII into the fourth ventricle had no effect.

Intraventricular injections. Intraventricular (IVT) injection of AII causes a marked pressor response in rats, cats, rabbits and goats (Rosendorff, Lowe,

Lavery & Cranston, 1970; Severs *et al.*, 1966; Severs, Summy-Long, Taylor & Connor, 1970); Andersson *et al.*, 1972; Eriksson, 1976). The pressor effect is dependent on sympathetic mechanisms almost entirely in cats, for it is lost after decerebration (Nashold, Mannarino & Wunderlich, 1962; Gildenberg, Ferrario & McCubbin, 1973). In rats, ganglion-blocking drugs only reduce the effect of IVT AII by half. The remainder of the effect appears to be mediated by ADH (Severs *et al.*, 1970).

There is some dispute as to the site from which sympathetic activation is achieved. In the cat there is good evidence that a portion of the peri-aqueductal gray (the subnucleus medialis) is the site of AII action. Certainly IVT AII has no pressor effect if this region is lesioned (Deuben & Buckley, 1970), or if the cerebral aqueduct is cannulated (Severs *et al.*, 1966), or if AII is added to the cerebrospinal fluid caudal to this region (Deuben & Buckley, 1970). In the cat too, it has been shown that the subnucleus medialis is the site of a relay in a pressor pathway originating in the hypothalamus (Enoch & Kerr, 1967*a,b*).

In rats, local application of AII to the region of the subnucleus medialis produces a pressor response comparable to IVT injection (S. Thornton, N. Sirett & J. Hubbard, unpublished observations). However, Hoffman & Phillips (1976*a*) claim that IVT injection is only effective if the anterior part of the third ventricle, not the posterior third or aqueduct, is perfused. The part of the anterior ventricle responding to AII is currently unknown.

Release of antidiuretic hormone

Plasma ADH levels increase in response to IVT administration of AII (Mouw, Bonjour, Malvin & Vander, 1971; Keil, Summy-Long & Severs, 1975) and this effect mediates some 50 % of the central pressor activity of AII in rats, as judged by the effects of lesions of either the supraoptic nuclei or the posterior pituitary (Severs *et al.*, 1970). Interestingly, in rats homozygous for hypothalamic diabetes insipidus the pressor response to IVT AII was almost absent (Hutchison, Schelling, Mohring & Ganten, 1976). IVT AII injection may affect the supraoptic or paraventricular nucleus, the latter being the closer of the two to the ventricles. Supraoptic neurones are excited by AII *in vivo* (Nicoll & Barker, 1971) and *in vitro* (Sakai, Marks, George & Koestner, 1974).

Intracarotid injection of AII was found to increase plasma ADH levels in conscious dogs (Mouw *et al.*, 1971) but this observation could not be confirmed on anaesthetised dogs (Shimazu, Shane & Claybaugh, 1973). The effects of intracarotid injection can plausibly be explained by direct action on the posterior pituitary since it is known that even a small intravenous or intracarotid dose of radioactive AII is concentrated in the pituitary (Osborne et al., 1971). In support it is found that spontaneous release of ADH from a rat pituitary in culture is accelerated by as little as 10^{-10}-M AII in the bathing medium and this effect is blocked by saralasin (Gagnon et al., 1973).

Release of adrenocorticotropic hormone

Intravenous AII infusion in cats and rats evokes ACTH release (Redgate, 1968; Campbell, Schmitz & Hskovitz, 1977). Intraventricular AII has a similar action (Reid & Jones, 1975). In rats with lesions in the median eminence, AII injection did not produce any increase in corticosteroid production. Nor did AII significantly increase ACTH release from anterior pituitary in vitro (De Wied, Witter, Versteeg & Mulder, 1969). A hypothalamic site of action could be inferred from this finding but it then becomes difficult to explain the effects of intravenous infusion, which would be most easily explained by a direct action on the anterior pituitary. It must be remembered too that many 'stressful' stimuli increase the release of ACTH and the list includes haemorrhage and pain which also increase the peripheral release of AII.

Drinking

Intravenous infusion of AII causes rats in water balance to drink more water (Fitzsimons & Simons, 1969). Injected directly into the brain, AII has the same effect in rats (Epstein et al., 1970) and every other species in which the procedure has been tried (Fitzsimons, 1972). In rats the regions of the brain most sensitive to AII injection (anterior hypothalamus, preoptic region and septum) are centred around the third and lateral ventricles.

While endeavouring to repeat these experiments with more physiological doses of AII, Johnson & Epstein (1975) found that AII injections only brought about drinking when the injection cannulas lay in or passed through the ventricles. If a cannula passed through a ventricle, AII was found to perfuse the ventricles. A periventricular site or sites of AII action thus appeared probable.

With a technique of plugging ventricles with Nivea cream injected through suitably placed cannulas, Hoffman & Phillips (1976a) showed that the posterior region of the third ventricle, the fourth ventricle and the lateral ventricle were not involved in the dipsogenic response to AII. This left the anterior third of the third ventricle which, if perfused with AII, always invoked drinking. Experiments in which plugs inserted in various parts of the anterior third ventricle were used to delineate the region sensitive to AII, showed that drinking responses occurred only when the ventral part of the anterior third ventricle was perfused with AII (Buggy & Fisher, 1976). In this region there are two structures outside the blood–brain barrier, the median eminence and the organum vasculosm of the lamina terminalis, which could mediate the intravenous effects of AII (Nicolaidis & Fitzsimons, 1975).

The subfornical organ. The subfornical organ (SFO) is a small nubbin of tissue lying on the anterior wall of the third ventricle between the interventricular foramina and dorsal to the anterior commissure. It lies outside the blood–brain barrier and thus could respond to intravenous AII and as it projects into the third ventricle it is accessible to IVT AII. Injections directly into this organ cause drinking, while its destruction causes IVT and intravenous injections of AII to become less effective (Simpson & Routtenberg, 1973, 1975). *In vivo* and *in vitro* neurones in the SFO are excited by AII and this action appears to be specific (Felix & Akert, 1974; Buranarugsa & Hubbard, 1976; Phillips & Felix, 1976). Some doubt that the SFO was the receptor for the dipsogenic effect of AII appeared, however, when it was found that its destruction did not completely abolish the dipsogenic effect of IVT or intravenous AII (Abdelaal, Assaf, Kucharczyk & Mogenson, 1974; Simpson & Routtenberg, 1975). Further systematic studies showed that in some rats at least, drinking in response to IVT AII injection continued in spite of 80–100 % destruction of the SFO. Furthermore, within 14 days animals began to drink again even if they had complete SFO lesions. It appeared that SFO lesions might block access to the ventral part of the anterior third ventricle, perhaps because of oedema and debris, for anterior third ventricular plugs could mimic the effect of SFO lesions (Buggy & Fisher, 1976; Hoffman & Phillips, 1976b).

It appears then that the SFO is neither necessary nor sufficient for the dipsogenic response to IVT AII (Buggy & Fisher, 1976). It may play some role in the dipsogenic response to systemic injections of AII. The finding that the threshold of drinking to AII was lower for injection in the SFO than for injection into any other brain locus, including the ventricles outside this organ, supports this idea, as does the finding that saralasin is more than 10 000 times as effective in blocking AII-induced drinking when injected into the SFO than when injected into the third ventricle (Simpson, Saad & Epstein, 1976).

REGIONAL LOCALISATION OF THE BRAIN
ISO-RENIN–ANGIOTENSIN SYSTEM

Enzyme studies

Studies of the regional localisation of various components of the system have so far shown very mixed results. Considering iso-renin activity, for instance, in puppies and dogs, all brain regions had much the same activity except for a localisation in the cerebellum (Ganten, Boucher & Genest, 1971a). Similarly in man all brain regions appeared to have equal activity except for the cerebellum which had an activity twice that of frontal cortex (Daul et al., 1975). In the rat brain more regional localisation was achieved (Haulica et al., 1975) so that giving the cortex concentration an arbitrary value of 1, the brainstem concentration was 1.8, the hypothalamus concentration 1.7 and the pituitary concentration 8.

Converting-enzyme activity likewise has been found to have different regional distributions in different studies. In the human brain, for instance, only the caudate nucleus stood out as a site of localisation (Poth, Heath & Ward, 1975) whereas working with the rat brain, Yang & Neff (1972, 1973) found regional localisations (cortex = 1), in the brainstem of 4.3, hypothalamus 1.5 and pituitary 10. The agreement with studies of renin activity in the rat brain (Haulica et al., 1975) is thus quite good, particularly in the indication of a pituitary localisation. In the Yang & Neff (1972) study significant localisations were also found in the cerebellum, striatum and hippocampus.

Receptor studies

Only studies of the rat and the calf brain have so far been published. In the rat brain there is agreement that the thalamus, hypothalamus, midbrain and brainstem contain the highest amounts of AII binding protein (McLean et al., 1975; Sirett et al., 1977a). In the calf brain (Bennett & Synder, 1976) the only significant regional binding of AII detected was in the cerebellar cortex (44 × cortex), cerebellar deep nuclei (7 × cortex) and the superior colliculus (1.4 × cortex). In our laboratory we have not so far been able to show significant localisation of AII binding in the cat and the monkey brain, using the techniques which worked successfully with the rat brain.

Fig. 1 shows our initial survey of specific AII binding activity in the rat (Sirett et al., 1977a). Only those regions – septum, thalamus, hypothalamus, midbrain, medulla – with significant localisations are shown. Not shown is the pituitary in which we also found significant binding activity. Other

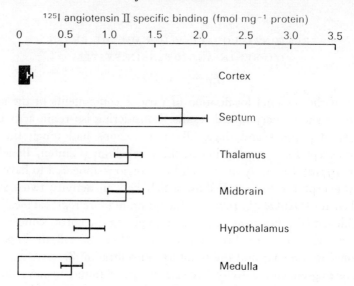

Fig. 1. The specific angiotensin II binding capacity of membranes from various regions of the rat brain. The number of brains were – cortex (black bar) six, septum eight, thalamus eight, midbrain five, hypothalamus seven, and medulla four. Bars represent ± the standard error. (Data from Sirett *et al.*, 1977*a,b*.)

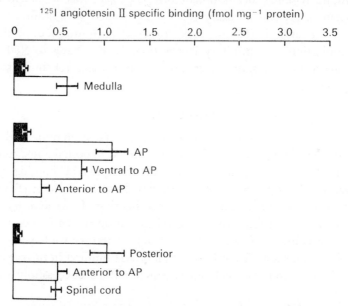

Fig. 2. The specific angiotensin II binding capacity of membranes from the spinal cord and regions of the medulla. In the first block, cortex (black) and medulla are compared. In the second, the parts of the medulla relative to the area postrema (AP), i.e. the area postrema, tissues ventral to the area postrema and tissues anterior to the area postrema, are compared. In the third block, binding in spinal cord, medulla anterior to area postrema and the remainder of medulla including area postrema (posterior) are compared. The number of rats was four or five. Bars represent ± the standard error. (Data from Sirett *et al.*, 1977*a,b*.)

large brain regions – cortex, striatum, hippocampus and cerebellum – were without significant activity in our hands. We have since turned our attention to finer subdivisions of those parts of the brain which had significant activity.

As Fig. 2 indicates the medulla proved extremely rewarding in this context. It contained significantly greater binding capacity than the cortex (Fig. 2, solid bar) and little or no binding activity was associated with the medulla anterior to the AP. The AP together with the medullary region ventral to it contained most of the AII binding activity of the organ. The spinal cord adjacent to this region again did not contain a significant localisation.

In the midbrain too, (Fig. 3) the AII binding activity was significantly greater than in the cortex. Subdivision of the midbrain into medial and lateral parts did not produce any marked localisation although both portions had significant binding activity. When, however, the division was dorsal/ventral, the dorsal quadrant containing the superior colliculi (and subnucleus medialis) proved to have a very high binding capacity (Sirett et al., 1977b).

An attempt was made to see whether there was a localisation of binding capacity in the tissue lining the ventral part of the anterior third ventricle. As Fig. 4 indicates the hypothalamus has significant binding capacity, certainly compatible with a physiological role of AII. Upon division into three parts the pre-optic region did not have a greater binding capacity than the anterior and posterior divisions of the hypothalamus. Medial/lateral division of the hypothalamus, however, did reveal a significant localisation of binding in the lateral region which is a region containing neurones excited by AII (Wayner, Ono & Nolley, 1973a,b). This may be accounted for by the presence of the lateral pressor pathway of Enoch & Kerr (1967b) which accompanies the median forebrain bundle in this region.

The subfornical region similarly did not prove to be a focus of AII binding despite the known effects of the peptide there. The septum was certainly a focus for binding activity but when the septum was divided (Fig. 5) anterior and posterior to the SFO, both portions had a high and roughly equal binding capacity. A medial/lateral division here, however, revealed a much greater binding capacity of the lateral septum, although the medial septum binding was, in itself, still significant. One source of the lateral pressor pathway of Enoch & Kerr (1967b) is of course the septum.

The thalamus, as Fig. 6 indicates, was again a region with a significant AII binding capacity. Attempts were made to section the thalamus into regions with a ventricular surface (medial) and with no ventricular surface (lateral). As Fig. 6 indicates, however, both sections had high and not

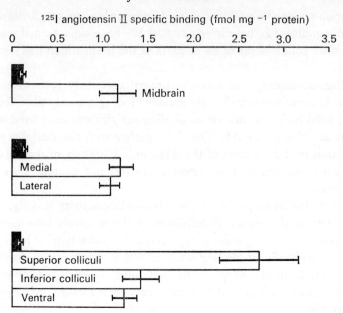

Fig. 3. The specific angiotensin II binding capacity of membranes from regions of the midbrain. The number of brains was four for the medial/lateral comparison and eight for the comparison of the colliculi. The value for the cortex is shown in black. Bars represent ± the standard error. (Data from Sirett *et al.*, 1977*a,b*.)

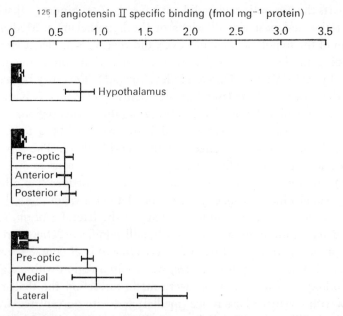

Fig. 4. The specific angiotensin II binding capacity of membranes from regions of the hypothalamus. The number of brains was four or five. The value for the cortex is shown in black. Bars represent ± the standard error. (Data from Sirett *et al.*, 1977*a,b*.)

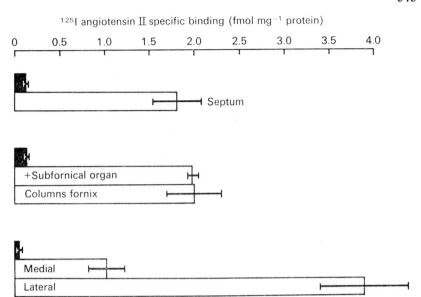

Fig. 5. The specific angiotensin II binding capacity of membranes from regions of the septum. The number of brains was four. The value for the cortex is shown in black. Bars represent ± the standard error. (Data from Sirett *et al.*, 1977*a,b*.)

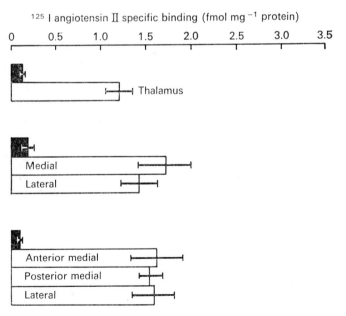

Fig. 6. The specific angiotensin II binding capacity of membranes from regions of the thalamus. The number of brains was four. The value for the cortex is shown in black. Bars represent ± the standard error. (Data from Sirett *et al.*, 1977*a,b*.)

significantly different binding capacities. Further subdivision of the medial portion into anterior and posterior parts again did not reveal differential binding so that we are unable to assume any particular specialisation of the thalamus.

SIGNIFICANCE OF ANGIOTENSIN II BRAIN RECEPTORS

Physiological significance

As the preceding section makes clear, specific AII receptor binding proteins have been found in a number of regions such as the AP (Fig. 2), superior colliculus (Fig. 3), septum, hypothalamus and thalamus which line the anterior third ventricle (Figs. 4, 5 and 6), and pituitary, where, as the evidence reviewed in earlier sections suggests, AII evokes its pressor, ADH- and ACTH-releasing and dipsogenic effects. The question naturally arises – what is the physiological role of these receptors? It has been suggested that AII is a transmitter (Barker, 1976), presumably in pathways involved in the control of blood pressure, hormone release and volume regulation. If this was so, we would expect to find AII receptor activity in the synaptosomal and microsomal fractions, where it is indeed concentrated (Sirett et al., 1977b).

Another criterion of transmitter action (Curtis & Johnson, 1974) is well established. AII certainly excites neurones in various areas of the brain (Nicoll & Barker, 1971; Wayner et al., 1973a,b; Felix & Akert, 1974; Sakai et al., 1974; Buranarugsa & Hubbard, 1976; Phillips & Felix, 1976). This excitatory action is dose-dependent and specific as judged by the effect of analogues and antagonists (Buranarugsa & Hubbard, 1976; Phillips & Felix, 1976; Schlegel & Felix, 1976). Of particular importance in this respect is the demonstration that AII does not excite all neurones in a preparation but only specific classes of neurones. For instance, in the SFO, three classes of neurones are found, excited by AII, carbachol and serotonin, respectively (P. Buranarugsa & J. Hubbard, unpublished observations).

The membrane action of AII has been explored only in a preliminary fashion. AII does depolarise spinal motoneurones in the frog. This action is calcium-dependent and tachyphylaxis was observed, a second dose of AII producing a small hyperpolarisation only, even after an interval of a few hours (Konishi & Otsuka, 1974).

The evidence that AII is formed within nerve terminals is again only preliminary. Both iso-renins (Ganten et al., 1976) and converting enzymes (Yang & Neff, 1972) are found in brain synaptosomal and microsomal fractions. The evidence that AII is within nerve terminals comes from immunohistochemistry (Fuxe, Ganten & Hokfelt, 1976). Encouragingly, some sites of localisation (hypothalamus and midbrain) contain high concentrations of putative AII receptors (Figs. 3 and 4).

The mode of destruction or removal of AII is not entirely clear although angiotensinases appear to be ubiquitous in the brain (Reid *et al.*, 1977). Again, evidence that AII is released from nerve terminals appears to be lacking although it may be inferred from the effectiveness of precursors in bringing about the same effects as AII when injected into brains (Epstein *et al.*, 1970; Cooling & Day, 1974; Reid & Ramsay, 1975).

Pathological significance

It is known that AII antagonists given to rats in sufficient dosage peripherally will antagonise the central actions of AII (Hoffman & Phillips, 1976c) and that the dipsogenic effects of AII given peripherally can be antagonised by IVT saralasin (Johnson & Schwab, 1975). Furthermore, all of the clinically active anti-hypertensive β adrenoreceptor antagonists produce a profound fall in blood pressure and heart rate after IVT administration in cats (Day & Roach, 1975). There is then a suspicion that effects of propranolol, saralasin and converting-enzyme inhibitors in normalising blood pressure in patients with essential hypertension (Laragh, Case, Wallace & Keim, 1976) may be exerted on the brain rather than on the peripheral circulation.

In this connection the two well-known strains of hypertensive rats – the genetic hypertensive (Phelan, Simpson & Smirk, 1976) and the Japanese spontaneously hypertensive (Yamori & Okamoto, 1976) – have been investigated. It has been claimed that for both types the hypertension results from a malfunction of the renin–angiotensin system in that their blood pressure can be normalised by IVT saralasin (Ganten, Hutchinson & Schelling, 1975b; Phillips, Phipps, Hoffman & Leavitt, 1975; Sweet, Columbo & Gaul, 1976). It appears, however, that the effect of saralasin is only pronounced when the blood pressure is elevated by stress (Phillips *et al.*, 1975) and we have found indeed (S. Thornton, N. Sirett & J. Hubbard, unpublished observations) that the effects of saralasin are much greater in the anaesthetised genetic hypertensive rat than in the conscious animal, which is hardly indicative of a pre-eminent role for the AII system in the genesis of this hypertension.

Research in the author's laboratory is supported by the Medical Research Council of New Zealand and the Neurological Foundation of New Zealand.

REFERENCES

Abdelaal, A. E., Assaf, S. Y., Kucharczyk, J. & Mogenson, J. (1974). Effect of ablation of the subfornical organ on water intake elicited by systemically administered angiotensin II. *Can. J. Physiol. Pharmac.* **52**, 1217–20.

Andersson, B., Eriksson, L., Fernandez, O., Kolinodin, C.-G. & Oltner, R. (1972). Centrally mediated effects of sodium and angiotensin II on arterial blood pressure and fluid balance. *Acta physiol. scand.* **85**, 398–407.

Barker, J. L. (1976). Peptides: roles in neuronal excitability. *Physiol. Rev.* **56**, 435–52.

Bennett, J. P. & Snyder, S. H. (1976). Angiotensin II binding to mammalian brain membranes. *J. biol. Chem.* **251**, 7423–30.

Bickerton, R. K. & Buckley, J. P. (1961). Evidence for a central mechanism in angiotensin-induced hypertension. *Proc. Soc. exp. Biol. Med.* **106**, 834–6.

Buckley, J. P. (1977). Central vasopressor actions of angiotensin. *Biochem. Pharmac.* **26**, 1–13.

Buckley, J. P. & Jandhyala, B. S. (1977). Central cardiovascular effects of angiotensin. *Life Sci.* **20**, 1485–94.

Buggy, T. & Fisher, A. E. (1976). Anterolateral third ventricle site of action for angiotensin induced thirst. *Pharmac. Biochem. Behav.* **4**, 651–60.

Buranarugsa, P. & Hubbard, J. I. (1976). Angiotensin receptors in rat subfornical organ *in vitro*. *Proc. Univ. Otago Med. Sch.* **54**, 3–4.

Campbell, W. B., Schmitz, J. M. & Hskovitz, H. D. (1977). (Des-Asp^1) Angiotensin I: a study of its pressor and steroidogenic activities in conscious rats. *Endocrinology* **100**, 46–51.

Cooling, M. J. & Day, M. D. (1974). Inhibition of renin–angiotensin induced drinking in the cat by enzyme inhibitors and by analogue antagonists of angiotensin II. *Clin. exp. Pharmac. Physiol.* **1**, 389–96.

Curtis, D. R. & Johnston, G. A. R. (1974). Amino acid transmitters in the mammalian central nervous system. *Ergebn. Physiol.* **69**, 98–142.

Daul, C. B., Heath, R. G. & Garey, R. E. (1975). Angiotensin-forming enzyme in human brain. *Neuropharmacology* **14**, 75–80.

Day, M. D. & Roach, A. G. (1975). The brain as a possible site for the cardiovascular effects of β adrenoceptor blocking agents in cats. *Clin. Sci. molec. Med.* **48**, 269–72S.

Day, R. P. & Reid, I. A. (1976). Renin activity in dog brain: enzymological similarity to cathepsin D. *Endocrinology* **99**, 93–100.

Deuben, R. R. & Buckley, J. P. (1970). Identification of a central site of action of angiotensin II. *J. Pharmac. exp. Ther.* **175**, 139–46.

De Wied, D., Witter, A., Versteeg, D. H. G. & Mulder, A. A. (1969). Release of ACTH by substances of central nervous system origin. *Endocrinology* **85**, 561–9.

Enoch, D. M. & Kerr, F. W. L. (1967a). Hypothalamic vasopressor and vesicopressor pathways. I. Functional studies. *Archs Neurol. Psychiat., Chicago* **16**, 290–306.

Enoch, D. M. & Kerr, W. L. (1967b). Hypothalamic vasopressor and vesicopressor pathways. II. Anatomic study of their course and connections. *Archs Neurol. Psychiat., Chicago* **16**, 307–20.

Epstein, A. N., Fitzsimons, J. T. & Rolls, B. J. (1970). Drinking induced by injection of angiotensin into the brain of the rat. *J. Physiol., Lond.* **210**, 457–74.

Eriksson, L. (1976). Sodium and angiotensin II in the central control of fluid balance. *Acta physiol. scand.* Supp. **444**, 1–36.

Felix, D. & Akert, K. (1974). The effect of angiotensin II on neurons of the cat subfornical organ. *Brain Res.* **76**, 350–3.

Ferrario, C. M., Gildenberg, P. L. & McCubbin, J. W. (1972). Cardiovascular effects of angiotensin mediated by the central nervous system. *Circulation Res.* **30**, 257–62.

Fischer-Ferraro, C., Nahmod, V. E., Goldstein, D. J. & Finkielman, S. (1971). Angiotensin and renin in rat and dog brain. *J. exp. Med.* **133**, 353–61.

Fitzsimons, J. T. (1971). The effect of drinking of peptide precursors and of shorter chain peptide fragments of angiotensin II injected into the rats' diencephalon. *J. Physiol., Lond.* **214**, 295–303.

Fitzsimons, J. T. (1972). Thirst. *Physiol. Rev.* **52**, 468–561.

Fitzsimons, J. T. & Simons, B. J. (1969). The effect on drinking in the rat of intravenous infusion of angiotensin, given alone or in combination with other stimuli of thirst. *J. Physiol., Lond.* **203**, 45–57.

Fuxe, K., Ganten, D. & Hokfelt, T. (1976). Immunohistochemical evidence for the existence of angiotensin II containing nerve terminals in the brain and spinal cord in rats. *Neurosci. Lett.* **2**, 229–34.

Gagnon, D. J., Cousineau, D. & Boucher, P. J. (1973). Release of vasopressin by angiotensin II and prostaglandin E_2 from the rat neuro-hypophysis *in vitro*. *Life Sci.* **12** (Part 1), 487–97.

Gagnon, D. J. & Sirois, P. (1976). Neurohypophyseal effects of angiotensin: further studies. *J. Pharm. Pharmac.* **28**, 777–8.

Ganong, W. F. (1976). The renin–angiotensin system and the central nervous system. *Fedn Proc. Fedn Am. Socs exp. Biol.* **36**, 1771–5.

Ganten, D., Boucher, R. & Genest, J. (1971*a*). Renin activity in brain tissue of puppies and adult dogs. *Brain Res.* **33**, 557–9.

Ganten, D., Ganten, U., Schelling, P., Boucher, R. & Genest, J. (1975*a*). The renin and iso-renin–angiotensin systems in rats with experimental pituitary tumours (38585). *Proc. Soc. exp. Biol. Med.* **148**, 568–72.

Ganten, D., Hutchinson, J. S. & Schelling, P. (1975*b*). The intrinsic brain iso-renin–angiotensin system in the rat: its possible role in central mechanisms of blood pressure regulation. *Clin. Sci. molec. Med.* **48**, 265–8*S*.

Ganten, D., Hutchinson, J. S., Schelling, P., Ganten, U. & Fischer, H. (1976). The iso-renin–angiotensin systems in extrarenal tissue. *Clin. exp. Pharmac. Physiol.* **3**, 103–26.

Ganten, D., Marquez-Julio, A., Granger, P., Hayduk, K., Karsunky, K. P., Boucher, R. & Genest, J. (1971*b*). Renin in dog brain. *Am. J. Physiol.* **221**, 1733–7.

Gildenberg, P. L. (1969). Localisation of a site of angiotensin vasopressor activity in the brain. *Physiologist, Wash.* **12**, 235.

Gildenberg, P. L., Ferrario, C. M., Alfidi, R. J. & McCubbin, J. W. (1971). Localisation of central nervous system vasopressor activity of angiotensin. In *Proc. 25th int. congr. physiol. sci.*, vol. 9, p. 203. Munich: German Physiol. Soc.

Gildenberg, P. L., Ferrario, C. M. & McCubbin, J. W. (1973). Two sites of cardiovascular action of angiotensin II in the brain cf the dog. *Clin. Sci.* **44**, 417–20.

Goldstein, D. J., Diaz, A., Finkielman, S., Nahmod, V. E. & Fischer-Ferraro, C. (1972). Angiotensinase activity in rat and dog brain. *J. Neurochem.* **19**, 2451–2.

Haulica, I., Branisteanu, D. D., Rosca, V., Stratone, A., Berbekeu, V., Balan, G. & Ionescu, L. (1975). A renin-like activity in pineal gland and hypophysis. *Endocrinology* **96**, 508–10.

Hoffman, W. E. & Phillips, M. I. (1976*a*). Regional study of cerebral ventricle sensitive sites to angiotensin II. *Brain Res.* **110**, 313–30.

Hoffman, W. E. & Phillips, M. I. (1976*b*). The effect of subfornical organ lesions and ventricular blockade on drinking induced by angiotensin II. *Brain Res.* **108**, 59–73.

Hoffman, W. E. & Phillips, M. I. (1976c). Evidence for Sar¹ Ala⁸ angiotensin crossing the blood cerebrospinal fluid barrier to antagonize central effects of AII. *Brain Res.* **109**, 541–52.

Horvath, J., Baxter, C., Furby, F. & Tiller, D. (1976). Endogenous angiotensin II levels in the central nervous system. *Clin. Res.* **24**, 223A.

Hutchinson, J. S. & Csicmann, J. (1977). Characteristics of immunoreactive angiotensin II in canine cerebrospinal fluid. *Proc. Aust. physiol. pharmac. Soc.* **8**, 66P.

Hutchinson, J. S., Schelling, P., Mohring, J. & Ganten, D. (1976). Pressor action of centrally perfused angiotensin II in rats with hereditary hypothalamic diabetes incipidus. *Endocrinology* **99**, 819–23.

Johnson, A. K. & Epstein, A. N. (1975). The cerebral ventricles as the avenue for dipsogenic action of intracranial angiotensin. *Brain Res.* **86**, 399–418.

Johnson, A. K. & Schwab, J. E. (1975). Cephalic angiotensin receptors mediating drinking to systemic angiotensin II. *Pharmac. Biochem. Behav.* **3**, 1077–84.

Joy, M. D. & Lowe, R. D. (1970). Evidence that the area postrema mediates the central cardiovascular response to angiotensin II. *Nature, Lond.* **228**, 1303–4.

Kaneko, Y., McCubbin, J. W. & Irvine, H. (1961). Ability of vasoconstrictor drugs to cause adrenal medullary discharge after 'sensitisation' by ganglion stimulating agents. *Circulation Res.* **9**, 1247–54.

Keil, L. C., Summy-Long, J. & Severs, W. B. (1975). Release of vasopressin by angiotensin II. *Endocrinology* **96**, 1063–5.

Konishi, S. & Otsuka, M. (1974). The effects of substance P and other peptides on spinal neurons of the frog. *Brain Res.* **65**, 397–410.

Laragh, J. H., Case, D. B., Wallace, J. M. & Keim, H. (1976). Blockade of renin or angiotensin for understanding human hypertension: a comparison of propranolol, saralasin and converting enzyme blockade. *Fedn Proc. Fedn Am. Socs exp. Biol.* **36**, 1781–7.

Laverty, R. (1960). The effect of drugs and vasoactive substances on the circulatory system of experimental animals with particular reference to the problems of experimental hypertension. Ph.D. thesis, University of New Zealand.

Laverty, R. (1963). A nervously-mediated action of angiotensin in anaesthetised rats. *J. Pharm. Pharmac.* **15**, 63–8.

Lewis, G. P. & Reit, E. (1965). The action of angiotensin and bradykinin on the superior cervical ganglion of the cat. *J. Physiol., Lond.* **179**, 538–53.

McLean, A. S., Sirett, N. E., Bray, J. J. & Hubbard, J. I. (1975). Regional distribution of angiotensin II receptors in the rat brain. *Proc. Univ. Otago Med. Sch.* **53**, 19–20.

Mouw, D., Bonjour, J.-P., Malvin, R. L. & Vander, A. (1971). Central action of angiotensin in stimulating ADH release. *Am. J. Physiol.* **220**, 239–42.

Nashold, B. S., Mannarino, E. & Wunderlich, M. (1962). Pressor–depressor blood pressure responses in the cat after intraventricular injections of drugs. *Nature, Lond.* **193**, 1297–8.

Nicolaidis, S. & Fitzsimons, J. T. (1975). La dépendance de la prise d'eau induite par l'angiotensin II envers la fonction vasomotrice cérébrale locale chez le rat. *C. r. hebd. Séanc. Acad. Sci., Paris, D* **281**, 1417–20.

Nicoll, R. A. & Barker, J. L. (1971). Excitation of supraoptic neurosecretory cells by angiotensin II. *Nature New Biol.* **233**, 172–4.

Onesti, G., Fernandes, M. & Kim, K. E. (1976). *Regulation of blood pressure by the central nervous system.* New York: Grune & Stratton.

Osborne, M. J., Pooters, N., D'Auraic, G. A., Epstein, A. N., Worth, M. & Meyer,

P. (1971). Metabolism of tritiated angiotensin II in anaesthetised rats. *Pflügers Arch. ges. Physiol.* **326**, 101–14.

Page, I. H. & Bumpus, F. M. (1974). *Angiotensin. Handbook of Experimental Pharmacology*, vol. 37. New York: Springer-Verlag.

Peach, M. J. (1977). Renin–angiotensin system: biochemistry and mechanisms of action. *Physiol. Rev.* **57**, 313–70.

Peters, G., Fitzsimons, J. F. & Peters-Haefli, L. (1975). *Control mechanisms of drinking.* New York: Springer-Verlag.

Phelan, E. L., Simpson, F. O. & Smirk, F. H. (1976). Characteristics of the New Zealand strain of genetically hypertensive (G-H) rats. *Clin. exp. Pharmac. Physiol.* Supp. **3**, 5–10.

Phillips, M. I. & Felix, D. (1976). Specific angiotensin II receptive neurons in the cat subfornical organ. *Brain Res.* **109**, 531–40.

Phillips, M. I., Phipps, J., Hoffman, W. E. & Leavitt, M. (1975). Reduction of blood pressure by intracranial injection of angiotensin blocker (P113) in spontaneously hypertensive rats. *Physiologist, Wash.* **18**, 350.

Poth, M. M., Heath, R. G. & Ward, M. (1975). Angiotensin-converting enzyme in human brain. *J. Neurochem.* **25**, 83–5.

Ramsay, D. J. & Reid, I. A. (1975). Some central mechanisms of thirst in the dog. *J. Physiol., Lond.* **253**, 517–25.

Redgate, E. S. (1968). Role of the baroreceptor reflexes and vasoactive polypeptides in the corticotropin release evoked by hypotension. *Endocrinology* **82**, 704–20.

Regoli, D., Park, W. L. & Rioux, F. (1974). Pharmacology of angiotensin. *Pharmac. Rev.* **26**, 69–123.

Reid, I. A., Day, R. P., Moffat, B. & Hughes, H. G. (1977). Apparent angiotensin immunoreactivity in dog brain resulting from angiotensinase. *J. Neurochem.* **28**, 435–8.

Reid, I. A. & Jones, A. (1975). Effect of angiotensin II and glucocorticoid on plasma angiotensinogen concentration in the dog. *Clin. Res.* **23**, 95*A*.

Reid, I. A. & Ramsay, D. J. (1975). The effects of intracerebro-ventricular administration of renin on drinking and blood pressure. *Endocrinology* **97**, 536–42.

Richardson, J. B. & Beaulnes, A. (1971). Cellular site of action of angiotensin. *J. cell. Biol.* **51**, 419–32.

Rosendorff, C., Lowe, R. D., Lavery, H. & Cranston, W. I. (1970). Cardiovascular effects of angiotensin mediated by the central nervous system of the rabbit. *Cardiovascular Res.* **4**, 36–43.

Roth, M., Weitzman, A. F. & Piquilloud, Y. (1969). Converting enzyme content of different tissues of the rat. *Experientia* **25**, 1247.

Sakai, K. K., Marks, B. H., George, J. & Koestner, A. (1974). Specific angiotensin II receptors in organ-cultured canine supra-optic nucleus cells. *Life Sci.* **14**, 1337–44.

Schlegel, W. & Felix, D. (1976). Structure–activity relations for angiotensin II action in subfornical organ. *Experientia* **32**, 761.

Severs, W. B. & Daniels-Severs, A. E. (1973). Effects of angiotensin on the central nervous system. *Pharmac. Rev.* **25**, 415–49.

Severs, W. B., Daniels, A. E., Smookler, H. H., Kinnard, W. J. & Buckley, J. P. (1966). Inter-relationship between angiotensin II and the sympathetic nervous system. *J. Pharmac. exp. Ther.* **153**, 530–7.

Severs, W. B., Summy-Long, J., Taylor, J. S. & Connor, J. D. (1970). A central effect of angiotensin: release of pituitary pressor material. *J. Pharmac. exp. Ther.* **174**, 27–34.

Shimazu, K., Shane, L. & Claybaugh, J. R. (1973). Potentiation by angiotensin II of the vasopressin response to an increasing plasma osmalality. *Endocrinology* **93**, 42–50.

Simpson, J. B. & Routtenberg, A. (1973). Subfornical organ: site of drinking elicitation by angiotensin II. *Science* **181**, 1172–5.

Simpson, J. B. & Routtenberg, A. (1975). Subfornical organ lesions reduce intravenous angiotensin-induced drinking. *Brain Res.* **88**, 154–61.

Simpson, J. B., Saad, A. W. & Epstein, A. N. (1976). The subfornical organ, the cerebrospinal fluid, and the dipsogenic action of angiotensin. In *Regulation of the blood pressure by the central nervous system*, ed. G. Onesti, M. Fernandes & E. K. Kim, pp. 191–202. New York: Grune & Stratton.

Sirett, N. E., Bray, J. J. & Hubbard, J. I. (1977*a*). Localisation of angiotensin II receptors in the superior colliculi. *NZ med. J.* **86**, 446.

Sirett, N. E., McLean, A. S., Bray, J. J. & Hubbard, J. I. (1977*b*). Distribution of angiotensin II receptors in rat brain. *Brain Res.* **122**, 299–312.

Sirois, P. & Gagnon, D. J. (1975). Increase in cyclic AMP levels and vasopressin release in response to angiotensin I in neurohypophyses: blockade following inhibition of the converting enzyme. *J. Neurochem.* **25**, 727–9.

Sweet, C. S., Columbo, J. C. & Gaul, S. L. (1976). Comparative antihypertensive effects of inhibitors of the renin–angiotensin system by central and peripheral administration in the malignant and spontaneously hypertensive rat. *Fedn Proc. Fedn Am. Socs exp. Biol.* **35**, 1056.

Ueda, H., Katayama, S. & Kato, R. (1972). Area postrema angiotensin-sensitive site in brain. *Adv. exp. Biol. Med.* **17**, 109–16.

Volicer, L. & Loew, C. G. (1971). Penetration of angiotensin II into the brain. *Neuropharmacology* **10**, 631–6.

Wayner, M. J., Ono, T. & Nolley, D. (1973*a*). Effects of angiotensin II applied electrophoretically on lateral hypothalamic neurons. *Pharmac. Biochem. Behav.* **1**, 223–6.

Wayner, M. J., Ono, T. & Nolley, D. (1973*b*). Effects of angiotensin II on central neurons. *Pharmac. Biochem. Behav.* **1**, 679–91.

Yamori, Y. & Okamoto, K. (1976). The Japanese spontaneously hypertensive rat (SHR). *Clin. exp. Pharmac. Physiol.* Supp. **3**, 1–4.

Yang, H.-Y. T. & Neff, N. H. (1972). Distribution and properties of angiotensin converting enzyme of rat brain. *J. Neurochem.* **19**, 2443–50.

Yang, H.-Y. T. & Neff, N. H. (1973). Differential distribution of angiotensin converting enzyme in the anterior and posterior lobe of the rat pituitary. *J. Neurochem.* **21**, 1035–6.

Yu, R. & Dickinson, C. J. (1965). Neurogenic effects of angiotensin. *Lancet* ii, 1276–7.

Zanchetti, A. & Stella, A. (1976). Renin release by central and reflex regulation. In *Regulation of blood pressure by the central nervous system*, ed. G. Onesti, M. Fernandes & K. E. Kim, pp. 235–50. New York: Grune & Stratton.

Fastigial (medial cerebellar) nuclear regulation of autonomic controls

C. B. B. DOWNMAN

Although Thomas Willis in 1664 reported that manipulation of the cerebellum stopped the heart (cited in Fulton, 1943) another 200 years elapsed before other evidence of cerebello–visceral relations appeared. Eckhard in 1872 and Balogh in 1876 (cited in Dow & Morruzzi, 1958) described alteration of heart rate of dogs during cerebellar stimulation, but this effect of electrical and mechanical stimulation could have been due to actions on the underlying medullary centres. Then in 1950, Morruzzi gave good evidence of circulatory changes during stimulation localized to the palaeocerebellum. These and other observations that cerebellar stimulation can affect the viscera are summarized by Morruzzi (1950) and Dow & Morruzzi (1958). Prior to 1969 there had been a few reports of visceral consequences of stimulating deeper in the cerebellum. One can list the following reports of responses during stimulation in or near the fastigial (medial cerebellar) nuclei: pupil constriction (Hare, Magoun & Ranson, 1937), micturition and pupil dilatation (Chambers, 1947), sham rage (Zanchetti & Zoccolini, 1954).

In 1969 there were reports from two laboratories that electrical stimulation of the fastigial nuclei can cause widespread cardiovascular changes (Achari & Downman, 1969; Miura & Reis, 1969). Since then there have been many more reports that fastigial stimulation not only elicits widespread cardiovascular and other visceral responses through mainly sympathetic pathways but that there is also inhibition or facilitation of autonomic reflex arcs at the brainstem and spinal levels.

Cardiovascular responses

Achari & Downman (1969, 1970), Miura & Reis (1969, 1970, 1971), Lisander & Martner (1971) and Doba & Reis (1972a) all agree that fastigial stimulation in the anaesthetized cat causes a marked and sustained rise of mean arterial pressure (BP) with increased pulse pressure and tachycardia. We have found that the tachycardia is more marked the slower the initial heart rate. BP and pulse pressure may double within 5–10 sec and the heart accelerates to about 260–280 beats min⁻¹ (Fig. 2). Cardiac contractility increases but, surprisingly in view of the great increase in pulse pressure, the stroke volume and cardiac output are little changed (Doba & Reis, 1972*a*). This is probably a consequence of the increased arterial pressure due to widespread peripheral vasoconstriction. Achari & Downman (1970) recorded marked decreases in the volumes of the paws, the limbs, the kidneys and loops of intestine during fastigial stimulation, even though BP was rising. Doba & Reis (1972*a*) recorded blood flows by flow transducers on the arteries; they found that while limb and abdominal visceral blood flows decreased as the regional vascular resistance increased, the flow in the common carotid arteries nearly doubled as BP rose without significant change of regional resistance.

Although the cardiovascular responses were first recorded in anaesthetized cats they are not an artefact of anaesthesia. Miura & Reis (1970) recorded the fastigial pressor response in decerebrate unanaesthetized cats. Achari, Al-Ubaidy & Downman (1973) and Lisander & Martner (1975) have recorded similar changes of BP and heart rate in unanaesthetized cats with indwelling electrodes and intra-arterial catheters. With stronger stimulations the cardiovascular changes are accompanied by slow posturing of the forelimbs, while Lisander & Martner noted a patterned behaviour of grooming and biting. However, the cardiovascular changes can be elicited below the threshold of somatic movements and the cats seem to be unaware of the stimulations. One may note here that the autonomic responses are best elicited by lower frequency stimulation (30–80 Hz) and somatic responses by higher frequency stimulation (200–300 Hz) (Miura & Reis, 1970).

Other fastigially induced responses

Dow & Morruzzi (1958) reviewed the literature of many autonomic responses to cerebellar stimulation, and some reports of various responses to fastigial stimulation have been mentioned above. During the last decade more reports have appeared with descriptions of changes of gastro-intestinal motility, piloerection, pupillary dilatation with retraction of nictitating membrane, bladder emptying, electrodermal responses and so on. There is now even evidence that fastigial stimulation inhibits antidiuretic hormone secretion induced by carotid occlusion (Hata & Miura, 1974). The general picture is of widespread autonomic discharge, the fastigial nuclei being the lowest threshold sites within the deep cerebellar nuclei for eliciting these and the cardiovascular responses described above.

Pathways activated by fastigial stimulation

The cardiac and peripheral vascular responses and also such responses as pupillary dilatation, retraction of the nictitating membranes and the electrodermal responses of the pads could all depend upon a widespread discharge through sympathetic pathways. This conclusion is supported by the abolition of responses by sympathetic receptor blockade. The chronotropic and inotropic responses of the heart, with consequent changes in pulse pressure, are all abolished by β-receptor blockade with propranolol. The residual pressor response is abolished by α-receptor blockade with phentolamine or tolazoline (Achari & Downman, 1970; Miura & Reis, 1972a). In our own experiments there was no alteration in the tachograph recording during fastigial stimulation after propranolol, 3 mg kg^{-1}, clearly indicating that there was no change in vagal tone during the stimulation.

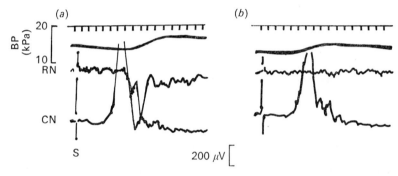

Fig. 1. Fastigial stimulation causing volleys in the inferior cardiac (CN) and renal nerves (RN). Volleys were recorded from central cut ends of these exposed nerves after unipolar stimulation (S) of the ipsilateral fastigial nucleus, 0.3 msec 1.0 V square pulse. Blood pressure (BP) recorded from femoral artery. Between recordings (a) and (b) the renal nerve was crushed proximal to the recording electrodes. Cat, anaesthetised chloralose 35 mg–urethane 700 mg kg^{-1} body weight. (N. K. Achari, S. Al-Ubaidy and C. B. B. Downman, unpublished.) Time marker interval = 10 msec.

More direct evidence of the sympathetic discharge was obtained by Achari, Al-Ubaidy & Downman (1971). In cats anaesthetized with chloralose–urethane the centrifugal activity was recorded in branches of various visceral nerves, i.e. splanchnic, renal and inferior cardiac. During stimulation of either an ipsilateral or a contralateral fastigial nucleus there was increased activity in the nerves. A short train of 3–5 stimulus pulses at 150 sec^{-1} through an insulated steel needle as unipolar electrode caused a single compact volley in the nerve; in some preparations a single shock was effective (Fig. 1). This volley was similar to the reflex volley due to a

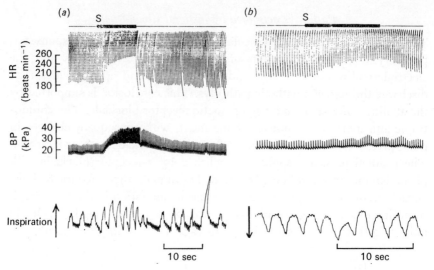

Fig. 2. Fastigial induction of cardiac acceleration after confirmed transection of the spinal cord between C5–C6 segments. Recordings in (a) show tachycardia with rise of blood pressure and pulse pressure. Recordings in (b) were made 5 min after transection, when repeating the fastigial stimulation still caused tachycardia. Throughout these recordings both common carotid arteries were occluded and both vagi were intact. Cat, anaesthetized chloralose 35 mg kg⁻¹ body weight after Althesin induction. Polygraph tracings from above downwards show: fastigial stimulation (S) marker; heart rate (HR) by Grass tachograph triggered by arterial pulsations; aortic arterial pressure (BP) from catheter inserted via femoral artery into abdominal aorta; breathing from intra-oesophageal balloon. Fastigial nucleus stimulated through dipolar insulated stainless steel wire electrode (tip separation 0.5 mm) 2.5 V r.m.s. 50-Hz sine wave current. (D. A.-H. Al-Senawi and C. B. B. Downman, 1976, unpublished.)

single shock stimulation of the central cut end of an intercostal nerve. In further experiments the discharge was defined as being sympathetic in origin by appropriate sections of nerve pathways and by its abolition by intravenous hexamethonium. When the same fastigial stimulation was continued for 5 sec or longer there was the expected rise of BP. Immediately after the stimulation there was a remarkably long silence of the nerve, about 6 sec, before the resting discharge returned.

All of the above evidence suggests that the cardiovascular responses could be explained as being due to fastigial activation of sympathetic pathways alone, with vagal pathways unaffected. However, an experiment with D. A.-H. Al-Senawi showed that some other pathway, possibly vagal, may also be involved in the tachycardia. In a cat under light chloralose anaesthesia a fastigial nucleus was stimulated before and after confirmed transection of the spinal cord between segments C5–C6, the vagi being intact and both common carotid arteries being occluded throughout. It is seen (Fig. 2) that the same cardiac acceleration occurred after severing the spinal sympathetic pathways. Supplementary evidence of extra-sympathetic acceleratory pathways was provided by the occurrence of fastigially induced tachycardia after the intravenous administration of bretylium tosylate 15 mg kg^{-1}; before blockade the heart accelerated during stimulation from 170 to 260 beats min^{-1}, and during blockade from 170 to 220 beats min^{-1}. It is possible, therefore, that fastigial regulation of cardiac vagal activity has to be sought in different conditions of anaesthesia and cardiovascular balance.

Fastigially induced cardiac dysrhythmia

In many of the polygraph recordings of BP, irregularities suggesting transient changes of cardiac rhythm were seen during prolonged fastigial stimulation. They were more obvious at higher paper speed. Recording the electrocardiogram (ECG) at the same time confirmed that the electrical activity of the heart showed a variety of bizarre changes (D. A.-H. Al-Senawi and C. B. B. Downman, 1976, unpublished). Using cats anaesthetized with chloralose–urethane or chloralose alone the ECG was recorded with classical limb leads from subcutaneous Ag–AgCl discs, or from similar electrodes attached to the chest. In about 50 % of our preparations, stimulation in pressor sites in a fastigial nucleus evoked marked dysrhythmia (Fig. 3). This did not begin until after many seconds of stimulation, and sometimes was delayed until after the 10–15-sec stimulation was ended. In some preparations the irregularity was confined to only one or a few beats, but in other cats the irregularity would continue for 1–1.5 min before spontaneous reversion to normal beating. Extra anaesthetic would reduce the duration of dysrhythmia or abolish it; conversely, the dysrhythmia could be provoked more frequently in lightly anaesthetized preparations. This susceptibility to anaesthesia recalls Miura & Reis' (1970) note that the fastigial pressor response itself is extremely sensitive to anaesthesia, being abolished by 20 mg kg^{-1} of sodium pentobarbitone or 80 mg kg^{-1} chloralose in decerebrate, unanaesthetized cats. We also found that the dysrhythmia was unpredictable, and the response might be lost during experiments for no obvious reason. The ECG showed a variety of changes, suggesting that multiple ectopic foci were activated at different times, sometimes ventricular, sometimes nodal; at other times sino-atrial inhibition and also shortened P-R with widening of QRS resembling the Wolff–Parkinson–White disorder in man. Since the dysrhythmia was prevented by bretylium tosylate, up to 15 mg kg^{-1}, the

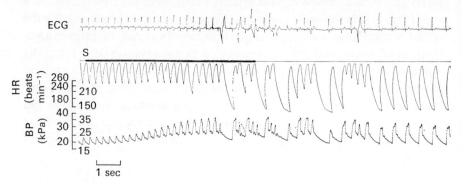

Fig. 3. Fastigial induction of cardiac dysrhythmia. During signal, S, the right fastigial nucleus was stimulated through a dipolar insulated stainless steel wire electrode,
5.0 V r.m.s. 50-Hz sine wave current. Polygraph tracings show from above downwards: electrocardiogram (ECG) lead, mid-sternum to right forearm; stimulus
marker; Grass tachograph recording of heart rate (HR) triggered from ECG;
arterial pressure (BP) from catheter inserted via femoral artery into abdominal
aorta. Cat, anaesthetised chloralose 35 mg kg^{-1} body weight intravenously, with
Althesin induction and supplements intravenously. This record also shows that
there may be no tachycardia when heart rate is high before fastigial stimulation.
(D. A.-H. Al-Senawi and C. B. B. Downman, 1976, unpublished.)

sympathetic innervation is probably involved. Since atropine, 0.6 mg,
prevented the dysrhythmia in some preparations, but was ineffective in
others, we deduce that the vagus innervation may also play some role.
These are preliminary findings and the mechanism requires further analysis.
They do, though, add another site to the list of places in the nervous
system which cause cardiac dysrhythmia on stimulation. Ectopic beats
with short P-R interval and wide QRS were reported by Weinberg &
Fuster (1960) after posterior hypothalamic stimulation. Korteweg, Boeles
& Cate (1957), Manning & Cotten (1962) and Melville, Blum, Shister &
Silver (1963) have also evoked cardiac dysrhythmia from diencephalic
stimulation.

Fastigial regulation of autonomic reflexes

Not only are autonomic pathways activated during fastigial stimulation but
autonomic reflex arcs are also inhibited. Achari & Downman (1970)
tested the effect of fastigial stimulation on the reflex bradycardia evoked by
carotid sinus distension, electrical stimulation of an exposed carotid sinus
nerve, noradrenaline injection, electrical stimulation of vagal afferents and
the chemoreflex bradycardia caused by injection of phenyl diguanide into
the right atrium. In all instances the reflex bradycardia was abolished.
Hockman, Livingston & Talesnik (1970) also showed that the reflex

bradycardia during stimulation of a carotid sinus nerve was abolished by fastigial stimulation in the encephale isolé preparation (cat) below the threshold of the BP pressor response. Lisander & Martner (1971) confirmed that fastigial stimulation reduced the responses to carotid sinus distension with a relatively greater reduction of the reflex action on the heart than on the vascular bed. Miura & Reis (1971) also found that the fastigial pressor response and the carotid sinus depressor responses were mutually inhibitory. Lisander & Martner (1975) have also shown that fastigial stimulation will cause increase of motility of the small intestine and colon by suppression of sympathetic influence on the gut, either when the latter is of spontaneous or reflex origin, i.e. intestino-intestinal inhibitory reflexes. Martner (1975) has shown that reflex vasodilatation of the colon is prevented by fastigial stimulation; since the vasodilatation occurred during adrenergic blockade with combined guanethidine and dibenzyline, the fastigial effect is presumably on a parasympathetic reflex pathway. Martner also reported inhibition of reflex bladder contraction and the defaecation reflex. These inhibitory responses were in both cases independent of the adrenergic sympathetic pathways; they were unaffected both by adrenergic nerve section and adrenergic receptor blockade but were not seen after pelvic nerve section. Facilitation of autonomic reflex action has also been described. Hultborn, Mori & Tsukahara (1973) found that fastigial stimulation caused bilateral monosynaptic discharges in the short ciliary nerves and increased the size of reflex volleys in the nerves from optic nerve stimulation.

The fastigial nuclei can both activate autonomic discharges and at the same time regulate autonomic reflex arcs.

DISCUSSION

All of the experimental evidence quoted above, without being a complete survey of the relevant literature, shows that the fastigial nuclei have particular and very effective connections with the autonomic efferent pathways. Most of the effects of fastigial stimulation depend upon activation of sympathetic pathways alone, without evidence of altered activity in parasympathetic pathways, but pelvic visceral regulations are mediated through the pelvic splanchnic nerves. The cardiovascular changes are marked and may include cardiac dysrhythmias. There is now direct evidence of this sympathetic discharge. The exception to this rule is the pelvic splanchnic outflow, and some evidence has been given above that the vagi may also be regulated by the fastigial nuclei. Not only will it be profitable to look more closely at the parasympathetic outflows but it may be also necessary to use newer anaesthetics. For example, the steroid anaesthetic Althesin (Glaxo: a mixture of alphaxolone and alphadolone acetate) is increasingly useful in investigations of autonomic function (Hilton, 1977).

The fastigial nuclei have very widespread connections throughout the nervous system, and are regulated by the vermian cerebellar cortex (Jansen & Brodal, 1954; Jansen & Jansen, 1955; Cohen, Chambers & Sprague, 1958; Voogd, 1964; Angaut & Bowsher, 1970; Bloedel, 1973; Eccles, Sabah & Tábořikova, 1974a,b). The efferents from the fastigial nuclei form the vast majority of fibres entering the inferior cerebellar peduncle. It is particularly relevant that the efferents from the rostral part of each nucleus, the part which is the source of the autonomic responses to stimulation, project mainly to the pons and medulla, with a contribution to the oculomotor nuclei. The crossed (uncinate) and direct fastigiobulbar tracts synapse in all the vestibular nuclei, in the dorsal pontine reticular substance, in the lateral and dorso-medial reticular substance of the medulla, including the paramedian nuclei. The fastigial nuclei receive a wide afferent input via spinocerebellar, cuneocerebellar, spino-olivo-cerebellar and spino-reticulo-cerebellar from the limbs and via direct afferents from the vestibular nuclei. The caudal part of each nucleus also receives an equally wide afferent input and also projects higher into the brainstem and ventrolateral and ventromedial nuclei of thalamus. Since the nuclei are so much a part of the organization of somatic movements one may speculate why they have such a particular role also in the control of autonomic, particularly sympathetic, outflow. Doba & Reis (1972a,b,c) have presented evidence that the fastigial nuclei may be concerned in cardiovascular adjustments to changes of posture. They point out that the overall pattern of cardiovascular responses to stimulation simulates the compensatory (orthostatic) responses which occur during the maintenance of the standing posture. This involves the direct fastigio–vestibular nuclear connections, and so the nuclei may be a pathway through which the visceral reactions are tied into a pattern of somatic behaviour. Elicitation of sham rage reactions (Zanchetti & Zoccolini, 1954) with their somatic and visceral components again links the fastigial control of sympathetic outflow with a pattern of somatic behaviour. In the experiments described above it is generally agreed that the most effective sites for stimulation lie within the rostral parts of the nuclei or just below them in the concentration of fastigial efferent fibres. Reports that neurones of the rostral parts of the fastigial nuclei respond to macular labyrinthine receptors and neck afferents (Erway, Gherladucci, Pompeiano & Stanojević, 1977a,b), and that there are direct fastigio-spinal pathways down to the neck segments of the cord (Wilson et al., 1977), again suggest that the fastigial nuclei have some particular role in coordinating postural somatic and visceral responses. More experiments need to be done using natural stimulations of the nuclei. The Purkinjě cell can only inhibit ongoing activity in the deep nuclei (Eccles, Ito & Szentagothai, 1967, Ito *et al.*, 1970) sustained through the collaterals of afferent fibres passing through the nuclei (Matsushita & Iwahori, 1971). Electrical stimulation will cause an excess of synchronous activity and as yet we have no evidence that any of the natural inputs can be equally successful in terms of quantity of output. Some of the more intense effects, e.g. cardiac dysrhythmia, may be of pathophysiological rather than physiological interest, but the fastigial nuclei could well be a site of integration and coordination of autonomic activity for visceral homeostasis during the somatic activities of posture and movement.

In all experiments on the fastigial nuclei I must gratefully acknowledge the co-operation and advice of N. K. Achari, D. A.-H. Al-Senawi and S. Al-Ubaidy.

REFERENCES

Achari, N. K., Al-Ubaidy, S. & Downman, C. B. B. (1971). Cerebellar initiation of discharges in sympathetic nerves. *J. Physiol., Lond.* **215**, 21–2.

Achari, N. K., Al-Ubaidy, S. & Downman, C. B. B. (1973). Cardiovascular responses elicited by fastigial and hypothalamic stimulation in conscious cats. *Brain Res* **60**, 439–47.

Achari, N. K. & Downman, C. B. B. (1969). Autonomic responses evoked by stimulation of fastigial nuclei in the anaesthetised cat. *J. Physiol., Lond.* **204**, 130.

Achari, N. K. & Downman, C. B. B. (1970). Autonomic effector responses to stimulation of nucleus fastigius. *J. Physiol., Lond.* **210**, 637–50.

Angaut, P. & Bowsher, D. (1970). Ascending projections of the medial cerebellar (fastigial) nucleus: an experimental study in the cat. *Brain Res.* **24**, 49–68.

Bloedel, J. R. (1973). Cerebellar afferent systems: a review. *Prog. Neurobiol.* **2**, 3–68.

Chambers, W. W. (1947). Electrical stimulation of the interior of the cerebellum in the cat. *Am. J. Anat.* **80**, 55–93.

Cohen, D. W., Chambers, W. W. & Sprague, J. M. (1958). Experimental study of the efferent projections from the cerebellar nuclei to the brainstem of the cat. *J. comp. Neurol.* **109**, 233–59.

Doba, N. & Reis, D. J. (1972*a*). Changes in regional blood flow and cardiodynamics evoked by electrical stimulation of the fastigial nucleus in the cat and their similarity to orthostatic reflexes. *J. Physiol., Lond.* **227**, 729–47.

Doba, N. & Reis, D. J. (1972*b*). Impairment of orthostatic reflexes by lesions of fastigial nucleus and vestibular nerves in cat. *Fedn Proc. Fedn Am. Socs exp. Biol.* **31**, 814.

Doba, N. & Reis, D. J. (1972*c*). Cerebellum: role in reflex adjustments to posture. *Brain Res.* **39**, 495–500.

Dow, R. S. & Morruzzi, G. (1958). *The physiology and pathology of the cerebellum.* Minneapolis: Univ. Minnesota Press.

Eccles, J. C., Ito, M. & Szentagothai, J. (1967). *The cerebellum as a neuronal machine.* Berlin, Heidelberg & New York: Springer-Verlag.

Eccles, J. C., Sabah, N. H. & Táboříková, H. (1974*a*). Excitatory and inhibitory responses of neurones of the cerebellar fastigial nucleus. *Exp. Brain Res.* **19**, 61–77.

Eccles, J. C., Sabah, N. H. & Tábořiková, H. (1974*b*). The pathways responsible for excitation and inhibition of fastigial neurones. *Exp. Brain Res.* **19**, 78–99.

Erway, L. C., Gherladucci, B., Pompeiano, O., Stanojević, M. (1977*a*). Crossed labyrinthine influences on cerebellar fastigial neurons originating from macular receptors. *Proc. Int. Congr. Physiol. Sci.* Abstr. 614.

Erway, L. C., Gherladucci, B., Pompeiano, O., Stanojević, M. (1977*b*). Responses of cerebellar fastigial neurones to stimulation of labyrinthine and neck muscle afferents. *Proc. Int. Congr. Physiol. Sci.* Abstr. 615.

Fulton, J. F. (1943). *Physiology of the nervous system.* London: Oxford Univ. Press.

Hare, W. K., Magoun, H. W. & Ranson, S. W. (1937). Localization within the cerebellum of reactions to faradic cerebellar stimulation. *J. comp. Neurol.* **67**, 145–82.

Hata, N. & Miura, M. (1974). The inhibitory effect of the cerebellar fastigial stimulation on ADH secretion. *J. Physiol., Lond.* **242**, 793–803.

Hilton, S. M. (1977). Influences on the circulation from the cerebral cortex. *Proc. Int. Congr. Physiol. Sci.* Abstr. **12**, 312.

Hockman, C. H., Livingston, K. E. & Talesnik, J. (1970). Cerebellar modulation of reflex vagal bradycardia. *Brain Res.* **23**, 101–4.

Hultborn, H., Mori, K. & Tsukahara, N. (1973). The neuronal pathway subserving the pupillary light reflex and its facilitation from cerebellar nuclei. *Brain Res.* **63**, 357–61.

Ito, M., Yoshida, M., Obata, K., Kawai, N. & Udo, M. (1970). Inhibitory control of intracerebellar nuclei by the Purkinje cell axons. *Exp. Brain Res.* **10**, 64–80.

Jansen, J. & Brodal, A. (1954). *Aspects of cerebellar anatomy.* Oslo: Johan Grundt Tanum Forlag.

Jansen, J. & Jansen, J. K. S. (1955). On the efferent fibres of the cerebellar nuclei in the cat. *J. comp. Neurol.* **102**, 607–23.

Korteweg, G. C. T., Boeles, J. T. F. & Cate, J. T. (1957). Influence of stimulation of some subcortical areas on electrocardiogram. *J. Neurophysiol.* **20**, 100–7.

Lisander, B. & Martner, J. (1971). Interaction between the fastigial pressor response and the baroreceptor reflex. *Acta physiol. scand.* Suppl. 351, 1–42.

Lisander, B. & Martner, J. (1975). Integrated somatomotor, cardiovascular and gastrointestinal adjustments induced from the cerebellar fastigial nucleus. *Acta physiol. scand.* **94**, 358–67.

Manning, J. W. & Cotten, M. (1962). Mechanism of cardiac arrhythmias induced by diencephalic stimulation. *Am. J. Physiol.* **203**, 1120–4.

Martner, J. (1975). Influences on the defecation and micturition reflexes by the cerebellar fastigial nucleus. *Acta physiol. scand.* **94**, 95–104.

Matsushita, M. & Iwahori, N. (1971). Structural organization of the fastigial nucleus. II. Afferent fibre systems. *Brain Res.* **25**, 611–24.

Melville, K. I., Blum, B., Shister, H. E. & Silver, M. D. (1963). Cardiac ischemic changes and arrhythmias induced by hypothalamic stimulation. *Am. J. Cardiol.* **12**, 781–91.

Mitra, J. & Snider, R. S. (1972). Nucleus fastigii influences on bloodflow and blood pressure. *Int. J. Neurosci.* **3**, 285–90.

Miura, M. & Reis, D. J. (1969). Cerebellum: a pressor response elicited from the fastigial nucleus and its efferent pathway in the brain stem. *Brain Res.* **13**, 595–9.

Miura, M. & Reis, D. J. (1970). A blood pressure response from fastigial nucleus and its relay path in the brainstem. *Am. J. Physiol.* **219**, 1300–36.

Miura, M. & Reis, D. J. (1971). The paramedian reticular nucleus: a site of inhibitory interaction between projections from fastigial nucleus and carotid sinus nerve acting on blood pressure. *J. Physiol., Lond.* **216**, 44–60.

Morruzzi, G. (1950). *Problems in cerebellar physiology.* Springfield, Illinois: Charles C. Thomas.

Voogd, J. (1964). *The cerebellum of the cat.* Assen: Van Gorcum.

Weinberg, S. J. & Fuster, J. M. (1960). Electrocardiographic changes produced by localized hypothalamic stimulations. *Ann. Int. Med.*, **53**, 332–41.

Wilson, V. J., Uchino, Y., Susswein, A., Maunz, R. A., Fukushima, K. (1977). Fastigiospinal neurons in the cat. *Proc. Int. Congr. Physiol. Sci.* Abstr. 2424.

Zanchetti, A. & Zoccolini, A. (1954). Autonomic hypothalamic outbursts elicited by cerebellar stimulation. *J. Neurophysiol.* **17**, 475–83.

The pathophysiology of central nerve fibres*

W. I. McDONALD

One of the outstanding developments in clinical neurology during the past decade has been the increased precision of diagnosis resulting from the application of physiological techniques to the investigation of patients. As a result of the systematic correlation of the histological changes of nerve fibres with their altered physiological properties in experimental and human disease, it is now possible not only to confirm the presence of central or peripheral lesions in patients with symptoms, but to detect abnormalities before they have produced a clinical deficit. This latter advance has proved to be invaluable in determining the pattern of peripheral neuropathies and in making an early diagnosis of multiple sclerosis. In addition, in the peripheral nervous system the characteristics of conduction can provide a useful guide to the pathological nature of the neuropathy, and thus play an important role in determining treatment and the prognosis.

In this chapter I want to trace the development of these advances which have arisen from the relationships of mutual dependence which exists between clinical and experimental neurology. My own special interest, initiated and nurtured by Archie McIntyre, is in the structure and function of nerve fibres and the way in which they react to disease. I shall consider a number of morphological and physiological questions hoping to show how the interaction of ideas originating in fundamental biology with those originating in clinical practice has led to specific advances in knowledge which are important to both disciplines.

The organisation of central myelin

How is central myelin organised? It has long been known that in the central nervous system, as in the peripheral nervous system, all but the smallest axons have myelin sheaths. After the introduction of electron microscopy it was quickly agreed that the sheaths in the two sites are formed in the

* Based on an Inaugural Lecture delivered at the Institute of Neurology on 2 October 1974.

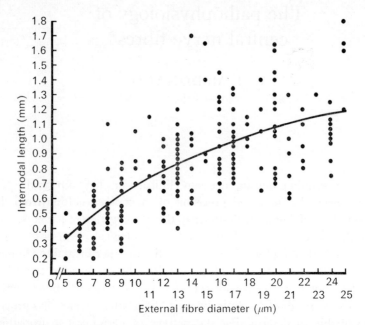

Fig. 1. Relationship between internodal length and external fibre diameter in 166 internodes on single fibres isolated from the spinal cord of three normal adult cats. Superimposed is the calculated regression line ($Y = 0.080 + 0.094X - 0.0018X^2$). (From Gledhill & McDonald, 1977.)

same general way by the elongation of the surface membrane of a specific satellite cell (oligodendrocyte or Schwann cell). The membrane is wrapped in a spiral around the axon, adjacent surfaces of the membrane subsequently becoming fused (Geren, 1954; Peters, 1960; Bunge, 1968).

The longitudinal organisation of central myelin was, however, the subject of long debate. One school, on the authority of Ranvier (1882), believed that there were continuous sheets of myelin and that the nodes seen on peripheral nerve fibres were absent. But as early as 1875, Tourneux & Le Goff had described nodes, and Porter (1890) illustrated them convincingly. Cajal (1909) described them and Bielschowsky (1928), having critically reviewed the literature, firmly stated that nodes occurred in the central nervous system. All this evidence was widely ignored. The problem was finally resolved by Hess & Young's (1949, 1952) systematic study with the light microscope in the rabbit, and by the electron microscopic studies later in the 1950s (see Bunge, 1968). It is now clear that central nerve fibres, like peripheral fibres, have nodes which divide the myelin into segments. The nodes are more or less regularly spaced, and the lengths of the segments are related to the diameter of the fibre.

The details of this relationship have been studied in our laboratory by Ohlrich and Gledhill in the cat and by Yogendran in the rat (McDonald & Ohlrich, 1971; Yogendran, 1976; Gledhill & McDonald, 1977). As Hess & Young (1949) first showed, the larger diameter fibres have longer internodal lengths (Fig. 1). In both the cat and the rat the curve relating fibre diameter to internodal length flattens as larger diameters are reached, this tendency being more marked in the rat.

Finally, electron microscopy has made it clear that although there are important resemblances between peripheral and central nerve fibres, there are significant differences. The outer cytoplasmic layer which is so prominent in peripheral fibres is reduced to a small 'tongue' in central fibres. As a result, the outer myelin layers of contiguous fibres come into contact and frequently fuse to form an intraperiod line, thus obliterating the extracellular space at these sites (Peters, 1960). Perhaps the difference with the greatest implications for the pathological reactions of central fibres is the relationship between the satellite cell and the axon. In the peripheral nervous system a single Schwann cell maintains a single internode. By contrast, in the central nervous system a single oligodendrocyte may form several internodes, the cell body being connected to each sheath by cytoplasmic processes (Bunge, 1968).

Pathological basis of paralysis and sensory loss

The second morphological question I want to consider arose directly out of clinical practice. What is the pathological basis of paralysis and sensory loss? In many situations loss of function is produced by complete degeneration of nerve fibres, both axons and myelin disintegrating completely. But as Jackson (1870) pointed out this is not the only mechanism of paralysis: 'Palsy depends on destruction of fibres . . . The word "destruction" is scarcely the correct word to use. By it is not meant that the nerve fibres are necessarily broken up, although they often are in palsy, but simply that there is a change in them which *destroys their function*' (Jackson's italics).

Perhaps the commonest morphological change of nerve fibres, short of total destruction, is demyelination, in which the myelin is destroyed leaving the axon in continuity. The first clear description of this change is contained in Charcot's (1868) account of multiple sclerosis, a disease in which demyelination is confined to the central nervous system. Demyelination was later shown to occur in the peripheral nervous system (Gombault, 1880) where the sequence of changes during normal organisation of myelin have been intensively studied. However, because the relationship between satellite myelin-forming cell and axon is so different on central and peripheral fibres, the results of the peripheral studies cannot be simply applied to the central nervous system. A direct examination is necessary.

The first to make such an examination was Babinski (1885) in his thesis on the histology of multiple sclerosis. He illustrated the myelin loss beautifully and considered the question of how the myelin sheath is disrupted. He showed that while in some regions there was extensive myelin loss, in others there was discontinuous myelin loss. He did not consider the question of relationship of demyelination to the nodes, since at that time orthodox teaching held that they did not exist. In the 1960s there were scattered reports of discontinuous demyelination in multiple sclerosis and experimental allergic encephalomyelitis. There had still not been, however, a systematic study of the pattern of demyelination in the central nervous system. A major reason for this was the lack of a suitable technique, adaptable to pathological material, which allowed the isolation of long lengths of single central nerve fibres with well-preserved myelin. It was this state of affairs which launched us on our studies of single nerve fibres in the central nervous system to which I have already referred. Ohlrich developed a suitable technique and the normal data accumulated by him and by Gledhill have provided the background against which we have studied pathological fibres. So far we have confined our attention to experimental lesions.

A consistent picture is beginning to emerge for demyelinating lesions produced in four quite different ways – acute compression of the spinal cord (Gledhill & McDonald, 1977; Harrison & McDonald, 1977) and a toxic lesion (Harrison, McDonald, Ochoa & Ohlrich, 1972), both in the cat, a naturally-occurring demyelinating lesion (Yogendran, 1976), and X-irradiation in the rat (Mastaglia, McDonald, Watson & Yogendran, 1976).

Demyelination begins in the paranodal region (Fig. 2). It may extend to involve long lengths of the fibre corresponding with one or more complete consecutive myelin segments. There has so far been no systematic study of the pattern of demyelination in human disease, although there are reports that demyelination may occur in relation to nodes in multiple sclerosis (Suzuki, Andrews, Waltz & Terry, 1969). From the variety of pathological mechanisms already shown to be capable of producing these changes, it seems likely that both paranodal and segmental demyelination will be found to occur in multiple sclerosis and probably other human demyelinating diseases as well.

Thus although the relationship between the myelin-forming cell and the sheath is quite different in the central and peripheral nervous systems, the sheaths seem to behave in a similar way when myelin is broken down during

demyelination. The destruction commences paranodally and whole segments of myelin may be lost.

I now wish to turn to some physiological questions and consider first the
mechanism by which demyelination produces a neurological deficit. As a
preliminary it is necessary to discuss the function of myelin itself.

Function of myelin

The significance of the myelin sheath did not much concern the early
anatomists and physiologists, but by the end of last century the matter
warranted short discussion in the standard text books of physiology.
Boruttau (1897) concluded that myelin made no essential difference to the
electrical properties of nerve fibres, and that there were quantitative
electrical differences only between myelinated and unmyelinated fibres.
Starling (1915) concluded that myelin had three functions.

(1) It acted as an insulator 'ensuring isolated conduction within any
given nerve fibre'.

(2) It was necessary for conduction in central nerve fibres, since embryological evidence suggested that central tracts did not appear to be functional before acquiring myelin sheaths.

(3) It had a trophic function for the axon.

These functions were re-affirmed by Gerard (1931), although he pointed
out, as indeed Starling had, that myelin did not seem to be essential as an
insulator since unmyelinated fibres appeared to function perfectly well
without it. There was also new evidence about foetal movements, and
Gerard concluded, unlike Starling, that myelin did not have a necessary
role in conduction, since certain late embryonic movements in the rat
occur before spinal cord tracts acquire their myelin sheaths. At that
time the possiblity that normal myelinated and unmyelinated fibres
might have significantly different properties seems not to have been
considered.

Gerard (1931) added an additional function to Starling's list – a relationship to conduction velocity. Lillie (1925) had proposed the hypothesis
of saltatory conduction as an analogy to the behaviour of his iron wire model
of nerve conduction. He suggested that the high electrical resistance of the
lipid-containing myelin would confine the excitation to focal regions of the
axon at the nodal gaps. The velocity of transmission would be increased
because the local electrical circuits set up at an active node would not have
to depolarise successively every portion of the whole length of the axon
but only the exposed portions at the nodes. This suggestion fitted with the

then known facts about the relative conduction velocities of myelinated and unmyelinated fibres.

Here matters rested until the decisive experiments of Huxley & Stämpfli (1949) on frog peripheral nerve established that the hypothesis of saltatory conduction was correct. In 1952, Tasaki established that conduction was saltatory in the frog spinal cord, and in 1969, BeMent & Ranck showed that the same was true for the mammalian spinal cord.

But is myelin essential to conduction in adult myelinated fibres? If one were to strip off the myelin would the fibre conduct like an unmyelinated fibre or would conduction stop altogether? These questions had been begged by the earlier investigators and the experiments of Huxley & Stämpfli were not designed to test this possibility. Tasaki (1955) placed a lipid solvent on isolated frog nerve fibres and showed that conduction ultimately ceased. It was not, however, clear whether conduction failure depended solely on damage to the myelin or whether the solvent had disrupted the axon membrane as well.

Some answers to these basic questions have been provided by studies which grew out of clinical questions about the mechanism of symptom production in demyelinating disease. The starting point for these studies is my fourth main question: How does myelin loss lead to paralysis?

Physiological basis of paralysis

The first evidence was provided by a remarkable and unjustly neglected paper of Gordon Holmes (1906) based on necropsies which he carried out on four paraplegic patients in the National Hospital, Queen Square – three had tuberculous caries of the spine and one had a glioma of the medulla. All had been paraplegic for up to eight and a half months before death. The essential pathological change was loss of myelin at the site of compression without evidence of secondary degeneration above and below the lesion. From this he concluded that the axons must have remained in continuity at the site of compression. He confirmed this conclusion by examining material from one of the cases using the Silver method for staining axons which had recently been introduced by Bielschowsky. In the discussion he makes the following remarkable statement:

In disseminated sclerosis, in which the essential and practically the only structural affection of the nervous tissues consists in the disappearance of the myeline sheaths, without necessarily producing secondary degeneration the loss of function may reach extreme degrees. From this fact alone it may be assumed that the presence of myeline sheaths is necessary for the functions of the tract fibres of the cord as conducting strands, and that when they are lost there may be a physiological block in the fibres. This condition, from the point of view of function, is equivalent to a structural break in the fibres, but differs in the fact that the anatomical integrity of the axis-cylinder and its trophic cell remain unaffected, and that reparative processes would be consequently possible.

The next piece of evidence came from experiments carried out by Denny-Brown & Brenner (1944*a,b*) during World War II in an attempt to understand more about the nature of peripheral nerve injuries. They produced paralysis in the hindlimbs of experimental animals by compressing the sciatic nerve. Subsequent electrical stimulation above the compressed region produced no contraction in the muscles. In some experiments, stimulation below the lesion produced full contraction. Microscopically the lesion in the latter animals consisted of selective destruction of myelin with preservation of axons. They concluded, as Holmes (1906) had, that demyelination produces conduction block.

The next step was to show this directly, and this is where my own interest in the pathophysiology of nerve fibres began. Archie McIntyre, who held the Chair of Physiology in Dunedin when I graudated, pointed out that there had been no direct experimental study of conduction in demyelinating lesions, and suggested that it would be interesting to examine this problem both because of its intrinsic interest to physiology, and because of its relevance to multiple sclerosis. I began with demyelination in peripheral nerve fibres because the problem was more accessible, and used parenteral diphtheria toxin which produced a convenient focal lesion in the dorsal root ganglion region of the cat (McDonald, 1963*a,b*). Since this chapter deals chiefly with the pathophysiology of central nerve fibres, I shall not describe those experiments in detail, but instead consider an analogous series later carried out on the central nervous system (McDonald & Sears, 1970). A focal demyelinating lesion was produced by the direct micro-injection of diphtheria toxin into the dorsal part of the cat spinal cord, and conduction examined two to four weeks after inducing demyelination. It was completely blocked at the level of large bilateral lesions, but persisted in the rostral and caudal histologically normal portions of the same tracts.

Conduction in fibres passing through smaller lesions persisted, but was abnormal in several respects. The conduction defect was analysed using a single fibre technique which permitted recording from individual undissected fibres (Sears & Stagg, 1967) and thus allowed a comparison of conduction in demyelinated and normal portions of the same single fibre. The ability to transmit long trains of impulses faithfully was impaired. The central portions of the intercostal muscle afferent fibres that we studied followed stimulation at 500–1000 Hz in normal animals. By contrast we observed intermittent conduction block down to frequencies of 70 Hz in demyelinated fibres. These findings were in keeping with the observed prolongation of the refractory period of transmission. This latter function provides a measure of the ability of an impulse *generated by a node* to traverse a damaged region (McDonald & Sears, 1970). It is likely that intermittent conduction failure occurs at frequencies lower than 70 Hz, since in peripheral demyelinated fibres we have observed conduction failure at 1 Hz (R. W. Gilliatt, W. I. McDonald and P. Rudge, unpublished observations). Prolonged stimulation of demyelinated fibres led to a progressive decline in the frequencies that could be transmitted, and reversible, complete conduction block finally occurred.

Comparison of the speed of conduction through the lesion with that through an adjacent normal region showed that the velocity was focally reduced by demyelination to as little as one sixth of the normal velocity.

These observations on central demyelinated fibres are closely similar to those obtained from comparable studies on peripheral fibres (see McDonald, 1974a). They have established that myelin is essential for normal conduction. Extensive demyelination results in complete conduction block, but smaller lesions permit intermittent conduction at a reduced velocity. The factors which determine whether block is persistent or intermittent have been analysed by computer simulation (Koles & Rasminsky, 1972) but have not yet been established experimentally, since it has so far proved impossible to study the same single fibre both morphologically and physiologically. Very recent experiments on peripheral fibres have shown that in certain circumstances saltatory conduction may be converted to continuous conduction in demyelinated fibres (Bostock & Sears, 1976). Again the factors governing this conversion have yet to be determined. Whether continuous conduction can develop in demyelinated central nerve fibres is unknown.

Human studies

To what extent are the findings in experimental disease applicable to man? Some answers to this question have come from the study of cerebral evoked potentials. Although the method is necessarily indirect, it is the only one at present regularly available for obtaining information about conduction in human central nerve fibres. Dr Martin Halliday and I began by examining the visual evoked potentials (VEP) in optic neuritis and multiple sclerosis. The most useful stimulus proved to be a chequer board pattern of black and white squares which was reversed at a frequency of 2 Hz. The response recorded with occipital scalp electrodes is dominated by a large positive wave with a latency of approximately 100 msec (Fig. 3). In different normal subjects there is a narrow range of latency, and in any

50 μm

Fig. 2. Portion of a single fibre isolated from the spinal cord of a rat with a naturally-occurring demyelin-ating disorder. The nodal gap is greatly widened beyond the normal upper limit (1 μ). A phagocyte closely related to the demyelinated portion of the axon contains intracellular myelin debris. (From Yogendran, 1976.)

25 μm

Fig. 5. Nerve fibres isolated from lesions produced by acute transient spinal cord compression in three cats, (a) five weeks, (b) three months and (c) six months previously. In each, an abnormally thin and short myelin segment is intercalated between sheaths of normal thickness. (From Gledhill & McDonald, 1977.)

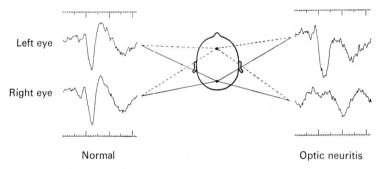

Left eye

Right eye

Normal   Optic neuritis

Fig. 3. Visual evoked responses recorded after pattern reversal stimulation at 2 Hz. Time scale 10, 50 and 100 msec. The records on the left are from a normal subject, and those on the right from a patient with a previous episode of acute right optic neuritis. (By courtesy of Dr A. M. Halliday.)

individual subject the latency from the two eyes is almost identical. We have now studied many hundreds of patients with demyelinating disease, and I want to describe two features which are important to the present discussion.

Delay. Demyelination produces a delay in the evoked response from the affected eye in 90% of patients (Halliday, McDonald & Mushin, 1972, 1973; see also Fig. 3). The mean delay in a series of 51 patients with multiple sclerosis was 43 msec but delays of up to 100 msec occur (Halliday *et al.*, 1973; Asselman, Chadwick & Marsden, 1975). The mechanism of these long delays is not yet clear and the contributing factors have been discussed elsewhere (McDonald, 1976, 1977). It is, however, clear from the experimental studies that slowed conduction must contribute.

Conduction block. Evidence for conduction block comes from patients with severe impairment of vision in the acute phase of optic neuritis. Fig. 4 shows the pairs of records from a patient with severe unilateral optic neuritis taken with an interval of four months between them. The lower records were taken after stimulation of the clinically unaffected eye. The upper left record was taken in the acute stages of the illness after stimulation of the affected eye, when the visual acuity was reduced to counting fingers. No pattern-evoked potential was recordable. The vision subsequently improved to 6/6 and a VEP with a substantial delay in the initial positive peak was present. The absence of the response in the early record and the presence of the delayed response in the later record implies that conduction was completely blocked in the majority of fibres in the early stage but returned, with a delay, as recovery proceeded. From the evoked potential study in man therefore, we conclude that both conduction block

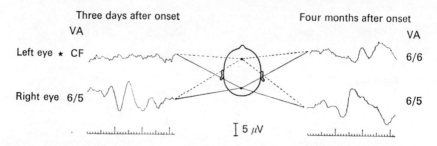

Fig. 4. Visual evoked responses recorded as in Fig. 3. The records on the left were made three days after the onset of acute left optic neuritis, when the visual acuity (VA) in the affected eye (*) was reduced to counting fingers (CF). The visual acuity in the right eye was normal (6/5). The records on the right were made after clinical recovery. (By courtesy of Dr A. M. Halliday.)

and slowing may occur in demyelinating lesions of central nerve fibres in man.

Relation of conduction defects to functional loss. We can now return to consider the question with which this section began, namely the physiological basis for paralysis and sensory loss in demyelinating disease. Complete conduction block in large numbers of nerve fibres is almost certainly the major factor in the production of severe clinical deficits. Less severe deficits probably arise from conduction block in a smaller proportion of the relevant groups of fibres and from the failure of a significant proportion of the impulses initiated in partially blocked fibres to reach their destinations. The accumulation of refractoriness of transmission, culminating in complete (though reversible) conduction block, is plausible as a contributory factor in the increasing weakness observed in sustained muscular contraction and during exercise. Slowing of conduction particularly when severe and of unequal degree in different fibres subserving the same function is likely to interfere with clinical tests – e.g. vibration sensibility – which depend on the delivery of synchronised bursts of impulses at particular sites in the nervous system. Irregular staggering of the velocities would be expected to distort the normal grouping of impulses.

These mechanisms are not necessarily exclusive. It has been suggested on the basis of experiments using tissue culture and the isolated frog spinal cord that there is a synaptic blocking factor in the serum of multiple sclerosis patients, but the effect has yet to be demonstrated *in vivo* (Bornstein & Crain, 1965; Cerf & Carels, 1966; Seil, Leiman & Kelly, 1976; Schauf *et al.*, 1976).

Clinical application

Have the various investigations I have described contributed to the management of patients? The serial study of our patients with optic neuritis and the patients with multiple sclerosis has shown that over 90 % of cases with clinically definite demyelinating disease have permanent delays in the VEP which may be present even when all the usual clinical signs of optic nerve damage are absent (Halliday et al., 1973). This observation led to an investigation of other types of optic nerve lesion and it has become clear that the technique is the most sensitive at present available for providing objective evidence of optic nerve damage. It is capable, for example, of demonstrating compression of the anterior visual pathways by pituitary tumour before there is field loss, a fact of considerable importance in planning treatment (Halliday et al., 1976). The widest application is, however, in the diagnosis of multiple sclerosis (McDonald & Halliday, 1977). The technique is particularly helpful in the assessment of the common clinical problem of the middle-aged patient with progressive spastic paraplegia who has no clinical evidence of an optic nerve lesion which, if present, would provide strong evidence for a diagnosis of multiple sclerosis. Approximately 50 % of such patients have strikingly abnormal VEPs (Halliday et al., 1974; Asselman et al., 1975) and in these cases we are now able to spare the patient the discomfort of myelography unless there are other compelling clinical reasons for suspecting spinal cord compression. It is important to stress that no VEP abnormality is specific to multiple sclerosis or to any other disease. The objective evidence of abnormality which evoked-potential studies provide has the status of an objective physical sign – taken in context it contributes to the formulation of a diagnosis and thus plays a useful role in determining the management of the patients. Recently, brainstem and spinal evoked responses have proved useful in demonstrating sub-clinical involvement of central white matter, and these techniques too promise to be helpful diagnostically (Kimura, 1975; Small, Beauchamp & Matthews, 1977 and unpublished observations; Robinson & Rudge, 1977).

Recovery from demyelinating lesions

Although one of the outstanding characteristics of demyelinating disease is the tendency to remission in the early stages, little is known about the mechanism of recovery following central nerve fibre damage. In principle, two fundamentally different processes might contribute. First, conduction might be restored in the damaged fibres themselves. Secondly, the functions of permanently blocked or degenerated fibres might be assumed by surviving fibres subserving the same general functions. There is no doubt from the recovery that occurs after spinal cord hemisection (e.g. Denny-Brown, 1966) or haemorrhagic capsular hemiplegia, that the second process occurs. Recently, evidence has been gained about its possible morphological and physiological substrata (see Wall, 1975). Our laboratory is currently concerned with the mechanism of conduction restoration in damaged fibres.

It is now clear that in certain circumstances demyelinated central nerve fibres can be remyelinated by oligodendrocytes, the normal myelin-forming cells. This fact was first established by Bunge and his colleagues in the early 1960s, studying a physically induced lesion in cats. Remyelination has since been demonstrated in allergic, compressive and a variety of toxic lesions (McDonald, 1974b). When the glial-limiting membrane is breached, Schwann cells from the nearby roots may migrate into the cord and re-myelinate central fibres (Blakemore, 1975). Recent work on experimental spinal cord compression has defined some of the characteristics of the newly formed central myelin (Gledhill, Harrison & McDonald, 1973a,b; Gledhill & McDonald, 1977; Harrison & McDonald, 1977). It is abnormally thin and although the thickness increases with time, adult dimensions are still not reached after 18 months. New nodes are formed but the internodal segments are abnormally short (Fig. 5). Fig. 6 compares the relationship between internodal length and external fibre diameter for 50 fibres from lesions three weeks to six months after compression with that for normal fibres. The filled circles represent data from normal animals, and open circles data from animals with compressive lesions. The diameter in the case of the experimental material is that of a thickly myelinated portion of the same fibre. There is a significant difference in length between the thin and thick internodes on fibres of the same diameter ($P < 0.05$).

Although the existence of central remyelination in experimental animals is now established, the evidence that it occurs in man is inconclusive (McDonald, 1974b), and it is unknown whether remyelination restores saltatory conduction in blocked central nerve fibres. While it has not yet been shown whether saltatory conduction can be converted to continuous conduction in demylinated central fibres, some form of surviving conduction in completely demyelinated human optic nerve fibres can be inferred from the following observations (Wiśniewski, Oppenheimer & McDonald, 1976).

(*Case report.* A 40-year-old woman with established multiple sclerosis was admitted to hospital four days before she died from a massive pulmonary

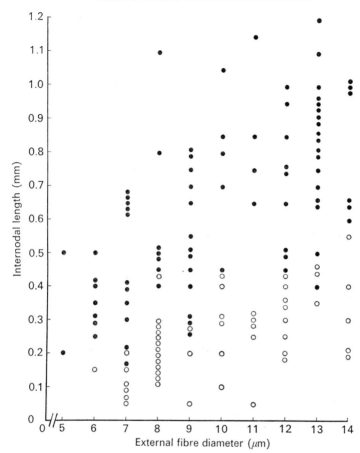

Fig. 6. Relationship between internodal length and external fibre diameter in the cat. ●, fibres from normal animals; ○, fibres from compression lesions. In the latter the internodal length is that of abnormally thin segments, and the diameter is the mean of the diameters of the adjacent segments on the same fibre with myelin of normal thickness. (From Gledhill & McDonald, 1977.)

embolus. Visual acuity was recorded by two separate examiners on admission. The patient was able to count fingers with each eye, and to read with telescopes. At necropsy both optic nerves were exhaustively examined by light and electron microscopy. Complete transverse sections were taken at many levels through the whole length of the nerves. Many areas in each nerve were examined by electron microscopy and at one level in one nerve a montage of a complete transverse section of the nerve was constructed from electron micrographs. Many intact axons were present in each nerve. Myelin sheaths were, however, completely absent in one nerve. In the other, myelin was present only over a length of 3 mm in a small group of fibres at

one edge of the nerve. The total length of demyelinated nerve was 30 mm
on one side and 27 mm on the other.)

Since this patient had some preserved visual function, it is clear that the
demyelinated axons were conducting. There is no other information about
conduction in this patient since she died before the introduction of VEP
techniques to clinical practice.

Thus at present, a number of phenomena have been identified which
might be relevant to recovery following central nerve fibre damage. Whether
they are relevant, in what way, and in what circumstances are the questions
with which we are much concerned at present.

CONCLUSION

If one looks back over the evolution of our knowledge of the pathophysiology of central nerve fibres, it is clear that the interests of basic scientists,
on the one hand, and of clinical neurologists, on the other, have led each to
formulate questions the answers to which have provided valuable data for
both. An examination of the background of the physicians who have
contributed fundamental observations to this field shows that each had had
a substantial period of laboratory training – Babinski with Charcot,
Bielschowsky with Weigert, Holmes with Edinger, Denny-Brown with
Sherrington. The conviction of the value of such experience in the training
of a physician is one of Archie McIntyre's fundamental tenets, and
through it he has had a wide influence on neurological science. Some of us
who came as medical students to his laboratory remained in basic disciplines,
while others returned to medicine not only with the skills he taught us, but
more importantly for the future, with his critical but at the same time
imaginative approach to the nervous system in order and in disorder. As a
result he has influenced not only the development of the field I have been
discussing, but in a very real way the practical management of patients
with neurological disease.

REFERENCES

Asselman, P., Chadwick, D. W. & Marsden, C. D. (1975). Visual evoked responses
    in the diagnosis and management of patients suspected of multiple sclerosis.
    Brain 98, 261–82.
Babinski, J. (1885). Recherches sur l'anatomie pathologique de la sclérose en
    plaques et étude comparative des diverses variétés de scléroses de la moelle.
    Archs Physiol. norm. pathol. Series 3, 5, 186–207.
BeMent, S. L. & Ranck, J. B. (1969). A quantitative study of electrical stimulation
    of central myelinated fibres. Exp. Neurol. 24, 147–70.

Bielschowsky, M. (1928). Zentrale Nervenfasern. In *Handbuch der Mikroskopischen Anatomie des Menschen*, ed. von Möllendorf, vol. IV/I, pp. 97–107. Berlin: Springer-Verlag.

Blakemore, W. F. (1975). Remyelination by Schwann cells of axons demyelinated by intraspinal injections of 6-amino-nicotinamide. *J. Neurocytol.* **4**, 745–57.

Bornstein, M. D. & Crain, S. M. (1965). Functional studies of cultured brain tissues as related to demyelinative disorders. *Science* **148**, 1242–4.

Boruttau, H. (1897). Der Elektrotonus und Phasischen Aktionsströme am Marklosen Cephalopodernerven. *Arch. ges. Physiol.* **66**, 285–307.

Bostock, H. & Sears, T. A. (1976). Continuous conduction in demyelinated mammalian nerve fibres. *Nature, Lond.* **263**, 786–7.

Bunge, R. P. (1968). Glial cells and the central myelin sheath. *Physiol. Rev.* **48**, 197–251.

Cajal, S. Ramon y. (1909). *Histologie du système nerveuxde l'homme et des vertébrés*, vol. 1. Paris: A. Maloine.

Cerf, J. A. & Carels, G. (1966). Multiple sclerosis: serum factor producing reversible alteration in bioelectrical responses. *Science* **152**, 1066–8.

Charcot, J. M. (1868). Histology de la sclerose en plaques. *Gaz. Hôp. civ. milit., Paris* **41**, 554–5, 557–8, 566.

Denny-Brown, D. (1966). *The cerebral control of movement.* Liverpool: Liverpool Univ. Press.

Denny-Brown, D. & Brenner, C. (1944a). Paralysis of nerve induced by direct pressure and by tourniquet. *Archs Neurol. Psychiat., Chicago* **51**, 1–26.

Denny-Brown, D. & Brenner, C. (1944b). Lesion in peripheral nerve resulting from compression by spring clip. *Archs Neurol. Psychiat., Chicago* **52**, 1–19.

Geren, B. B. (1954). The formation from the Schwann cell surface of myelin in the peripheral nerves of chick embryos. *Exp. Cell Res.* **7**, 558–62.

Gerard, R. W. (1931). Nerve conduction in relation to nerve structure. *Q. Rev. Biol.* **6**, 59–83.

Gledhill, R. F., Harrison, B. M. & McDonald, W. I. (1973a). Demyelination and remyelination after acute spinal cord compression. *Exp. Neurol.* **38**, 472–87.

Gledhill, R. F., Harrison, B. M. & McDonald, W. I. (1973b). Pattern of remyelination in the central nervous system. *Nature, Lond.* **244**, 443–4.

Gledhill, R. F. & McDonald, W. I. (1977). Morphological characteristics of central demyelination and remyelination. A single fiber study. *Ann. Neurol.* **1**, 552–60.

Gombault, A. (1880). Contribution a l'étude anatomique de la névrite paranchymateuse subaiguë et chronique – névrite segmentaire péri-axile. *Archs Neurol., Paris* **1**, 11–38.

Halliday, A. M., Halliday, E., Kriss, A., McDonald, W. I. & Mushin, J. (1976). The pattern-evoked potential in compression of the anterior visual pathways. *Brain* **99**, 357–74.

Halliday, A. M., McDonald, W. I. & Mushin, J. (1972). Delayed visual evoked response in optic neuritis. *Lancet* i, 982–5.

Halliday, A. M., McDonald, W. I. & Mushin, J. (1973). Visual evoked response in diagnosis of multiple sclerosis. *Br. med. J.* **4**, 661–4.

Halliday, A. M., McDonald, W. I. & Mushin, J. (1974). Delayed pattern evoked responses in progressive spastic paraplegia. *Neurology* **24**, 360–1 (abstract).

Harrison, B. M. & McDonald, W. I. (1977). Remyelination after transient experimental compression of the spinal cord. *Ann. Neurol.* **1**, 542–51.

Harrison, B. M., McDonald, W. I., Ochoa, J. & Ohlrich, G. D. (1972). Paranodal demyelination in the central nervous system. *J. neurol. Sci.* **16**, 489–94.

Hess, A. & Young, J. Z. (1949). Correlation of internodal length and fibre diameter in the central nervous system. *Nature, Lond.* **164**, 490–1.

Hess, A. & Young, J. Z. (1952). The nodes of Ranvier. *Proc. R. Soc.*, B **140**, 301–20.

Holmes, G. (1906). On the relation between loss of function and structural change in focal lesions of the central nervous system, with special reference to secondary degeneration. *Brain* **29**, 514–23.

Huxley, A. F. & Stämpfli, R. (1949). Evidence for saltatory conduction in peripheral myelinated nerve fibres. *J. Physiol., Lond.* **108**, 315–39.

Jackson, J. H. (1870). A study of convulsions. *Transactions of St Andrew's Medical Graduates Association.* **3**. Reprinted in 1931 in *Selected writings of John Hughlings Jackson*, ed. J. Taylor, vol. 1, p. 8. London: Hodder & Staughton.

Kimura, J. (1975). Electrically elicited blink reflex in diagnosis of multiple sclerosis – review of 260 patients over a seven year period. *Brain* **98**, 413–26.

Koles, Z. J. & Rasminsky, M. (1972). A computer simulation of conduction in demyelinated nerve fibres. *J. Physiol., Lond.* **227**, 351–64.

Lillie, R. S. (1925). Factors affecting transmission and recovery in the passive iron nerve model. *J. gen. Physiol.* **7**, 473–502.

Mastaglia, F. L., McDonald, W. I., Watson, J. V. & Yogendran, K. (1976). Effects of X-radiation on the spinal cord: an experimental study of the morphological changes in central nerve fibres. *Brain* **99**, 101–22.

McDonald, W. I. (1963a). The effects of experimental demyelination on conduction in peripheral nerve: a histological and electrophysiological study. I. Clinical and histological observations. *Brain* **86**, 481–500.

McDonald, W. I. (1963b). The effects of experimental demyelination on conduction in peripheral nerve: a histological and electrophysiological study. II. Electrophysiological observations. *Brain* **86**, 501–24.

McDonald, W. I. (1974a). Pathophysiology in multiple sclerosis. *Brain* **97**, 243–60.

McDonald, W. I. (1974b). Remyelination in relation to clinical lesions of the central nervous system. *Br. med. Bull.* **30**, 186–9.

McDonald, W. I. (1976). Conduction in the optic nerve. *Trans. ophthal. Soc. U.K.* **96**, 352–4.

McDonald, W. I. (1977). Pathophysiology of conduction in central nerve fibres. In *New developments in visual evoked potentials in the human brain*, ed. J. E. Desmedt, pp. 427–37. London: Oxford Univ. Press.

McDonald, W. I. & Halliday, A. M. (1977). Diagnosis and classification of multiple sclerosis. *Br. med. Bull.* **33**, 4–9.

McDonald, W. I. & Ohlrich, G. D. (1971). Quantitative anatomical measurements on single isolated fibres from the cat spinal cord. *J. Anat.* **110**, 191–202.

McDonald, W. I. & Sears, T. A. (1970). The effects of experimental demyelination on conduction in the central nervous system. *Brain* **93**, 583–98.

Peters, A. (1960). The structure of the myelin sheaths in the central nervous system of xenopus laevis (Daudin). *J. biophys. biochem. Cytol.* **7**, 121–6.

Porter, W. T. (1890). The presence of Ranvier's constrictions in the spinal cord of vertebrates. *Q. Jl microsc. Sci.* **31**, 91–8.

Ranvier, L. (1882). Des modifications de structure qu'éprouvent les tubes nerveux en passant des racines spinales dans la moelle épinière. *C. r. Hebd. Séanc. Acad. Sci., Paris* **95**, 1066–9.

Robinson, K. & Rudge, P. (1977). Abnormalities of the auditory evoked potentials in patients with multiple sclerosis. *Brain* **100**, 19–40.

Schauf, C. L., Davis, F. A., Sack, D. A., Reed, B. J. & Kessler, R. L. (1976). Neuro-electric blocking factors in human and animal sera evaluated using the isolated frog spinal cord. *J. Neurol. Neurosurg. Psychiat.* **39**, 682–5.

Sears, T. A. & Stagg, D. (1967). Methods used for the recording and subsequent analysis of spontaneous efferent and afferent activity in intact intercostal nerve filaments. *J. Physiol., Lond.* **191**, 108–9P.

Seil, F. J., Leiman, A. L. & Kelly, J. M. (1976). Neuro-electric blocking factors in multiple sclerosis and normal human sera. *Archs Neurol., Chicago* **33**, 418–22.

Small, D. G., Beauchamp, M. & Matthews, W. B. (1977). Spinal evoked potentials in multiple sclerosis. *Electroenceph. clin. Neurophysiol.* **42**, 141 (abstract).

Starling, E. H. (1915). *Principles of human physiology.* London: J. & A. Churchill.

Suzuki, K., Andrews, J. M., Waltz, J. M. & Terry, R. D. (1969). Ultrastructural studies of multiple sclerosis. *Lab. Invest.* **20**, 444–54.

Tasaki, I. (1952). Properties of myelinated fibres in frog sciatic nerve and in spinal cord as examined with micro-electrodes. *Jap. J. Physiol.* **3**, 73–94.

Tasaki, I. (1955). New measurements of the capacity and the resistance of the myelin sheath in the nodal membrane of the isolated frog nerve fibre. *Am. J. Physiol.* **181**, 639–50.

Tourneux, F. & Le Goff, R. (1875). Note sur les étranglements de tubes nerveux de la moelle épinière. *J. Anat. Physiol., Paris* **11**, 403–4.

Wall, P. (1975). Signs of plasticity and reconnection in spinal cord damage. In *Outcome of severe damage to the central nervous system.* Ciba Foundation Symposium 34 (new series). Amsterdam: Elsevier.

Wiśniewski, H., Oppenheimer, D. & McDonald, W. I. (1976). Relation between myelination and function in MS and EAE. *J. Neuropathol. exp. Neurol.* **35**, 327, (abstract).

Yogendran, K. (1976). *Isolation of single nerve fibres from the rat spinal cord.* F.I.S.T. thesis, Institute of Science Technology, London.

Self-organizing systems in psychobiology

C. TREVARTHEN

Whatever the exact nature of the physical forces which mould the developing tissues into organs, they must certainly be organized into self-regulating systems.
[Waddington, *The nature of life*, 1961.]

Let us note that perception and acquired behaviour alike, whether elementary or higher, do, in fact, contain autoregulatory processes . . . On the perceptual level, one of the most remarkable is that which regulates constancy of size, shape, etc. In the case of size constancy, the zoologist von Holst has admitted the existence of a hereditary system of feedbacks . . . It is obvious that all learning by trial and error (a groping) presupposes feedback structures . . . Operations of thought and especially those of elementary logico-mathematical thought can be considered as a vast autoregulatory system which gives autonomy and coherence to the process of thought . . . It goes without saying that these regulatory mechanisms, in knowledge at all levels, raise the problem of their relationship with organic regulations.
[Piaget, *Biology and knowledge*, 1971.]

The doctrine which I am maintaining is that the whole concept of materialism only applies to very abstract entities, the products of logical discernment. The concrete enduring entities are organisms, so that the plan of the whole influences the very characters of the various subordinate organisms which enter into it. In the case of an animal, the mental states enter into the plan of the total organism and thus modify the plans of the successive subordinate organisms until the ultimate smallest organisms, such as electrons, are reached.

There are thus two sides to the machinery involved in the development of nature. On the one side there is a given environment with organisms adapting themselves to it . . . The other side of the evolutionary machinery, the neglected side, is expressed by the word creativeness. The organisms can create their own environment. For this purpose the single organism is almost helpless. The adequate forces require societies of cooperating organisms. But with such cooperation and in proportion to the effort put forward, the environment has a plasticity which alters the whole ethical aspect of evolution.
[A. A. Whitehead, *Science and the modern world*, 1925.]

My chapter begins in a memory I keep of a lecture on homeostatic mechanisms given by Archie to a little class of dedicated students of physiological science many years ago. That opened my eyes for good to the power of the brain to keep ambitious goals intact, and to profit from the forces that try to push its function this way and that. I have never stopped trying to understand that lecture.

The Author.

Is a mind biology possible?

Some bridges between disciplines in science promise more than they deliver. They conceal gaps of knowledge and confuse hard-won understanding about structure at different levels in nature. On the whole, 'physiological psychology', for example, has been a somewhat unsatisfactory structure, mainly because the psychology end had reductionist, antimentalist weaknesses and the physiology was often summary and inaccurate.

The word 'psychobiology' may strike a physiologist as one more name for an optimistic bridge of thought without proper credentials. He may feel, as Pavlov did, that psychology and brain physiology make best sense when kept distinct, each working with the knowledge it can control well, as the more 'fundamental' science steadily encroaches on the other. Nevertheless, in an evolutionary perspective, 'psychobiology' is quite as logical as 'biophysics' or 'biochemistry'. In principle, a realm of science exists in which biology and psychology are inseparable.

Let us define an ideal psychobiology as the science of structures and processes in mental life. The proper aim of this science is to discover biological principles in the behaviour of animals and in the conceptions and experiences that guide what animals do. It should, with one set of principles, encompass consciousness in man and all the feelings, emotions, intuitions and intentions that motivate the various adaptive forms of human understanding. There need be no reduction of psychology to a subordinate biology. Just as physiological systems gain a permanent independence of their foundations in inorganic physics and chemistry, psychological systems become free from the 'vegetative' biology of organisms that have action by growth and form, but no behaviour. We need the right level of physiology and anatomy and the right psychology.

Recently the names 'neurobiology' and 'brain science' have become emblems for many scientists. Presumably the fields so labelled aim to escape from certain limitations in traditional neurophysiology. There is simultaneously a prodigious activity in 'objective' psychological science, the experiments leading to recognition of powerful cognitive systems. In a few limited areas there seem to be real psychophysiological fusions as, for example, between psychophysics of the visual periphery and unit neurophysiology of the retina, optic pathways and striate cortex. But the gap between the abstract central processes psychologists aspire to understand and the structures and functions biologists see is still awe-inspiring. To bridge the gap we will need to form new perceptions of the systems that organize genuinely intelligent mental actions. Only then will the Cartesian distinction between matter (*res extensa*) and mind (*res cogitans*) have finally lost meaning, as the scientific aims of physiology logically require it should.

If we are quickly out of touch with normal life when we attempt to break psychology down to more elementary 'material' processes of nerve cell physiology and neuroendocrinology, it is equally misguided to look for structures in the brain that map directly onto abstract thought patterns of a highly rational and deductive psychology; one that, while being philosophically sophisticated, or at least complexly argued, is highly artificial, remote from the activities and experiences of normal conscious living. To establish basic principles for psychobiological research we need to guard, then, against both biological reductionism and a high-flown philosophy of discourse in texts – what Whitehead (1925) calls 'the products of logical discernment'.

Biology itself, we should recall, has long been in battle between the extremes of mechanism and 'objective vitalism' (Waddington, 1961). The problem of how to choose the right level of explanation is not new. As biology has had to defend its own physics and chemistry, so psychobiology must clearly state its own concepts of form, self-regulation and organization analogous to those that have proved essential in evolutionary theory, in genetics, in embryology, in the study of metabolism, in whole-brain physiology – indeed, in all areas where biology has tried to obtain more direct perception of the systems it wants to understand. These notions of organization are not 'elaborate' in biological explanations. They are elementary and fundamental. It is, as Whitehead (1925) explained, getting things upside down, a misplacing of the idea of what is the concrete entity one knows about, to believe that the smaller parts of self-organizing 'enduring' things are more real than the wholes of their existence. The integrations may be more real than the matter passing through them. It all depends what one is trying to comprehend, the forest or the trees. Forests are not just aggregates of trees. Some trees are inseparable bits of integral forests.

I wish, in this paper, to illustrate how two classical techniques of biological enquiry, dissection of anatomical patterns and the study of development, have together made discoveries in the past 10 years that bring the known brain systems and mental processes into more harmonious relation. I believe that what has been found requires a reappraisal of certain basic ideas about genetics. The facts certainly fly in the face of the dominant mechanistic doctrine of recent psychology. They require a much more elaborate nativistic or, rather, organismic approach in this subject. But, first let me explain my understanding of the continuum of biological

functions that has led from morphogenesis of organisms to creation of mind
and understanding.

DEVELOPMENT

In the formation of an organism from a fertilized egg biologists distinguish
differentiation, by which the organic matter becomes more varied and
complicated, and a counteracting process of reintegration that restrains
microscopic forms, pulling them in to one macroscopic system. Differentiations and reintegrations result from reactions between self-replicating
elements. The mutually exciting structures inside cytoplasm, cells and
tissues have a feedback-controlled instability which generates patterns of
growth in elaborate hierarchies of biological matter. All large organisms are
built by a process of sorting and selecting in which large numbers of cells of
the same genetic constitution interact together and negotiate their increasing
differences of form and constitution. Eventually the process of epigenesis
leads beyond the relations of cell to cell and organ to organ, to start a
unique and controlled two-way involvement of one species of organism and
its environment. This life activity or 'habit' opens up new prospects for
regulation of exchanges inside the total life form of the organism.

Nervous life forms

Nerve cell networks revolutionize all the processes of development. In
spite of their miraculous capacity for controlled creation of forms, and of
forms of forms, plants, because they do not have neuromotor systems, have
spatially more limited and slower changing command over the conditions
of their existence. Nerve networks boost that control to a new level. A
skeleto-muscular machinery is excited by nerves to act rapidly and it can
make large changes in the environment. Instinctive movements coordinated
by a patterned nerve net may take an animal to another place and a different
state of existence. Movements differentiated out of whole body locomotion
can regulate a new efficiency in feeding or mating. Receptor cells and
perception circuits respond to stimulus information about changes of
condition or layout in the environment, and excite delicately formed
patterns imaging the world in the central nervous system. Motor impulse
patterns automatically create sensory reflections of the changing situation
of the organism, and this 'reafference' permits the central nervous system
to detect the world radiating out from the body and so to modify the
patterns of movement adaptively with reference to goals and appetites
already set within the system. Finally, sensory perception of the forms of
action of *other* beings transforms social interactions between individuals

and leads to cooperative behaviours that greatly multiply the power of control of the species over the environment. How do all these mechanisms develop?

Nerve membrane impulses are, we now know, freely translatable into other forms of cell activity such as uptake of metabolites, protein synthesis, gene action, growth and differentiation. In the embryo, nerve cells are potentially in fine and influential communication with differentiating tissues before they are capable of transmitting excitatory effects to muscles. Indeed, nervous systems have probably evolved out of the communication systems that integrate cells in all large tissue aggregates and that control the development of any organism. The neuroblasts of embryos are indistinguishable from other tightly joined ectodermal cells. Later, close junctions, which are lost or greatly modified in other tissues, are retained as the highly specialized synapses of the nerve net.

The embryos of all intelligent animals still pass through a pre-behavioural stage in which intricate sensory and motor structures arise in and around the nervous system in anticipation of an environment to be acted on and perceived (Trevarthen, 1973, 1974a). Like gene products which automatically segregate into cell types and tissues, the sensory integrative and motor structures of the embryo central nervous system are pre-set to generate habits of adaptive behaviour in a suitably supporting environment. A fate map of functions is formed in the embryo nerve net, mirroring the body (Benzer, 1973). It is capable of formulating adaptations of body function and behaviour that come into effect only at a remote time when the body moves about testing the field of experience. Part of this cerebrogenesis or brain epigenetics is by intercellular negotiations prior to the conduction of action potentials. Then, growth at every link in the neurone network and in the receptors and musculo-skeletal machinery may be controlled by the ion-moving and membrane-distorting propagation of nerve potentials. At first, spontaneous impulses assist selection from among a vast excess of cell-to-cell connections in the net (Willshaw & von der Malsburg, 1976). Later, trains of action potentials retain images of patterns of stimuli, and hence of physical order in the body and in the outside world. A history of assimilated effects and images is created among neurones as a continuation of embryogenesis.

In this continuum of developments, psychological functions arise where metabolism, acts of movement and the feedback effects of movements on the stimulation from surroundings meet. Thus the original patterns of motivation state, of intention to act, and of experience emerge in pre-established relationships out of the tissue differentiations and reintegrations of the embryo.

Embryology of the field of behaviour

The classical experiments of Spemann (1938) showed that the polarity and symmetry of the embryo body depend upon one locus on the surface of the zygote. If this organizer is cut from one early embryo and transplanted to another, it can preside over forces that form a new body added to the host one. Cells in the body normally move, adhere and differentiate relative to the organizer locus. Subsequent developments depend on autonomous interactions between cells locally, and a complex pattern of territories is set up, but the organizer is the anchor or reference origin responsible for fixing the polarity and symmetry of the body. The embryo nervous system has its regulated place in the body. It is formed in register with the set of relations that determine body form. Later the body becomes segmented and branched and a chain or tree of subordinate organizer regions and growth points is formed. In the same way the nervous system develops a bilaterally symmetric array of parts, including lobes that contribute to rudiments for organs of special sense (eyes, ears, nostrils) and fields of the central nervous system to which these later project.

Recent research in neuroembryology has brought to light a simple but powerful case of individuation in formation of the nerve cell arrays that anticipate the space for behaviour. Surgical rearrangement tests with the eye primordia of lower vertebrates and with the visual projection fields in the midbrain roof have demonstrated that a replica of the body's organizer is active in the diencephalon just above the tip of the notochord (Chung & Cooke, 1975). This appears to be a *brain organizer*, giving the primordial centres information to establish polar coordinates equivalent to those of the body. In this way the eye and visual field in the brain have the same code for anterior and posterior, left and right and up and down as those for the body. When a horde of nerve axons and dendrites grow out in the late embryo stage, they use the main polarity markers and their offspring as anchor posts or poles of reference (Hunt, 1975).

In morphogenesis of the body, cell sheets and clusters interact, compete and mutually adjust. When nerve cells grow axons to establish a network of contacts they also interact, compete and mutually adjust. Rearrangement tests, in which eyes or midbrain visual areas of embryo amphibia are cut out or transplanted and duplicated or rotated, show how these adjustments, eventually under control of nerve conduction and its chemical consequences, are capable of ordering the nerve projections (Sperry, 1965; Gaze, 1970). Between field polarization by organizers, on the one hand, and sorting to ordered series by communication between neighbouring nerve terminals in dynamic competitive interaction, on the other, are specified an incredibly refined ordering of excitatory projections from eye to midbrain, one that sets them in correct orientation with respect to other parts that relate to them. Moreover, it appears as if the different modalities of sensory reception all gain their capacity to represent events in an orderly space relative to the body by the same epigenetic mechanism. As a result of this process of mapping between fields of affinity, and with added differentiations that refine and separate the degrees of affinity, all the modalities of sense are inherently equivalent in representing this space. The organism has an 'embryogenic egocentre' to which all sense modalities refer just as it has an anterior pole. In consequence, the various brain fields communicate in one and the same space–time code.

The implications of this pattern of cerebral embryology and of its great generative potentialities are revolutionary for psychology. Previous arguments of philosophers and psychologists about the *acquisition* of a unity of experience, the integration of elementary sensations and the equivalence of modes of experience have been highly artificial and speculative and they have begun with the facts of the environment alone. Now our theories may be brought towards a less uncertain plane of biological reality based on knowledge of the growth of the organism and its regulated response to the environment.

Work on the motor side shows that projection of integrative patterns of nerve connection from brain to muscle units of the trunk, neck, limbs and receptors of head and extremities is governed by processes that match those for sensory systems (Gaze, 1970). This, in the end, assures orderly mapping from stimulus to brain to motor act. It governs the primary correspondence of an act with other acts (motor coordination or motor equivalence) and the matching of act to perception. The latter is the basis for all instinct and all intentionality.

In theory, these psychomorphogenetic principles can be extended to explain innate features of higher-order mappings of acts into habits and skills, and of modalities of perception into object concepts and representational thinking to which Piaget refers in the quotation at the beginning of this paper (Trevarthen, 1978*a*). But the implications of the embryological process behind formation of interneuronal and neuromuscular projections do not stop here. The language of impulse transmission in the embryo, transmitter substances at intercellular junctions, is the same as that by which the mature nervous system regulates behaviour and responds to its consequences. An extended brain embryogenesis creates a nervous system that can adapt and remember (Mark, 1974). New patterns of experience that relate to patterns of intention in particular, advantageous ways can be incorporated into unfinished synaptic arrays as memories.

Experiments with amphibian neuromuscular systems and with the cortical visual circuits of kittens and newborn hamsters, rabbits and monkeys show that competition between overlapping nerve terminals of different origin for effective control over the cells they reach can be influenced by stimulation as well as by pre-established codes of cell-to-cell interaction (Trevarthen, 1978b). At present the power of experience to modify brain circuits seems less than was expected. It is tightly constrained by the morphogenesis. But it is equally clear that the adaptive refinements of perceptual mechanisms as well as the mechanisms of skillful movement cannot mature without the patterns of excitation to which they are adapted. A wonderful fitness appears between the structure-forming power of the brain and the opportunities of the environment. The growing brain actively calibrates itself to experience reality and to act on it precisely. It can set out psychological activities ahead of experience, and it may even mould the environment to pre-establish ends, as when a bird builds a nest or a spider spins a web. At the same time, within limits set by evolution, the memory process can image events never before encountered and in the higher forms a seemingly endless adaptability is assured. That seems to be the unique achievement of mental organisms. They have great plastic response to experience that grows out of an equally impressive power to make the universe plastic to their will.

NEUROPSYCHOLOGY

Anatomies of imagery and intention

The life history of a tree, an organism which develops by choice among a large excess of buds, shows up in its morphology when it is fully grown. The same is true for a brain, in spite of a vast amount of cell loss by 'editing' in the process of its growth. Some of the anatomy of the mind becomes visible by systematic correlation of brain form with psychological functions Luria, 1973).

Unlike any other single form of intervention, surgical disconnection of the cerebral hemispheres (cerebral commissurotomy) brings out features of the brain's organization as the constantly changing but highly determined mechanism of mental control (Sperry, 1968, 1977). The findings by Sperry's split-brain technique have been complemented by a revolutionary increase in our knowledge of both fine-grain anatomy (as reported, for example, in the *Journal of Comparative Neurology* over the past 15 years) and general topography of sensory, motor and integrative systems of the brain. A delicate and staggeringly complex functional anatomy is being discovered. On the one hand, this anatomy is clearly adapted to govern the flow of

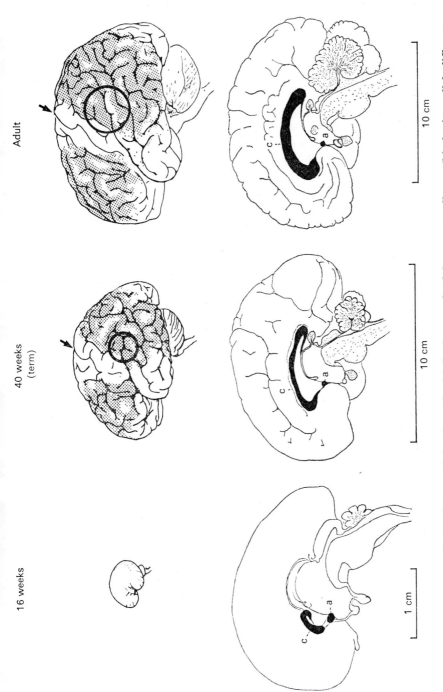

16 weeks   40 weeks   Adult
(term)

Fig. 1. As the brain forms in the foetus and develops after birth, the growth of the corpus callosum (black, c) parallels differentiation of the association cortex (stipple). Circles mark the approximate location of speech mechanisms that develop in the left hemisphere. The arrows mark the central sulcus; 'a' is the anterior commissure.

experience in relation to the hierarchy of forms of movement, and on the other, it bears a clear record of the embryological processes by which it was formed before experience began.

The story of split-brain research is one of steady progress away from behaviourism and Pavlovian sensory–motor psychology (Sperry, 1975). Beginning as an inspired attempt to discover the pathways of perception and learning, the research effort has led to analysis of the anatomical basis of intentions, the commanding role of consciousness and the different forms of creative understanding in human life. No single research project has been more influential in establishing a psychobiological view of man. The mechanism for voluntary control of the experience of objects, and for extending awareness in fruitful directions, has been found to have far more elaborate organic creativity than supposed by the experimental analysts of human performance and their theories of man's capacity for processing of sensory information in various 'channels'.

The effects of commissurotomy, brought out by appropriate tests, show the self-regulating generative mechanisms of knowing and intending (Trevarthen, 1974b, 1975; Sperry, 1977). These mechanisms are not related in a simple way to the classical projection pathways for the different modalities of sensation. They relate best to inherent psychological motives and modes of performance in an active subject (Trevarthen, 1978a).

From split brain to split mind

Myers (1956) showed that new visual memories of a cat could be located in the hemispheres. Perception and learning in a training situation were brought under total surgical control. At the same time, the corpus callosum was proved capable of transmitting experience from one side of the brain to the other mirror half and of establishing perceptual equivalence between stimuli presented to one eye at a time (Myers, 1962). Under conditions of one-eyed training and testing, callosal fibres were necessary for perceptual equivalence and memory transfer, and, at least for easy kinds of discrimination, they were sufficient for complete equivalence or transfer.

A further striking discovery, confirming findings that puzzled earlier workers, was that a cat with completely split mechanisms for experiencing shapes by seeing them or touching them with the pads of the forepaws was still a well-coordinated motor cat. Either hemisphere could read out experiences into well-integrated actions of any part of the body. The linkage of experience and motor patterning seemed to be duplicate and the motor brain undivided (Sperry, 1961).

In my experiments with monkeys, I found that separate experiences could be created and remembered even when both halves of the divided brain were open to experiences at the same time (Trevarthen, 1962). That meant that the mirror nerve mechanisms that responded distinctively to the forms which the monkeys were looking at were capable of acting independently and simultaneously while the monkey as a whole was looking at one place, at one time. However, two new effects came out of these consciousness-duplicating experiments that changed the direction of my analysis.

First I noticed that the split-brain monkeys rarely performed double symmetric acts of response, pushing with two hands at once. Usually they spontaneously preferred one hand to touch the response panels. Secondly, I found that there were some visual discriminations where acting and remembering with one half-brain either blocked learning of the reverse choice on the other side or even transferred to it a trace of experience. In the classical view of monkey visual anatomy such interocular transfer of perception after optic chiasm and corpus callosum were cut was ruled out. But it happened, and it meant that a pathway existed to project information from one retina to a territory accessible to the other retina.

Visual information must have been passed through the brainstem. This led me to the hypothesis that the extra-geniculate midbrain visual territories which were bilaterally integrated by commissures were either capable of perceiving and memorizing some kinds of stimuli correctly on their own, or they were sufficiently discriminating avenues of input to perceptions and memories integrated in the cerebral hemispheres. Phylogenetic and ontogenetic considerations added weight to the idea that the midbrain roof of a monkey could play a vital part in the seeing of space and some details of form or colour and not just integrate oculomotor reflexes. There was reason to believe that brainstem vision can transmit some data about forms of object to the more critical mechanisms of perception in the cortex.

I jumped from my suggestive data with two split-brain monkeys that were completely doubled for shape discrimination, and only partly so for brightness and some colour discriminations, to the idea that the midbrain would be adapted to see in a functionally different, probably less elaborate and less refined, way than the cerebrum (Trevarthen, 1965). Considering the known anatomical relations of midbrain and forebrain with the eye and with the motor system, and encouraged by the results of an elegant double dissociation experiment by Schneider (1967) with the hamster, I later decided that there was sufficient evidence to propose a two-stage vision in the primate brain. Two principal levels of the visually informed brain, midbrain and striate cortex, corresponded with two levels of moving and seeing. I called one *ambient* and the other *focal*. The perceptual mechanism had been divided not simply in two by the chiasm–callosum operation, but into three, two focal and one ambient. The psychological functions of attending and orienting in the egocentric array of experiences throughout the ambient visual field were evidently in close two-way communication with the image forming recognitive mechanisms of the cortex. The latter obtained information via the high-resolution foveal part of the visual projection by oculomotor sampling, while ambient vision depended on the combined effects of excitations at less finely distinguished loci scattered more widely in the field.

In recent years, behavioural work with brain-operated cats by Sprague and Berlucchi and physiological studies by Rizolatti and others on the cat and Wurtz and Goldberg and others on the monkey have greatly increased our knowledge of how intimately midbrain and cortical visual systems interact (Ingle & Sprague, 1975). I believe the original concept of two principal levels of structure and function in seeing still holds firm.

About 1958, Downer and I working independently became curious as to how the split-brain monkey behaved when unable to indulge in an obvious preference for working with the hand contralateral to the cortex that was engaged in making visual discriminations (Downer, 1959; Trevarthen, 1962). Unlike the split-brain cat, the split-brain monkey, though still agile in free locomotion still a walking, jumping and orienting creature, had divided intentional coordination of reaching and manipulation. When forced for the first time to touch and push or pick up with the hand on the same side as the seeing cortex, by having the preferred contralateral hand tied down or blocked off from the response panels, the operated monkeys became apraxic. They misreached, fumbled and, quite reasonably, showed great reluctance to use their hand (Trevarthen, 1965). Soon they overcame the misreaching and were able to pick up peanuts or pull objects apart with dexterity, but they remained clumsy in preprogramming how to use their fingers on visual input alone.

These disorders of voluntary actions have been brilliantly analysed by Brinkman & Kuypers (1973) and brought by Kuypers (1973) into relation with a general scheme for the motor pathways of the brain. This seems closely parallel to that proposed for visual structures. Brinkman, by detailed film analysis of picking up movements of split-brain monkeys seeking to extract food pellets jammed in a form board, showed that each cortex, when disconnected from its partner, had visual control for fine grasping movements of the opposite hand only. Touch was confused by knobs in the form board so the subjects had to guide their fingers to extract the food with focal vision. The ability to do this with the ipsilateral (non-preferred) hand–eye pairs never recovers even after forced training. The tests showed how each nearly blind hand could immediately compensate for the approximate visual guidance by taking up touch feedback.

These experiments bring out beautifully how the brain is designed to obtain experience in complementary modes of local percepto-motor action in a common reaching space. The main levels of function are related to the way the subject uses sense data rather than the physical nature of these data. That is, they are morphogenetically created correspondences between movement and experience (Trevarthen, 1974c).

By making lesions in one hemisphere before or after cutting the corpus callosum, Myers (1962) had tried to find the areas necessary in one hemisphere to transfer memory to the other hemisphere and to find the regions in which memories of the two hemispheres were combined when the corpus callosum was intact. Sperry did similar tests to find the minimal amount of tissue for storing a memory of a touch or vision experience in one of the disconnected hemispheres (Sperry, 1961). The cat's touch system appeared compact, within the anterior part of the brain, but the visual mechanism involved more than the whole posterior half of the cortex. Mishkin (unpublished) has carried out a painstaking analysis by the lesion technique of the touch system of the monkey and it seems to be like that for vision in both the cat and the monkey (Mishkin, 1972). Perceptual identification of objects is gradually built up by confluence of information from primary projection fields towards areas at the base of the temporal lobe where object recognition is integrated (secondary somesthetic area and the infero-temporal cortex). Then the major memory trace for the evaluated identity of the object is built up in the medial temporal and ventral frontal cortices. Callosal connections interconnect the secondary perceptual and the memory-storing regions but not the primary sensory fields. They are part of the system that relates experience in the different modalities to integrated plans for action and to motivation.

In Professor Paillard's laboratory at Marseille I was able to show that when eyes and both hands of baboons work together in mastering a complex mobile structure, a puzzle box that opened only by an intricate sequence of hand movements, the two hemispheres also work in close coordination (Trevarthen, 1978d). Commissurotomy after learning seriously disrupted this coordination and revealed that the functional relations between the hemispheres tend in such a task to be asymmetric (Trevarthen, 1965, 1978c). Just as most normal monkeys show manual dominance or strong spontaneous preferences for using one hand in visually guided manipulations, so the memory of how to open a puzzle box to get food was laid down more in one half of the undivided cerebrum of my baboons. I believe this lopsided retention of skill in the intact brain is due to a prewired, innate asymmetry in the schemata for hand use which is differentiated further by practice of a particular skill. The monkey brain is not as symmetric as it appears to be.

Two kinds of thought in one head

All the findings with split-brain monkeys have been greatly extended now by an inspired series of studies of commissurotomy patients in Los Angeles by Sperry, Bogen, Gazzangia, Milner and Taylor, Nebes, Gordon, Levy, D. and E. Zaidel and others. The patients, under care of the neuro-surgeons Vogel and Bogen who performed commissurotomy to achieve control over mentally crippling epilepsy, give detailed, subjective information about the effects of this surgery on people. They have collaborated, at the California Institute of Technology, in intricate tests of attention perceiving, motor coordination at all levels, emotional reaction, memory, intellectual comprehension, judgment and imagination.

The findings of this remarkable research involving cooperation of surgeons, patients and psychologists are available in a number of reviews (Levy, 1974; Gazzaniga, 1970; Sperry, 1968, 1970, 1974; Sperry, Gazzaniga & Bogen, 1969). I can only sample these findings here to give an impression of how surgical transection of the biggest nerve tract in the brain with some 200 million fibres affects the human mind.

A few hours after surgery a commissurotomy patient may be alert and able to discuss his or her experiences and feelings. He or she looks and sounds no worse than someone who has had a major operation somewhere outside the brain. At first sight the mind is perhaps a little tired or confused, but otherwise it seems normal. However, tests which carefully control orientations show a clear separation of two different kinds of mind, one for each hand. Bogen (1969) has administered such tests soon after the operation. Objects felt in the right hand out of sight are identified readily, but if they are placed in the left hand and explored by the patient out of sight and without emitting any distinctive sound the patient can say nothing to identify them. It is as if the object were held in the hand of another person. In contrast, if required to copy drawings of even quite simple geometric forms, the patient uses the right hand poorly and is more skilled with the left.

A large series of simple but powerful tests has been evolved for these patients in Sperry's laboratory at the California Institute of Technology out of those given there to split-brain cats and monkeys, taking proper account for the first time of the anatomical segregation of inputs from each half of the visual field to the striate cortex of the opposite brain half (Sperry et al., 1969). These tests have demonstrated that commissurotomy

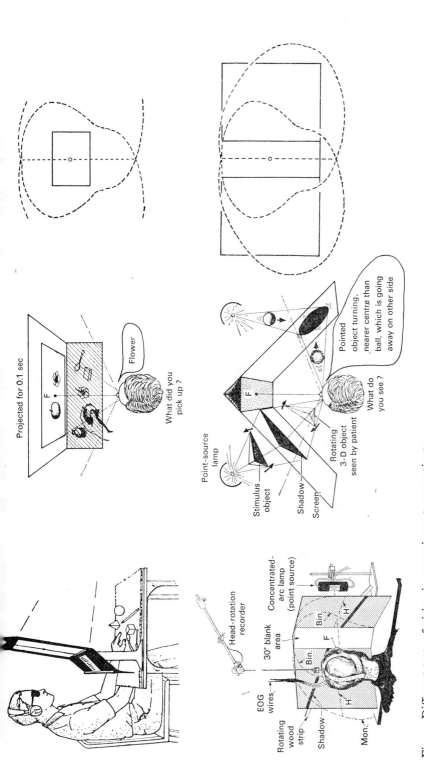

395

Fig. 2. Different tests of vision in commissurotomy patients.

Above: brief pictures in central vision are seen separately by the hemispheres in focal vision. Only stimuli to the right of the fixation point (F) can be spoken about.

Below: large dynamic shadow stimuli excite an undivided ambient vision. EOG, electro-oculographic; Bin, edge of binocular field; Mon., edge of monocular field; H, horizontal meridian of the visual field.

Right: visual fields and size of stimuli.

396

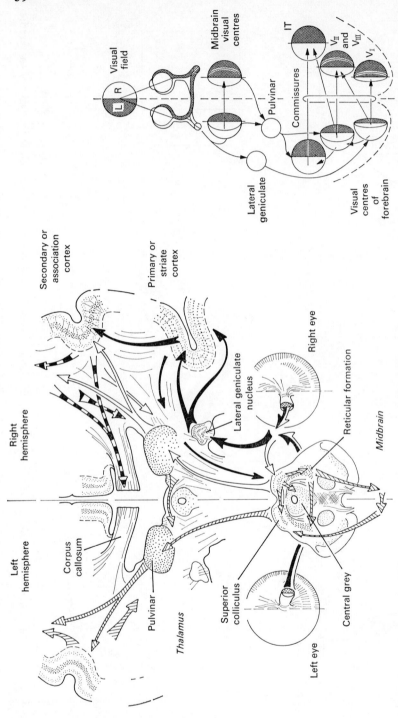

Fig. 3. Visual anatomy in man. Ambient vision involves midbrain centers (white arrows). Focal vision depends on the geniculo-striate system (black arrows). There are many maps of the visual field (right). V_I, V_{II}, V_{III} are successive visual fields of the cortex; IT, infero-temporal cortex. Black and white arrows: integrated visual processes. Cross-hatched arrows: pathways for subcortical cross-over of visual information include the reticular formation and cerebellum.

leaves a person with two separate visual awarenesses as well as two separately aware hands (Fig. 2). Only one consciousness, that of the left hemisphere, gaining information from the right half of the visual field and the right hand, is able to talk and so give a full account of things perceived. The other right hemisphere consciousness is mute. Nevertheless highly elaborate patterns of intention and explorations of experience can be sustained by each of the hemispheres. These are brought to light by testing them separately.

As with the monkey, each cerebral cortex can command movements of postural orientation of looking with the eyes and reaching with either hand on its own. It can set goals for the integrated mechanisms of the brainstem, cerebellum and spinal cord. But more subtle articulations of each hand involving guidance by vision, touch or audition are confined to the contralateral hemisphere. Movements of speaking and those of writing are almost totally controlled by the left brain. This hemisphere alone can give an account of itself in speech or writing.

Sperry and I have explored vision with large moving stimuli invading the far periphery of the visual field which is important in the control of posture, locomotion and reaching (Figs. 2, 3 and 4). We found that ambient vision of space at large is undivided (Trevarthen & Sperry, 1973). This means that the idea of two levels of vision derived from the monkey tests was correct for man too. Sub-hemispheric mechanisms that regulate the body as a whole motor structure and coordinate its parts in relation to the external layouts in space are capable of participating in organizing a general context of space awareness. More discriminating focal vision may be deployed in this context by fixation eye movements.

Kinsbourne and I used briefly flashed pictures to show that both hemispheres may perceive a whole shape in spite of receiving information about only half of it to one side of the vertical meridian of the field (Trevarthen, 1974b). This means that consciousness of something on the fixation point and extending to left and right of it can be synthesized out of the one-sided data in each striate cortex. But refined tests of stereognosis by Levy (1974) and Nebes (1974) have shown that touch recognition of shapes and sense of a whole form completed from pieces of information in vision or touch is best performed in the right hemisphere. Evidently the right unspeaking hemisphere has more of the mechanism that generates forms in immediate awareness out of incomplete samples of sensory information picked up at one moment by eye or hand.

With Dr Sperry, Levy and I have developed a method of rapidly exploring the interplay and relative strength of different kinds of visual experience in the two hemispheres, using what we call bilateral chimeras –

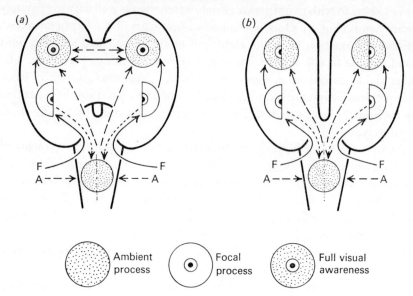

Fig. 4. Ambient (A) and focal (F) processes (*a*) before and (*b*) after commissurotomy.

stimuli made by joining the left half of one stimulus to the right half of another (Levy, Trevarthen & Sperry, 1972; Levy & Trevarthen, 1976, 1977). When a split-brain subject sees the chimera briefly with the join between the stimuli centred on his fixation point and lined up with the vertical meridian of his vision, each half stimulus excites an experience in the opposite hemisphere only (Fig. 5). The two experiences remain independent until by means of some act of response that reflects reafferent information back to both brain halves, or by picking up additional information in other ways, the subject is able to repair the division of his mind.

It is certainly correct to speak of a divided awareness and divided memory when describing the condition of these people. But it is not at all clear that intentions or attentions are so completely dissociated. It was thought at one point that the integral coordinated behaviour of a commissurotomy patient, which makes him capable of ordinary acts like any other person, was entirely due to feedback of reafference between the two brain halves. However, more careful tests show that there is a high degree of central unity of purpose and of motor command in generation of direction of interest and of orientation or prehension. Indeed, the undivided mechanisms of the brain are capable of exercising powerful directive influence over the awareness and memory of the hemispheres.

Levy and I have used the chimeric stimuli to demonstrate that each

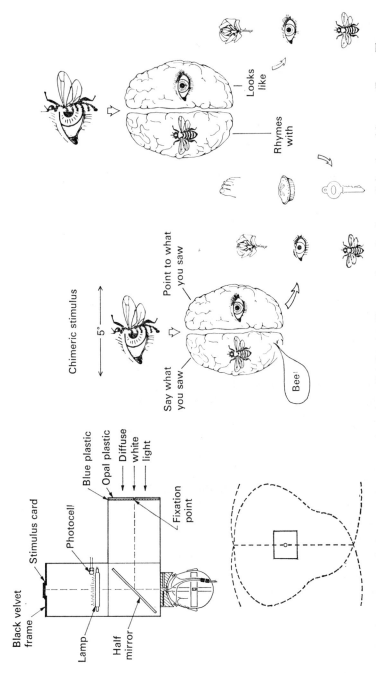

Fig. 5. Tachistoscopic (150 msec) presentation of small (5°) stimuli centred on the fovea. Stimulus chimeras (Levy, Trevarthen & Sperry, 1972; Levy & Trevarthen, 1977) facilitate tests of hemispheric differences in cognitive processes. Only the left hemisphere can generate the phonological image of the name of an object seen, to say what it is, or to guide pointing (without speaking) to a picture of a rhyming object (e.g. 'bee' matches 'key'). Direct visual matching is done more efficiently by the right hemisphere. Lower left: visual field and size of stimulus.

hemisphere has both a superiority and a preference for carrying out one class of cognitive process, matching of appearances in the right hemisphere and analysing functional significance in the left hemisphere, but that a metacontrol process, presumably involving instinctive brainstem functions of limited cognitive potential, can override the process of adaptation and generate an inappropriate lateralization of brain activity. When this happens the subject makes poor choices and cannot comprehend what he is asked to do (Levy & Trevarthen, 1976).

This test and related ones on visual search and vigilance which are impaired in commissurotomy subjects indicate that concepts of information processing based on a linear or cascade sequence of steps in coding and categorizing sensory information cannot be applied without severe qualification to the brain mechanisms of perception and cognition. Intrinsic organizations and spontaneously generated schemata widespread in the brain appear to have a decisive influence over the way patterns of stimulation are experienced. The split-brain subject does not have two entirely separate perceptual processing mechanisms nor double capacity for information uptake.

The innate principle that distinguishes mental activities of the hemispheres still remains imperfectly understood. Recent experiments by Zaidel (1978) using a contact lens device to keep vision to the left- or right-half field while the subject makes inspection eye movements over a text or picture, and chimeric tests by Levy & Trevarthen (1977) of elementary language processes support the idea that unilateral generation of auditory phonological images of words to be uttered is the key factor confining vocal expression to the left half of the divided brain. The disconnected right brain cannot 'imagine' the sound of a word seen in print or the name of an object perceived, even though the name seen or heard is recognized by that hemisphere as an attribute identifying the object as might the colour of butter or the form of a matchbox. Similarly it would appear as if the immediate and precise visual guidance of the dominant hand is due to a unique mechanism in the left hemisphere. Touch guidance of finger movements in the imagining of an extended image of a form to be handled is more strongly carried out by the left hand and right brain. But other results are paradoxical. There is no simple identification of a hemisphere with special abilities in one modality of sense.

The left hemisphere is weaker in guiding the right hand to draw what the subject has seen freely, the right hemisphere is weaker at controlling writing. The former seems to reflect a superior shape-conceiving mechanism in the right and the latter may depend on assistance normally given to writing by an interior phonological image of the words in the left. The left hemisphere cannot perceive and/or remember faces of strangers or other nameless shapes as well as the right, but it appears to be better at systematically exploring a form by a tactic that relates separate familiar elements – 'features' or 'units of structure'. Underlying these differences must be principles of intention built into the genetics for brain growth because about 90 % of people have the right hand dominant and control of speech and language in the left hemisphere (Levy, 1976).

A particularly interesting recent finding concerns the geometric talents of left and right hemispheres of commissurotomy patients who have only very moderate education (Franco & Sperry, 1977). Franco has found that the hemispheres differ in their recognition of geometric similarity related to the mathematical rules for space by which forms are constructed. All forms are grasped or understood better by the right hemisphere, but there was a gradient for this lateralization. The more rule-bound, more verbalizable Euclidean forms are best understood by the left hemisphere and forms in topological space are most strongly apprehended by the right brain. Forms relating to projective and affine groups produce intermediate results. Evidently these different mathematical groups engage differences in the analytical capacities of the hemispheres. These capacity differences are intuitive, not inculcated by the education of these patients who had had no lessons in the relevant geometry.

Normal duplicity of brains

The sensational and very clear findings with commissurotomy patients have helped stimulate experimental studies of normal subjects of different ages and of patients with lateralized cerebral lesions or pathological degenerations in the brain. The experiments attempt to quantify the principles by which the human brain is built asymmetrically. They confirm that people with normal intact brains have lateralized half-brain functions hidden inside their apparently unified and freely oriented intelligence (White, 1969). Large populations of subjects permit inferences about inherent differences in brain growth in the two sexes, between left and right handers, and changes in the degree and quality of the lateralized mental processes in the course of childhood (Corballis & Beale, 1976). They illuminate the causes of abnormal mental conditions such as dyslexia which clash with any uniform educational practice.

The studies with children prove that the lateralized functions of the brain, first formed in the foetus, undergo slow maturation. This correlates well with the known cycle of maturation of the corpus callosum and of areas of the association cortex richly cross-connected by the commissures (Trevarthen, 1974a, 1978b). There is evidence that if the corpus callosum fails to form in the early foetus, then the hemispheres do not differentiate. If the brain growth is disturbed by emotional disturbance or illness of the mother when the hemispheres are actively differentiating, this may affect the cerebral dominance of the infant. Evidently the commissure-association system astride the gap between the mirror symmetric sets of projections from the sense organs and to the motor system differentiates into a progressively more diversified generator for cognitive and intentional schemata that segregate more and more strongly between the hemispheres of a child.

Although it takes more than a decade for the brain's asymmetry to mature, there are large differences in the hemispheres well before birth. The left hemisphere of most foetal brains has enlargements in the regions that years later come to control speaking and hearing language (Wada, 1969, cited in Geschwind, 1974; Le May & Culebras, 1972; Witelson & Pallie, 1973)(Fig.1). Complementary enlargements in the right hemisphere, of unknown function, probably relate to its special visuo-spatial and constructive abilities. There is evidence that other congenital anatomical features correlate with the various forms of handedness that are only indirectly related to the lateralization of language (Levy, 1976). These interacting elements in the brain together produce many different forms of mind that together aid each other in human social achievements. Education, job training, etc. furthers the specialization of the different minds that contribute to the success of the community as a whole. The analogy with differentiations and reintegrations of a developing organism is strikingly close.

Because of this psychobiology of split brains, or at least helped by it, there is a shift of opinion in psychology away from both behaviourist and idealist explanations of human faculties and differences of personality. Philosophers argue about the meaning to be given to an operation that can divide consciousness. Sperry (1976, 1977) has been led to assert that his work proves that consciousness is not dependent on language and communition with others to achieve knowledge of self because the right hemisphere has a profound and far-ranging consciousness that cannot talk, and that conscious processes, innate in principle to the brain, have a primary determining role in brain function, overruling processes at the level of neurones or reflex networks. He rejects the generally held view that human awareness is an emergent or epiphenomenal outcome of impulses running through a nerve net. It is a consequence of the form of organs of consciousness in the nerve net and that, while innate in basic plan, grow with the aid of experience.

In man there seem to be many kinds of purpose as there are many forms of action to gain experience. These consciousness-making acts are controlled by cerebral hemispheres that have developed complementary forms of purposiveness and complementary ways of attending to the world.

The attempts currently being made to explain cerebral hemispheric asymmetry have been led in a most intriguing direction. Apparently the evolution of man has been based on the development of new powers of seeing purpose in others. Attentive, purposeful action by one conscious agent may reveal itself to another. This is the basis of communication and of cooperative intelligence in animals and it is carried to its greatest achievements in man. What one human does another understands and humans can gesture and speak to transfer what they know. It seems that the special brain mechanisms for this transfer of awareness by transmission of new, highly informative effects of moving have become segregated from those for gaining experience directly from the world. Thus the left hemisphere is better in most persons at monitoring expressive trains of movement and at detecting them in others. Cerebral asymmetry in humans probably evolved, therefore, in relation to a unique development of cooperative intelligence. The main factor at the start seems to have been a bias of sub-humans to choose one hand for gestures specially adapted to communication, and this has led to one-sided control for speech and for sequencing all signal movements in speech or gesture, or in playing a musical instrument. Asymmetries recently found in the brain anatomy and behaviour of apes seem to fit the hypothesis that it was watching to see what another is doing or expressing that set evolution on the way towards near consistent one-sidedness in the structures for language.

EVIDENCE FROM INFANCY

Intending to know

The human body is completed in a rudimentary but distinctly human form in eight weeks. The main regions of the nervous system are also differentiated in the embryo. Yet it takes a further 32 weeks, four times longer, for a human being to be ready for birth. In the foetus most development is in the tissues of the brain, especially in the cerebral hemispheres and cerebellum (Fig. 1). It is impossible for a biologist to observe the growth of the brain in the foetus and not conclude that major functions of the human mind are sketched out before birth. What is constructed then determines what a human infant will do and how it will begin responding to and regulating experience of the outside world.

Remarkably, this conclusion has had little authority in psychological thinking. Some powerful doubt or taboo has ruled to exclude the possibility that the human infant is born wise in principle to the opportunities of the world and to the human and animal agents that inhabit it. Now this is changing, thanks to a renaissance of careful imaginative study of what infants actually do to regulate their experience. A deliberate effort is at last being made as well to observe how human interpersonal communication comes about. Thus a bridge is being built between the psychological processes of human social life and the biology of the brain.

Piaget stimulated modern research in the psychology of infants by taking the position that concepts of objects and representations in the mind of the identities and functions of objects could emerge in infancy before language and before thinking based on speech. His remarkably detailed and ingenious examinations of how his own infants would act to control experience and solve conceptual puzzles showed him how the uptake of experience was continually under control of acts (Piaget, 1953). Looking at and tracking objects to observe their motions and changes, and reaching to and grasping them to handle, see, hear, mouth and taste them were inborn strategies of the infant which led inexorably to formation of representational schemata in the mind.

Piaget proposed that infants less than nine months of age were reflex or 'sensori-motor' organisms lacking intentional structures. He felt that the development of reaching for a seen object, for example, required discovery by chance of the hand within the reflex-orienting field of the eyes. The orienting reflexes of the forelimb directed by touch, especially on the hand itself, became associated with the visual reflexes by chance encounters of eyes with hand. Images of objects could not exist in the child's mind until these responses to immediate circumstances had left 'internalized' prediction of their joint action.

In the last 10 years or so, with the aid of film and television and with carefully controlled experiments, it has been found that Piaget's inspired concepts of mental development underestimated the newborn human brain. By carefully recording adjustments of orientations of eyes and head to anticipate experiences or to choose between consequences, Bower (1974), for example, has shown that even in the third or fourth month a baby is 'aware' of the three-dimensionality of objects and their movements in space (Fig. 6). The brain transforms the patterns of retinal stimulation into a 3-D representation of the nearby world and objects are distinguished from their backgrounds, at least when they are in motion, before the infant is capable of grasping and manipulating them.

A newborn is able to coordinate visual explorations in auditory space.

Within a few weeks objects are perceived to have some continued existence when they are concealed by going behind a screen. A two-month-old baby shows various signs of surprise or heightened curiosity if an object changes appearance radically as it moves behind a screen, but not if a thing simply turns about by rotating. To achieve such predictions and recognitions the baby must carry, at least for a short time, some image of the object's essential features and a notion of what changes of appearance distinguish a real change in identity from mere displacement in space.

Conditioning tests show that learning begins immediately after birth and it must play a big role in development of the models for reality in the mind from the start. But no learning theory can explain the intuitive readiness of infants to coordinate experience by various parts of the body.

Within a month of birth babies show regular prototypes of the intended movements of adults. A four-week-old cannot walk, sit up to look about or reach and grasp an object, but all of these show highly differentiated precursor patterns in the spontaneous movements a baby makes when awake but quiet (Fig. 6). Looking with two eyes in precise conjugation to track an object or to select places for fixation by central (foveal) vision are present at birth, paced in time like adult-looking movements and already coordinated with weakly developed aiming movements of the head (Trevarthen, 1974c). Films of neonates show outlines of stepping and reaching and grasping and these movements are already coordinated in some measure both spatially and temporally with orienting to look at or listen for (Trevarthen, 1974d).

We must conclude that a general field of space and time for acting with all parts of the body together is provided for in the brain before there is any outside world to perceive. Knowing objects are out there to be reached to or walked to and then put in the mouth or manipulated has a powerful basis in these motor patterns. Apparently the strategy of mental development requires that a rudimentary power of motor prediction or intentionality has to be laid down in the cerebral structures before perception can develop its images of reality (Trevarthen, 1978b).

The infant as a person

Of all the behaviours of infants, those attracting most attention at present are the ones directed adaptively towards persons and to communicating with them (Trevarthen, 1974e). Piaget has never studied social behaviour specifically and he treats the infant as a cognizing or knowing agent developing schemata for things and events in a social milieu but out of his own efforts. According to him there is no language-like communication before infants develop representational thought for themselves in the second year. Piagetian symbolic functions are made by socializing the mental representations developing in the child's brain in the first two years and beyond. Most current educational practice accepts this approach as basic.

However, observation of infants in the first year suggests that of all the preadaptations to voluntary life those for communication as a person with other persons are the most elaborate and the most precociously manifested in infancy. Infants have astonishing capacities for perceiving other persons when very young. They are primed to make selective responses to the odour, warmth, rhythmical movements, voice sounds and face configurations of a woman minutes after birth. They 'learn' highly specific features of their own caretaker, almost always their biological mother, beginning within hours. By two or three months they show clear predispositions to watch the eyes mouth and hands of persons who address them, and controlled experiments prove, what most mothers believe, that babies have refined powers of perceptual discrimination for differently articulated sounds of speech (Trevarthen, 1978c).

Now that adequate recordings have been made it is obvious that human infants are endowed with an extremely complex set of actions preadapted to communication with others (Trevarthen, 1974d,e, 1978c). They make, even as premature newborns, almost the full range of face movements for emotional and other kinds of expression, jumbled but remarkably differentiated. We have found that movements of lips and tongue specifically adapted to speaking become common when cooing and gestures emerge within the rich repertoire of spontaneous emissions that two-month-olds make. These socially useful movements are not mature. They are not used to give specific messages with reference to the world the baby shares with people. But they are already sensitive to what people do when trying to communicate. An extraordinary finding is that a one-month-old may imitate a visually perceived face movement or a hand movement. That means that the face and hands are represented in the brain of the baby by a mechanism that can generate a movement to match one seen. The perceptual image and the motor image are linked. The anatomical basis for this must have great power in establishing the infant's rapport with others and in setting the foundations for a self-concept. It cannot possibly be learned though learning will automatically contribute to its differentiation.

All the expressive movements of a two-month-old may be regulated by actions of others (Fig. 7). Films show that the talking, moving, touching of a mother clearly interest and influence what the infant does, but the infant is still spontaneously generating smiles, pre-speech, coos, hand waving or pointing movements and a wide variety of associated postural displays in

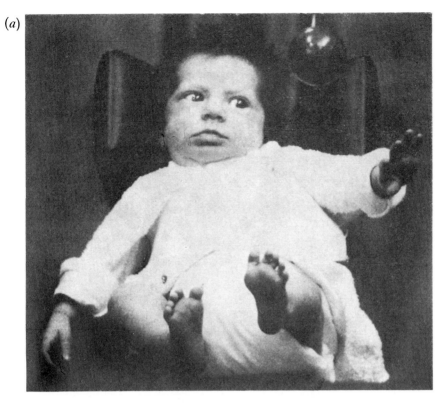

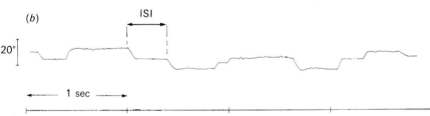

Fig. 6. (a) A one-month-old girl focuses on and tracks a moving ball, moving as if to reach for it and step on it.
(b) Electro-oculogram of a nine-day-old shows conjugate saccades almost identical with adult eye movements and with essentially the same periodicity (intersaccade interval, ISI).

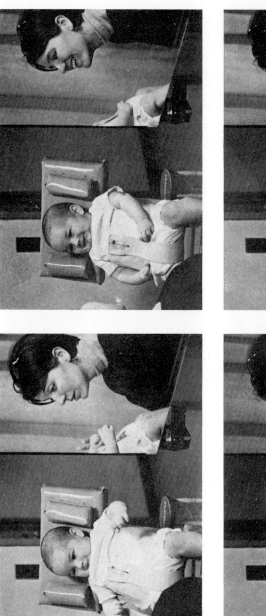

Fig. 7. Person–person play at two and a half months. A complex, highly regulated 'dialogue' is generated with his attentive mother.

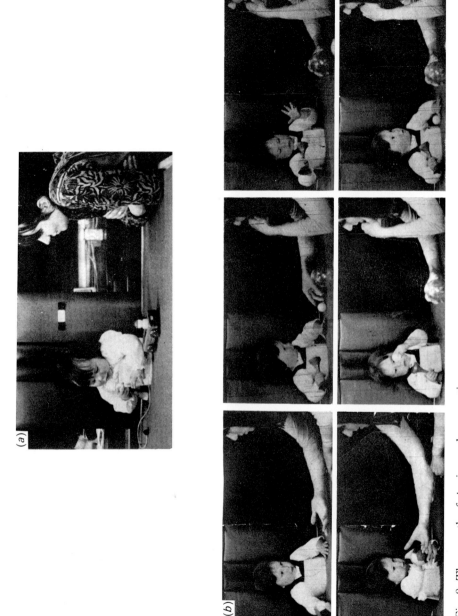

Fig. 8. The growth of sharing and cooperation.
(a) At eight months a little girl, happily engrossed in play, is watched by her mother. (b) At ten months she has begun to give when asked, to take a suggestion, to share a joke.

'reply' to the mother (Stern, 1974). To achieve this kind of interaction which may appear to be a prototype of adult conversational behaviour with both partners generating 'utterances', it is necessary for the adult to adapt expression in a special way. We find that the 'baby talk' of mothers and all her movements of attentive, solicitous and playful action constitute a subtle adaptation by the mother to infant perceptions of human messages. It seems certain that, like the infant, the mother has an innate aptitude for developing her part in this highly asymmetric form of human interaction. The result of success in interpersonal contact established so early in life is a unique relationship in which all sorts of communicative skills may be practised by both partners, but the basis is certainly not learned.

Making culture happen

We have been following the detailed development of infant communicative powers in the first year (Trevarthen, 1977a; Trevarthen & Hubley, 1978). There are complex adjustments between intentions towards objects and those towards people in the infant's brain. The developments appear strongly regulated in the infant's mind so that they follow a predictable timetable. I shall single out here one remarkable change which appears to show the inherent foundations for being taught and for becoming part of a human cultural achievement. We seem to have found a psychobiological basis for the transmission of knowledge and technique by education, which is the fundamental self-maintaining process of human life in organized societies (Fig. 8).

We have never seen an infant under about nine months of age give an object to another person with a smile to show the act was intended as a giving, nor ask a partner to help in doing something with an object. A six-month-old may take an object when it is offered and will join in a game in which a toy is a vehicle of play being animated by the playmate. The early form of communication seen at two or three months, which proves that infants see people as possible partners in a dialogue-like exchange of expressions and utterances, is complicated after four months by development of a heightened awareness of the possibilities for experiencing physical things in surroundings. Often the infant shows a marked unwillingness at six months for mixing interest in persons with interest in things. Nevertheless the more complex powers of predicting and recognizing can be exercised in games and a six-month-old is able to express complex feelings of joy, satisfaction, dismay, annoyance, refusal within the course of a game with another person. Games regulate the integration of control by the baby of intentions to experience alone in certain ways with his control of intentions to share with others the feelings that accompany those experiences.

At nine or ten months something happens in the infant's mind which transforms this highly dynamic balance of interests. For the first time the infant can combine a project related to use of an object with an exchange of feelings or intentions with another person. As a consequence he or she will for the first time follow directions how to combine or move objects, simultaneously beginning to respond appropriately and with evident pleasure to the names of familiar objects or acts, and the mother can thenceforth direct the baby's interests with a word alone. The vocalizations and other expressions of the baby encompass the new understanding of how to communicate about things or goals. The linguist Halliday calls the ability to express intentions this way 'protolanguage' because it follows basic grammatical forms.

These findings appear to show that it is a development in the mind of the child that opens the way towards predicative communication and co-operating in the use and exchange of knowledge about the world. A baby under one year can engage in a form of cooperative intelligence beyond any sub-human primate, but related to the extraordinary powers of chimps for sharing knowledge of the world (Menzel, 1971). The fact that a baby human can be taught by another how to use elementary objects of his culture so early and that he shows spontaneously the necessary interest to do so at a definite stage of development is, I think, evidence that human culture is not just a learned acquisition of man – it is something he is born to explore and extend as he grows through childhood in partnership with his society (Trevarthen, 1977b).

Developments of tissue in the brain gain the power through behaviour to take part in a cooperative mastery of the environment, a mastery that seems to have no limit. Cultural development could go on for ever, but it must remain dependent on the innate functions that control its generation in the brain.

CONCLUSION

Servo-mechanisms are recognized to be essential in all the levels of biological processes. Gene mechanisms in the nucleus, metabolic organelles in the cytoplasm, organs, bodies, brains, conscious ideas, concepts, social systems, all are homeostatically regulated in a coherent hierarchy. But observing this does not explain directed change; it does not explain development. There has to be some push to keep the self-regulations moving and combining, and there has to be some frame of boundary states to give masses of interacting elements pattern or shape. Evidently developmental change arises by a principle or mechanism that sets goals beyond what is presently attainable. Forward-looking organizations among elements created by growth keep the motivations going for increasingly organized systems.

A baby is born with world-imaging eyes connected to a brain that can turn a visual image of the world in light into a plan for movements. Both eye and brain were laid out by intercellular negotiations months before. Human awareness and the impulse to act on it and talk about it are anticipated in a double cerebrum. The rudiments of immensely prepotent categories of thought, like language and topological understanding of form, are segregated before experience begins.

But the most remarkable of all anticipations, the hardest to believe, is the mental anticipation of human culture and its evolution in society. No baby is born knowing what human history has been or will be like, but that there will be a history of discoveries, conventions, institutions and co-operative achievements is evidently a certainty inside every baby's head. All the passions and motives, pure and impure, that vitalize the life of humans together seem to be outlined as embryonic rudiments in the response of a one-year-old to other persons.

When we consider these facts there appears to be no reason, except our ignorance or the limits of our scientific imagination, to explain separately the biological and the mental.

REFERENCES

Benzer, S. (1973). Genetic dissection of behaviour. *Scien. Am.* **229**, 14–37.

Bogen, J. (1969). The other side of the brain. I: Dyscographia and dyscopia following cerebral commissurotomy. *Bull. Los Ang. neurol. Soc.* **34**, 73–105.

Bower, T. G. R. (1975). *Development in infancy.* San Francisco: Freeman.

Brinkman, J. & Kuypers, H. G. J. M. (1973). Cerebral control of contralateral and ipsilateral arm, hand and finger movements in the split-brain rhesus monkey. *Brain* **96**, 653–74.

Chung, S. H. & Cooke, J. (1975). Polarity of structure and of ordered nerve connections in the developing amphibian brain. *Nature, Lond.* **258**, 126–32.

Corballis, M. C. & Beale, I. L. (1976). *The psychology of left and right.* Hillsdale, N.J.: Erlbaum.

Downer, J. L. de C. (1959). Changes in visually guided behaviour following midsagittal division of optic chiasm and corpus callosum in monkey (*Macaca mulatta*). *Brain* **82**, 251–9.

Franco, L. & Sperry, R. W. S. (1977). Hemisphere lateralization for cognitive processing of geometry. *Neuropsychologia* **15**, 107–14.

Gaze, R. M. (1970). *Formation of nerve cell connections.* New York: Academic Press.

Gazzaniga, M. S. (1970). *The bisected brain.* New York: Appleton-Century-Crofts.

Geschwind, N. (1974). The anatomical basis of hemispheric differentiation. In *Hemispheric function in the human brain*, ed. S. J. Dimond & J. G. Beaumont, pp. 7–24. London: Paul Elek.

Halliday, M. A. K. (1975). *Learning how to mean: explorations in the development of language*. London: Eward Arnold.

Hunt, R. K. (1975). Developmental programming for retinotectal patterns. In *Ciba foundation symposium on cell patterning new series*, vol. 29, pp. 131–59. New York: American Elsevier.

Ingle, D. & Sprague, J. M. (eds) (1975). *Sensorimotor function of the mid-brain tactum*, Neurosciences Research Program Bulletin, vol. 13, pp. 169–288. Cambridge, Massachussetts: Neurosciences Research Program.

Kuypers, H. G. J. M. (1973). The anatomical organization of the descending pathways and their contributions of motor control, especially in primates. In *New developments in electromyography and clinical neurophysiology*, ed. T. E. Desmedt, vol. 3, pp. 38–68. Basel: Karger.

Le May, M. & Culebras, A. (1972). Human brain – morphological differences in the hemispheres demonstrable by carotid arteriography. *New Engl. J. Med.* **287**, 168–70.

Levy, J. (1974). Psychobiological implications of bilateral asymmetry. In *Hemisphere function in the human brain*, ed. S. Dimond & J. G. Beaumont, pp. 121–83. London: Paul Elek.

Levy, J. (1976). A review of evidence for a genetic component in the determination of handedness. *Behav. Genet.* **6**, 429–53.

Levy, J. & Trevarthen, C. (1976). Metacontrol of hemispheric function in human split-brain patients. *J. exp. Psychol. Hum. Percept. Perform.* **2**, 299–312.

Levy, J. & Trevarthen, C. (1977). Perceptual, semantic and phonetic aspects of elementary language processes in split-brain patients. *Brain* **100**, 105–18.

Levy, J., Trevarthen, C. & Sperry, R. W. (1972). Perception of bilateral chimeric figures following hemispheric deconnection. *Brain* **95**, 61–78.

Luria, A. R. (1973). *The working brain*. Harmondsworth: Penguin Paperbacks.

Mark, R. F. (1974). *Memory and nerve cell connections*. London: Oxford Univ. Press.

Menzel, E. W. (1971). Communication about the environment in a group of young chimpanzees. *Folia Primatol.* **15**, 220–32.

Mishkin, M. (1972). Cortical visual mechanisms and their interaction. In *Brain and human behaviour*, ed. A. G. Karczmar & J. C. Eccles, pp. 187–208. New York: Springer-Verlag.

Myers, R. E. (1956). Function of corpus callosum in interocular transfer. *Brain* **79**, 358–63.

Myers, R. E. (1962). Transmission of visual information within and between the hemispheres. In *Interhemispheric relations and cerebral dominance*, ed. V. B. Mountcastle, pp. 51–73. Baltimore: Johns Hopkins Press.

Nebes, R. D. (1974). Hemispheric specialization in commissurotomized man. *Psychol. Bull.* **81**, 1–14.

Paiget, J. (1953). *The origins of intelligence in children*. London: Routledge & Kegan Paul. (Original French edition, 1936.)

Paiget, J. (1962). *Play, dreams and imitation in childhood*. New York: Basic Books.

Piaget, J. (1971). *Biology and knowledge*. Edinburgh: Edinburgh Univ. Press.

Schneider, G. E. (1967). Contrasting visuomotor functions of tectum and cortex in the golden hamster. *Psychol. Forsch.* **31**, 52–62.

Spemann, H. (1938). *Embryonic development and induction*. New Haven: Yale Univ. Press.

Sperry, R. W. (1961). Cerebral organization and behaviour. *Science* **133**, 1749–57.

Sperry, R. W. (1965). Embryogenesis of behavioral nerve nets. In *Organogenesis*, ed. R. L. De Haan & H. Ursprung, pp. 161–86. New York: Holt.

Sperry, R. W. (1968). *Mental unity following surgical disconnection of the cerebral hemispheres.* The Harvey Lectures. New York: Academic Press.

Sperry, R. W. (1970). Perception in absence of the neocortical commissures. *Ass. Res. Nerv. Ment. Dis.* 48, 123–38.

Sperry, R. W. (1974). Lateral specialization in the surgically separated hemisphere. In *The neurosciences: third study program,* ed. F. O. Schmitt & F. G. Warden. Cambridge, Massachussetts: MIT Press.

Sperry, R. W. (1975). In search of psyche. In *The neurosciences: paths of discovery,* eds. F. G. Warden, J. P. Swazey & G. Adelman, pp. 425–34. Cambridge, Massachussetts: MIT Press.

Sperry, R. W. (1976). Mental phenomena as causal determinants in brain function. In *Consciousness and the brain,* ed. G. Globus, G. Maxwell & I. Savodnik. New York: Plenum.

Sperry, R. W. (1977). Forebrain commissurotomy and conscious awareness. *J. Med. Phil.* 2, 101–26.

Sperry, R. W., Gazzaniga, M. S. & Bogen, J. E. (1969). Interhemispheric relationships: the neocortical commissures; syndromes of hemisphere deconnection. In *Handbook of clinical neurology,* ed. P. J. Vinken & G. W. Bruyn, vol. 4, pp. 273–90. Amsterdam: North-Holland.

Stern, D. N. (1974). Mother and infant at play: the dyadic interaction involving facial, vocal and gaze behaviours. In *The effect of the infant on its caregiver,* ed. M. Lewis & L. Rosenblum. New York: Wiley.

Trevarthen, C. (1962). Double visual learning in split-brain monkeys. *Science* 136, 258–9.

Trevarthen, C. (1965). Functional interactions between the cerebral hemispheres of the split-brain monkey. In *Functions of the corpus callosum,* ed. E. G. Ettlinger, pp. 24–40. Ciba Foundation Study Group No. 20. London: Churchill.

Trevarthen, C. (1968). Two mechanisms of vision in primates. *Psychol. Forsch.* 31, 299–337.

Trevarthen, C. (1973). Behavioural embryology. In *Handbook of perception,* vol. 3, ed. E. C. Carterette & M. P. Friedman, pp. 89–117. New York: Academic Press.

Trevarthen, C. (1974a). Cerebral embryology and the split brain. In *Hemispheric disconnection and cerebral function,* ed. M. Kinsbourne & W. L. Smith, pp. 208–36. Springfield, Illinois: Charles C. Thomas.

Trevarthen, C. (1974b). Analysis of cerebral activities that generate and regulate consciousness in commissurotomy patients. In *Hemisphere function in the human brain,* ed. S. J. Dimond & J. G. Beaumont, pp. 235–63. London: Paul Elek (Scientific Books) Ltd.

Trevarthen, C. (1974c). L'action dans l'espace et la perception de l'espace: mecanismes cerebraux de base. In *De l'espace corporel a l'espace edologique,* ed. F. Bresson *et al.,* pp. 65–80. Paris: Presses Universitaires de France.

Trevarthen, C. (1974d). The psychobiology of speech development. In *Language and brain: developmental aspects,* Neurosciences Research Program Bulletin, vol. 12, ed. E. H. Lenneberg, pp. 570–85.

Trevarthen, C. (1974e). Conversations with a two-month old. *New Scien.* 62, 230–5.

Trevarthen, C. (1975). Psychological activities after forebrain commissurotomy in man. Concepts and methodological hurdles in testing. In *Les syndromes de disconnection calleuse chez l'homme,* ed. F. Michel & B. Schott, pp. 181–210. Lyon: Hopital Neurologique.

Trevarthen, C. (1977a). Descriptive analyses of infant communication behaviour. In *Studies in mother–infant interaction: the Loch Lomond symposium,* pp. 227–70. London: Academic Press.

Trevarthen, C. (1977*b*). Instincts for human understanding and for cultural co-operation: their development in infancy. Paper presented at the Werner-Reimer-Stiftung Symposium on Human Ethology, Bad Homburg, West Germany, 25–29 October 1977. (To be published.)

Trevarthen, C. (1978*a*). Modes of perceiving and modes of acting. In *Psychological modes of perceiving and processing information*, ed. H. J. Pick, pp. 99–136. Hillsdale, N.J.: Erlbaum.

Trevarthen, C. (1978*b*). Neuroembryology and the development of perception. In *Human growth: a comprehensive treatise*, vol. III. New York: Plenum. (In Press.)

Trevarthen, C. (1978*c*). Communication and cooperation in early infancy. A description of primary intersubjectivity. In *Before speech: the beginnings of human communication*, ed. M. Bullowa. London: Cambridge Univ. Press. (In Press.)

Trevarthen, C. (1978*d*). Manipulative strategies of baboons and origins of cerebral asymmetry. In *Asymmetrical function of the brain*, ed. M. Kinsbourne. London: Cambridge Univ. Press. (In Press.)

Trevarthen, C. & Hubley, P. (1978). Secondary intersubjectivity: confidence, confiding and acts of meaning in the first year. In *Action gesture and symbol: the emergence of language*, ed. A. Lock. London: Academic Press. (In Press.)

Trevarthen, C. & Sperry, R. W. (1973). Perceptual unity of the ambient visual field in human commissurotomy patients. *Brain* 96, 547–70.

Waddington, C. H. (1961). *The nature of life*. London: Unwin.

White, M. J. (1969). Laterality differences in perception: a review. *Psychol. Bull.* 72, 387–405.

Whitehead, A. N. (1925). *Science and the modern world*. New York & London: Macmillan.

Witelson, S. F. & Pallie, W. (1973). Left hemisphere specialization for language in the newborn: neuroanatomical evidence of asymmetry. *Brain* 96, 641–7.

Willshaw, D. J. & von der Malsburg, C. (1976). How patterned neural connections can be set up by self-organization. *Proc. R. Soc., B* 194, 431–45.

Zaidel, E. (1978). Auditory language comprehension in the right hemisphere following cerebral commissurotomy and hemispherectomy: a comparison with child language and aphasia. In *The acquisition and breakdown of language: parallels and divergences*, ed. E. Zurif & A. Caramazza. Baltimore: Johns Hopkins Press. (In Press.)

Claude Bernard's internal environment revisited

J. R. ROBINSON

Next year we have the centenary of Claude Bernard's death. When he died on 10 February 1878 he was correcting the proofs of his book on the phenomena of life common to animals and plants. The book appeared later in the same year, and it contained the famous statement '*La fixité du milieu intérieur est la condition de la vie libre, indépendente*'. The constancy of the internal environment is the condition for free and independent life. Bernard went on 'All the vital mechanisms, varied as they are, have only one object, that of preserving constant the conditions of life in the internal environment.' This principle of the constancy, or the stability, of the internal environment has become so much a part of the fabric of physiology that it is usually taken for granted nowadays. It seems, then, not too bad an idea to try to take stock, to see how Claude Bernard's century-old ideas have worn, and whether they need patching or extending.

We usually present the internal environment to our students as a sort of water bath or chemostat comprising the blood plasma and the other extracellular fluids. These provide a secluded habitat in which the cells of a higher animal can live out their own existence untroubled by the vicissitudes of a changeable and sometimes hostile external world. I interpret free and independent life to mean that multicellular animals can live, and men can work, and even engage in athletic contests, all over the world, including some pretty inhospitable places, only because their cells do not have to cope with the local climate. A mammal's cells live in a relatively small and very overcrowded pond of extracellular fluid. Many of them are highly specialized and very sensitive to changes in their immediate surroundings. They are therefore very exacting in the demands they make on the internal environment, for food and oxygen; for stability of osmotic pressure, reaction and temperature; and for the removal of their waste products. The vital mechanisms really have their work cut out to keep the internal environment fit for the cells to live in! W. B. Cannon coined the term 'homeostasis' for maintenance of stability and described the mechanisms which work for it as 'homeostatic'. His (1932) book, *The wisdom of the body*, had chapters on: the fluid matrix; thirst and hunger; the water content and the salt content of the blood; the blood sugar, the blood protein, the blood fat, the blood calcium; oxygen supply, neutrality, body temperature; and the role of the sympathico-adrenal system in homeostasis. A great deal of physiology can be presented as the story of the mechanisms which maintain the constancy of the internal environment. This is a good way for students to organize their knowledge and to arrive at an integrated picture and understanding of the subject.

Although the principle of the constancy of the internal environment was not clearly enunciated until 1878 its beginnings were much earlier. Claude Bernard introduced the idea of an internal environment into his teaching of general physiology in the 1850s. He referred to the cells as 'living molecules' or as the 'histological elements of the tissues'. He taught that these were the primary living units, and also that they had to live in a watery medium. Most of them are buried deep inside the bodies of higher animals; yet they all need things like water, foodstuffs and oxygen, which exist in the world outside the animal. Bernard saw that the blood offers a means of bringing things from the external environment within reach of the cells, most of which are fixed and need to have things carried to and from them in solution. He taught that it is only by having a liquid internal medium for their cells to live in that higher animals can exist in the atmosphere.

Claude Bernard was also a fine experimental biologist; and one of his chief aims was to bring physical and chemical measurements to bear on the phenomena of life. He believed that cells were made of ordinary substances obeying the general laws of physics and chemistry. But because chemical analysis inevitably killed and destroyed cells, he regarded the interior of the cells as inaccessible to direct experiment. However, he also firmly believed that the life of the cells was a product of continual interaction with their immediate surroundings. 'The vital properties of the tissues', he said 'are manifested through the physico-chemical properties of their milieu.' If this milieu was a liquid that could be removed and analysed, at least one aspect of the life of the cells would be accessible to investigation (cf. Holmes, 1963).

Bernard was struck by the fact that living things do not respond to external changes as non-living things do. This made sense because, as he put it, 'the living organism does not really exist in the *milieu extérieur*, but in the liquid internal medium formed by the circulating organic fluid which surrounds and bathes all the tissue elements'. Hence living things should not respond directly to external changes – only to changes in the internal environment in contact with the cells. And he was struck by the fact that the blood, which was perhaps the most accessible part of this internal environment, had rather constant properties. He had measured the temperature of the blood inside the vessels, and found that it became a little warmer going through the liver, or through the gut during digestion; and a little cooler going through the lungs. But these were small differences. The average temperature of the blood was remarkably constant.

The chemical composition of the blood was also beginning to be investigated, and he was impressed that this, too, was remarkably constant. So gradually the blood, and the associated extracellular fluids in closer contact with the tissue elements, came to appear as a bland, rather uniform medium which could nourish the cells and protect them from changes in the world outside. There was a possible contradiction here, because the protecting, insulating, internal medium also had to serve as a line of communication with the external world (Holmes, 1967). But although it brought the external environment into the animal, it was itself not part of the external environment. Bernard himself had discovered that the blood contained a specific sugar, glucose, which did not come from the diet, but was made in the liver. And so he developed the idea that the internal environment was not just external fluid that had got trapped inside the animal's body, but was, as he put it in 1865 in his *Introduction to the study of experimental medicine*, 'a true product of the organism which preserves the necessary relations of exchange and equilibrium with the external cosmic environment'. Sir Charles Sherrington paraphrased this in *Man on his nature*: 'Life overcame the obstacles of unfavourable environment by manufacturing a piece of environment suited to itself and carrying it about with it.'

Bernard placed most stress on the constancy of the internal environment just before he died, in his lectures on the common phenomena of life. He defined three modes of life: latent, as in seeds, where life is not evident; oscillating, when the manifestations of life vary with external conditions, as in coldblooded animals; and constant, when life is relatively independent of external changes. He claimed that he was the first to urge that animals really live in two environments, an external one in which they are placed, and an internal one in which their cells live: and that the mechanisms which enable life to be free and independent of external changes are just those that ensure that the internal environment provides the conditions necessary for the life of the cells. He specified these conditions as water, oxygen, heat and nutritive chemical substances or reserves. He added 'These are the same conditions as are necessary for life in simple organisms; but in the perfected animal whose existence is independent, the nervous system is called upon to regulate the harmony which exists between all these conditions.'

Looking back over the 20 or 30 years during which Claude Bernard's thoughts were maturing, Holmes (1967) concluded that his great principle that the constancy of the internal environment is the primary condition for free and independent existence just grew out of his whole experience. He arrived at it intuitively rather than by any formal process of logic. Perhaps this is a genuine, and triumphant, example of induction! Paul Bert (1878) said of Bernard that no-one ever made discoveries more simply; and quoted one of his students; 'He seemed to have eyes all round his head. In the course of an experiment, students were stupefied when they saw him point out quite evident phenomena which no-one but himself had seen. He discovered as others breathed.' L. J. Henderson (1927) remarked that there was no better illustration of Claude Bernard's penetrating intelligence than the way in which 'a few scattered observations on the composition of the blood served to justify, in his opinion, the assertion that the constancy of the internal environment is the condition of free and independent life.'

This principle has been generally accepted; though Sir Joseph Barcroft, whom I had the good fortune to have as my teacher in Cambridge, once asked what purpose was served by so carefully preserving the constancy of the internal environment, and what dire consequences would befall if it were not preserved. The answer seems to be that though gross alterations are lethal, lesser variations show up first as disturbances of central nervous or mental function, like the distortions of consciousness that come with fever, water intoxication, hypoxia or hypoglycaemia; the mental confusion and coma that occur if the blood gets too acid, or the tetany that occurs if it is too alkaline. Barcroft (1934) concluded that the constancy of the internal environment is most necessary for the central nervous functioning of higher animals. Fine homeostatic controls had first to be evolved to set the stage before higher forms of nervous function, culminating in human consciousness, could emerge. 'To look for a high intellectual development in a milieu whose properties have not become stabilized is', he said, 'to seek music amongst the crashings of a rudimentary wireless or ripple patterns on the surface of a stormy Atlantic.'

Some recent experimental evidence suggests that the internal environment may actually be better controlled for the cells of the brain than for those of other tissues. Micro-electrodes have been applied to measure concentrations of some ions and of oxygen at pericellular sites. And the microenvironment of the cells has seemed more constant in the brain than in other organs (Silver, 1975).

There are also mechanical devices that help to stabilize the brain's environment. Thus the arrangement of the circulation helps to protect the mammalian nervous system from fluctuations in the composition of the blood. Blood from active tissues is mixed with blood from resting tissues and from the liver, and this mixed venous blood is passed through the lungs before it reaches the brain. The extent to which depletion of oxygen and loading with carbon dioxide have been corrected is monitored by chemoreceptors on the way to the brain and within the nervous system; and the output from these receptors serves to match respiratory exchange to metabolic needs. So the arterial blood distributed to the body at large, is, as it were, scrutinized and controlled to meet the brain's demands. The arrangement of the cardiovascular system also protects the brain against changes in temperature. Close apposition of their arteries and veins helps to keep heat in busy muscles, so that they can work a little above the general temperature; warmed blood from active tissues is mixed with venous blood from resting tissues and cooled blood from the skin before passing through the lungs to the brain. And temperature is regulated by neurones centrally placed in the hypothalamus. However, if the brain is especially favoured, it is fair to say that the organization of the circulation ensures that the rest of the cells can enjoy the tune that the brain's cells call.

The cells themselves help to maintain the constancy of their common internal environment. This is perhaps most evident in cells which guard the body's surfaces and control traffic across frontiers with the external environment. In aquatic vertebrates, cells lining the skin, gills, gut, renal tubules, and, in amphibia, the urinary bladder too, do much of the work of sustaining large differences in osmotic pressure and ionic concentration between the body fluids and the aquatic habitat, so that cells inside the body are less concerned with osmoregulation. Exchanges across the membranes of cells within the body are less obvious but no less important. Potassium-rich cells in a sodium-rich environment must work continuously to retain their stores of potassium and to avoid taking in sodium. Though the osmotic pressure is set for them, they must still carry out their own ionic regulation and control their volume. What they do to the internal environment is the mirror image of what they do for themselves. Their active accumulation of potassium keeps the concentration low outside. Just how important this is can be seen from the fact that the loss of a few per cent of a mammal's intracellular potassium could raise the concentration in the extracellular fluid enough to stop the heart. The cells cannot just bask in their constant internal environment. They have to work for it! Even what we call resting cells, which do not appear to be doing anything in particular, require uninterrupted supplies of metabolic energy.

One question which arises from discoveries since Bernard's time is whether the many circadian and other more or less periodic changes we know stand in contradiction to his principle of the constancy of the internal environment. Should we, as Janet Harker (1964) suggested, speak rather of a 'controlled' internal environment of which some features show periodic fluctuations? Claude Bernard was well aware of diurnal variations in the temperature of the body. They were of plus or minus around 1 °C, and he characterized them as 'rather feeble fluctuations'. About an average of 310 °K, they amount to no more than plus or minus a third of 1 %, and in no way invalidate his conclusion that the temperature of the body is remarkably constant. Other prominent rhythms, like those of pulse rate and the excretion of water, Na^+, K^+ and H^+, do not of course necessarily imply parallel changes in concentrations in the extracellular fluid, though daily variations in the renal excretion of phosphate do follow the concentration in the plasma (Mills, 1966). The largest documented variations in concentration in the blood are probably those of adrenocortical and other hormones, which Bernard could not have known about. These variations appear to be partly endogenous, so that even if they share the timing of events like the sequence of day and night it is not certain that they are directly determined by changes in the external environment. They could be related to patterns of feeding or other habitual behaviour. The concentrations of hormones are most likely to reflect their role in the homeostatic mechanisms which control the volume of the extracellular fluid and the concentrations of its major constituents. Hence fluctuations in the concentrations of hormones might be classed with the nervous impulses which Claude Bernard regarded as regulating the harmony and preserving the conditions of life in the internal environment. They help the organism to adapt to its surroundings and to maintain its integrity and independence. This must frequently require actions to compensate for changes in the external environment.

Indeed, by the end of his life Claude Bernard had come to realize that the maintenance of a stable internal environment was an active, dynamic process, in which, as he said in 1878, 'external variations are at each instant compensated for and equilibrated. Therefore, far from being indifferent to the external world, the higher animal is, on the contrary, constrained in a close and masterful (or wise) relation with it, of such fashion that its equilibrium results from a continuous and delicate compensation, established as if by the most delicate of balances.' There is nothing static about this idea of fixity. Bernard seems to have realized that homeostatic regulation implies variation in the quantity that is controlled; this variation is required to trigger the controlling mechanism. Once it is conceded that the independence of the organism is not achieved by mere isolation or indifference but by a controlled interaction with the external environment, the contradiction between overall stability and incidental fluctuations ceases to be important.

I would now like to consider a possible extension to Claude Bernard's doctrine of the internal environment. For him the cell was the indivisible unit of living matter. We might be tempted to locate the seat of vital activity inside the cell, at the surfaces of enzymes, or specific receptors, or organelles that electron microscopy has revealed to us. These sites are no more in direct contact with Claude Bernard's *milieu intérieur* than the cells are in contact with his external cosmic environment. They are shut off by selectively permeable cell membranes and bathed in intracellular fluids which are middlemen for exchanges within cells just as extracellular fluid is the middleman for exchanges between cells. The intracellular fluids, not Claude Bernard's *milieu intérieur*, form the immediate environment of intracellular enzymes, mitochondria and other organelles. Eugene Robin (Robin & Bromberg, 1959) suggested that we might speak of an *external* internal environment, which is the extracellular fluid, and an *internal* internal environment which each cell possesses within its own membrane. Claude Bernard's *milieu intérieur* is internal to the animal but external to its cells, a public environment shared by all the cells and especially important for communication and integration within the body. Robin's *internal* internal environment is private to each cell, and intimately concerned in many cellular functions. For one example, muscle cells contract because their myofibrils contract. The myofibrils do not exist in Claude Bernard's *milieu intérieur*, but in the intracellular fluid which contains far more potassium and far less sodium and calcium. The concentration of calcium has to be several orders of magnitude less than that in the extracellular fluid; and its changes control the sequence of contraction and relaxation. Thus the concentration of calcium can *not* be constant in the internal internal environment of the muscle cell. But are there bulk constituents whose concentrations are kept constant; and how far does Claude Bernard's principle of constancy extend to Robin's internal internal environment?

There do seem to be a few properties which should be approximately uniform through all the body fluids. Cells can hardly differ appreciably in temperature or in osmotic pressure from the fluids around them. Hence the internal internal environment should share the thermal and the osmotic stability of Claude Bernard's *milieu intérieur*. Diffusible solutes not subjected to active transport, like urea, and oxygen and carbon dioxide, should have similar concentrations in cell water and in the extracellular fluid, within the limits of the gradients needed to move them across cell membranes and within cells by diffusion. Other solutes, including most ions and many metabolites, have very different concentrations on opposite sides of cell membranes. There is no *a priori* reason why the concentrations of these solutes should be the same in all cells. They might be expected to be similar in cells of the same kind and state of activity. There is nonetheless a rather remarkable uniformity in the sense that cells generally are rich in potassium and contain little sodium and chloride. But this composition is maintained dynamically as a steady state by the cells' own activity, not primarily by the mechanisms which stabilize the *milieu intérieur*. Free-living cells, like Protozoa, which live in Claude Bernard's external, cosmic environment, are similarly rich in potassium and poor in sodium.

The internal internal environment cannot be a single homogeneous phase like the *milieu intérieur*. It exists in as many separate portions as there are cells. And even apart from their minute size, it is difficult to make reliable estimates of the concentrations in the fluid inside cells. It cannot be assumed that all the ions found by analysing tissue extracts were ever uniformly dissolved in the cell water. Although the sodium and potassium of some invertebrates' giant nerve fibres seem to be entirely in solution and free to diffuse, this may not be so in mammalian cells with their more complex ultrastructure and extensive intracellular membranes. One consequence of the vast areas of membrane in many cells is that no part of the cell water can be far from a surface. Bernal (1965) suggested that intracellular surfaces are covered by layers of water about half a dozen molecules thick which behave more like ice than liquid water. Individual molecules are relatively immobilized, and ions may be excluded from solution. If any substantial part of the water in cells differs in such ways from water in bulk, concentrations cannot be expected to be uniform throughout the cellular water. (Indeed if the thermodynamic activities of the solutes are to be uniform, the concentrations will have to vary.)

Another consequence of the minute subdivision of water in cells was pointed out by Sir Rudolph Peters in 1949 in an address to the British Association for the Advancement of Science. The reaction inside many cells is believed to be near pH 7. That means there is about 1 mol of H^+ in each 10 million litres of water; and since a litre of water contains 55 mol, there should be about one H^+ for each 550000000 water molecules. Peters added that one can *see* 550000000 water molecules. They would fill a cube with a side measuring 0.25 μm. There are spaces in cells of this order of size, for example between the cristae of mitochondria. If pH 7 means there is one H^+ in such a region, pH 6.7 means there are two, and pH 6.4 four. But what about pH 7.3 and 7.6? Can you have half an H^+, or a quarter of an H^+? In less than a few thousand million water molecules pH can only be a continuous variable if it means something like a time average or the probability of finding an H^+ there. In the 0.25-μm cube, pH 7.3 could mean there was an H^+ there half the time. It may be that the ordinary laws of physics and chemistry, on which Claude Bernard pinned his faith, do not operate in quite the same way inside cells after all. The working rules of everyday chemistry are statistical laws which give precise predictions in test tubes and vats, but they may not do so where molecules are few enough to be handled one by one, as enzyme systems in effect do handle molecules. But maybe we should not be too worried. H^+ ions are an extreme example because their absolute concentration is very small, and from Drabkin's (1975) estimates of their volumes there should be around 4000 H^+ ions in an erythrocyte and 150000 in a liver cell.

With intracellular pH so hard even to define it is not surprising that attempts to measure it have led to discrepancies which are still not resolved. Measurements of the distribution of weak acids or bases between cells and interstitial fluid have indicated that the interior of mammalian muscle cells is close to pH 7, while direct measurements with the micro-electrodes used by Carter, Rector, Campion & Seldin, (1967) give results close to pH 6. The discrepancy may mean simply that the methods measure quite different things. A micro-electrode presumably samples a minute portion of intracellular fluid near its tip, while analytical methods give averages over all the phases in large numbers of cells. The uncertainty is tantalizing, because cell fluids constitute the bulk of the body's total fluids, and the body's homeostasis of H^+ is fantastically precise. Forty litres of human body water at pH 7 would contain 0.004 mmol of H^+. This is ordinarily kept constant within perhaps 10% despite the daily production of 50 to 80 mmol of strong inorganic acid by metabolism. The daily variation in total body H^+ is about half of one second's production. This is the accuracy of a watch that keeps time better than to one second a day: and it implies that at least one important concentration must be well stabilized in the internal internal environment! How is this done?

Cells generally seem to be freely permeable to the unchanged molecules of carbon dioxide, and so they should have at least as great a pCO_2 as the extracellular fluids. They have, however, about half the concentration of bicarbonate, so they should be more acid (or less alkaline). The membranes do not seem to be so freely permeable to HCO_3^-, so that the concentration of bicarbonate in the cells does not immediately reflect changes in external (bicarbonate). It seems to change only slowly, over hours or days, while intracellular pCO_2 follows external changes in seconds or minutes. Because of this, the pH in the cells and the pH in the extracellular fluids may sometimes change in opposite directions (Robin & Bromberg, 1959). Respiratory disturbances that alter pCO_2 shift intracellular and extracellular pH in the

same direction. But acidification of the blood stimulates breathing by way
of the carotid body chemoreceptors and this lowers pCO_2 so that the cells
are overprotected and become more alkaline. Similarly, alkalinization of
the blood leads to retention of carbon dioxide and this should make the
cells more acid. This behaviour suggests that the cell fluids do form a sort
of internal internal environment which is partly independent of Claude
Bernard's *milieu intérieur*.

We can write an approximate Henderson–Hasselbalch equation for cell
fluid;

$$pH_{cell} = 6.1 + \log \frac{(HCO_3^-)_{cell}}{0.03 \ pCO_{2_{plasma}}} \tag{1}$$

The cells would control the numerator while the mechanisms that fix
pCO_2 in the *milieu intérieur* control the denominator. The concentration of
carbon dioxide in Claude Bernard's internal environment is therefore an
important part of the mechanism that sets the reaction in the cells, and we
might expect there to be some way of controlling it to suit their needs.

There is one important collection of extracellular fluid which is separated
off from the general body of extracellular fluid in the sense that it is accessible only by crossing cellular membranes. This is the cerebrospinal fluid
(CSF), isolated by the blood–brain barrier and the choroid plexuses, and
often characterized as a 'transcellular fluid'. It resembles the intracellular
fluids in that its pCO_2 equilibrates quickly with the blood, while its concentration of bicarbonate is independent, and takes hours or days rather than
minutes to adjust. Consequently the pH of the CSF responds to a number
of circumstances very much as the cell fluids do. Retention of carbon
dioxide makes the CSF and cells, as well as the blood, more acid, and loss of
carbon dioxide makes all three fluids more alkaline. But if addition of a
non-carbonic acid makes the blood acid and lowers pCO_2, this should make
both the cells and the CSF more alkaline. Likewise, retention of carbon
dioxide by reduction of pulmonary ventilation when alkali is added to the
blood should make both the CSF and the cells more acid. In these acute
disturbances, CSF pH varies in the same way as cell pH, no matter whether
this is in the same direction as extracellular fluid pH or opposite. Hence the
pH of the CSF should reflect the intracellular pH over the body as a whole.
Moreover, since, unlike blood and cell fluids, the CSF contains practically
no buffer base other than bicarbonate, the pH of the CSF should be more
sensitive than the pH of the cell fluids to alterations in pCO_2, and should
provide an amplified index of changes in pH within the cells. This makes
the control of the pH of the CSF unusually interesting, for the mechanisms
which control the pH of CSF could also be controlling pH in the internal
internal environment inside the cells.

An approximate Henderson-Hasselbalch equation for the CSF is very like that for the cells:

$$pH_{csf} = 6.1 + \log \frac{(HCO_3^-)_{csf}}{0.03 \ pCO_{2_{plasma}}} \tag{2}$$

The pH is determined by the concentration of bicarbonate in the CSF and the pCO_2 transmitted across the blood–brain barrier. There is a very efficient short-term regulation which keeps the reaction within 0.1 of a unit of pH 7.32 in man by the respiratory control of pCO_2. The pH of the CSF is monitored by intracranial chemoreceptors on or near the surface of the medulla which act to increase pulmonary ventilation if the CSF becomes too acid and to decrease ventilation if the CSF becomes too alkaline. When pCO_2 increases throughout the body (as when production increases during exercise) the intracranial chemoreceptors reinforce the peripheral ones in the carotid bodies, and pulmonary ventilation increases briskly. But when the reaction in the cells and the CSF shifts in an opposite direction to that in the blood, the intracranial chemoreceptors can act in opposition to the peripheral ones and protect the reaction of the CSF, the brain, and the cells, in preference to that of the blood. A greater sensitivity of the intracranial chemoreceptor reflexes due to the absence of non-bicarbonate buffers from the CSF may give Robin's internal internal environment a higher priority than Claude Bernard's *milieu intérieur*.

Longer-term control depends upon slower adjustments over a few hours to a day or so, as new CSF is secreted with a concentration of bicarbonate that is roughly proportional to the partial pressure of carbon dioxide in the blood perfusing the choroid plexuses. Consequently if an alteration in pCO_2 is prolonged, the initial change in pH of the CSF and the effect of this on respiration both fade. There may be corresponding slow adjustments in the cells, but we do not know much about these as yet. In the meantime it seems not unreasonable to think of the intracranial chemo-receptors as indirectly monitoring pH in the internal internal environment in the cells. The peripheral chemoreceptors in the carotid bodies seem concerned rather with the reaction of Claude Bernard's *milieu intérieur*; but more importantly with its pO_2 and the supply of oxygen to the cells.

Some recent work on poikilotherms throws new light on the control of pH. Rahn and Robin found that as a number of animals, snapping turtles, carp, toads and bullfrogs, became adapted to temperatures from 40 °C down towards 0 °C, the pH of the blood did not remain constant. It rose, as the pH of pure, neutral water rises, when the temperature falls and the ionization of the water decreases. The turtles' blood changed from pH 7·6 at 35 °C to pH 8 at 10 °C, and so was about 0.7 pH unit on the alkaline side of neutrality at all temperatures. Constancy of the internal environment might suggest keeping the concentration of H^+, or pH, constant. But instead of keeping the pH constant when the temperature fell, these animals lowered the concentration of H^+ and kept the ratio of (OH^-) to (H^+) constant, at a little over 20 to 1. The turtles did this actively, by maintaining a constant rate of respiratory ventilation while the metabolic rate, the oxygen consumption and the production of carbon dioxide all fell to about half for each 10 °C drop in temperature (Rahn, 1974). This reduced pCO_2, so that the blood, and presumably the cells also, became alkaline as the temperature was lowered. The change in the cells has recently been confirmed for skeletal and heart muscle of dogfish (Heisler, Weitz & Weitz, 1976).

Rahn puzzled over the question why pH should be constrained to vary with temperature in this way, and noticed the curious fact that the imidazole groups of histidine residues in proteins have nearly the same temperature coefficient of ionization as has water. Consequently a system which causes pH to vary parallel with the neutral point will maintain a constant state of ionization of imidazole groups, which are about half ionized at prevailing intracellular pH, and so provide good buffering to hold the intracellular fluids near neutrality. This is likely to be important for the working of intracellular enzymes. It could also be important for receptors controlling respiration which set the partial pressure of carbon dioxide throughout the body fluids. This is an exciting idea, because receptors activated by the state of ionization of intracellular imidazole groups would place an important factor of the *milieu intérieur* under the control of a primary factor in an internal internal environment that Claude Bernard had not envisaged. (The argument cannot be applied directly to mammals, because they fix their body temperatures. Consequently it is not clear whether they are preserving primarily the pH or the ratio of (OH^-) to (H^+) in their body fluids. In either event they must still be fixing the degree of ionization of their imidazole groups.)

Rahn, Reeves & Howell (1975) emphasized the advantages of an intracellular pH near neutrality. Davis (1958) had noted that almost all of several hundreds of compounds taking part in intermediary metabolism are either carboxylated or phosphorylated, or, in fewer instances, have ionizable basic groups. Consequently these compounds are all fully ionized at pHs close to neutrality. Since cell membranes generally are far less permeable to ions than to uncharged molecules, a neutral pH keeps all the intermediates in a form in which they are effectively trapped inside the cells, whereas products of fermentation and waste products like urea are typically uncharged and can readily escape. Extracellular fluids seem usually to be kept a little more alkaline than the cells, and this makes them effective sinks for carbon dioxide and other acidic waste products. The ability to retain essential metabolites in the cells in small concentrations is important. These substances are so numerous that their individual concentrations must be kept small if they are to be accommodated together with important bulk ions like potassium within the limit set by the osmotic pressure of body fluids. Atkinson (1969) remarked that 'Conservation of low concentrations of metabolites, both individually and collectively, is one of the most fundamental requisites for a viable metabolizing system.' It seems that 'good' metabolites have been evoked which react rapidly, so that their concentrations remain low, and which are ionized, so that they are not lost from the cells.

It remains to consider whether Claude Bernard's concept of a stable internal environment as the basis for independence of life needs to be modified or extended. The stable *milieu intérieur* appears to have been evolved for the benefit of the cells. Departures from constancy upset cellular functions, and it would be appropriate if cellular changes initiated adjustments to restore and preserve optimal conditions. Robin's internal internal environment appears to share with the extracellular fluids three intensive properties which are transmitted through cell membranes – temperature, osmotic pressure and the partial pressure of carbon dioxide. These variables affect all cells throughout the body, but the adjustments called for to hold them constant are the responsibility of a relatively small number of cells in the central nervous system or in peripheral receptors. Thus changes in osmotic pressure in the *milieu intérieur* make all cells swell or shrink; but the adjustments of water balance required to correct disturbances of osmolality and restore cellular volume depend upon hypothalamic osmoreceptors which act through neurones signalling thirst, and through neurosecretory cells releasing antidiuretic hormone. Similarly, all cells are affected when the temperature of the body changes, but corrective adjustments depend mainly upon a relatively small number of cells strategically placed in the hypothalamus. Peripheral receptors in the carotid body and receptors in the CSF (which functions as an extension of the mass of cell fluids) control the partial pressure of carbon dioxide. This in turn is doubly important for controlling reaction throughout the body, for it appears directly in the denominator of the Henderson–Hasselbach equations for blood, CSF and intracellular fluids; and it also controls the numerator through its effect upon the secretion of bicarbonate into the plasma by the renal tubular epithelium and into the CSF by the choroid plexus (cf. Robinson, 1975).

Hence for temperature and the concentrations of water and carbon dioxide it seems reasonable to accept Robin's proposal of an internal internal environment as an extension of Claude Bernard's *milieu intérieur*. The same is probably true for any 'innermost' internal environments that might be postulated inside mitochondria or other intracellular organelles. But for substances to which cell membranes are not readily permeable, or which pumps keep in different concentrations on the two sides of the membrane, we cannot really speak of any common internal internal environment. Each cell is a world of its own, partly a law unto itself, as when by altering its concentration of bicarbonate it adjusts its own pH on the background of a partial pressure of carbon dioxide transmitted from the external fluid. But we can admire the way in which cells have acquired control over the constancy of their fluid matrix. Claude Bernard's statement that the nervous system regulates the harmony between the conditions in the internal environment has turned out to be prophetic. So long as we still accept the cell as the smallest unit that is fully alive we do not need to go far beyond his grand conception. His *milieu intérieur* remains the one essentially constant, public internal environment, and the stable foundation upon which each cell can build its own pattern of activity. The common medium protects the cells and supplies their needs without demanding any stultifying uniformity of individual behaviour. Our lives would be duller if our cells were all identical and had always to do the same things at the same time.

REFERENCES

Atkinson, D. E. (1969). Limitation of metabolite concentrations and the conservation of solvent capacity in the living cell. *Curr. Top. cell. Regul.* **1**, 29–42.

Barcroft, J. (1934). *Features in the architecture of physiological function.* London: Cambridge Univ. Press.

Bernal, D. J. (1965). The structure of water and its biological implications. *Symp. Soc. exp. Biol.* **19**, 17–32.

Bernard, C. (1865). *Introduction a l'étude de la médicine expérimentale.* Paris: Baillière et Fils.

Bernard, C. (1878–79). *Leçons sur les phénomènes de la vie communs aux animaux et végétaux.* Paris: Baillière et Fils.

Bert, P. (1878). Claude Bernard. Obituary notice in *An introduction to the study of experimental medicine* by Claude Bernard, translated by H. C. Greene, pp. xiii–xix. New York: Macmillan (1927).

Cannon, W. B. (1932). *The wisdom of the body.* London: Kegan Paul, Trench, Trubner & Co.

Carter, N. W., Rector, F. C., Campion, D. S. & Seldin, D. W. (1967). Measurement of intracellular pH of skeletal muscle with pH sensitive glass microelectrodes. *J. clin. Invest.* **46**, 920–33.

Davis, B. D. (1958). The importance of being ionized. *Arch. of Biochem. Biophys.* **78**, 497–509.

Drabkin, D. L. (1975). The environment of function of liver and red blood cells. *Ann. N.Y. Acad. Sci.* **244**, 603–23.

Harker, J. E. (1964). Diurnal rhythms and homeostatic mechanisms. *Symp. Soc. exp. Biol.* **18**, 283–300.

Heisler, N., Weitz, H. & Weitz, A. M. (1976). Extracellular and intracellular pH with changes of temperature in the dogfish *Scyliorhinus stellaris. Resp. Physiol.* **26**, 249–63.

Henderson, L. J. (1927). Introduction, in *An introduction to the study of experimental medicine*, by Claude Bernard, translated by H. C. Greene, pp. v–xii. New York: Macmillan.

Holmes, F. L. (1963). The milieu intérieur and the cell theory. *Bull. Hist. Med.* **37**, 315–35.

Holmes, F. L. (1967). Origins of the concept of the milieu intérieur. In *Claude Bernard and experimental medicine*, ed. F. Grande & M. B. Vischer, pp. 179–91. Cambridge, Massachusetts: Schenkman Publishing Company.

Mills, J. N. (1966). Human circadian rhythms. *Physiol. Rev.* **46**, 128–71.

Rahn, H. (1974). PCO_2, pH and body temperature. In *Carbon dioxide and metabolic regulations*, ed. G. Nahas & K. E. Schafer, pp. 152–62. New York: Springer-Verlag.

Rahn, H., Reeves, R. B. & Howell, B. J. (1975). Hydrogen ion regulation, temperature and evolution. *Am. Rev. resp. Dis.* **112**, 165–72.

Robin, E. D. & Bromberg, P. A. (1959). Editorial: Claude Bernard's milieu intérieur extended: intracellular acid–base relationships. *Am. J. Med.* **27**, 689–92.

Robinson, J. R. (1975). *Fundamentals of acid–base regulation*, 5th ed. Oxford: Blackwell.

Sherrington, C. S. (1940). *Man on his nature.* Cambridge: Cambridge Univ. Press.

Silver, I. A. (1975). Measurement of pH and ionic composition of pericellular sites. *Phil. Trans. R. Soc.*, B **271**, 261–72.

Index

ATP: released with ACh at neuromuscular junction, and with NA at sympathetic nerve terminals, 39; suggested as transmitter at inhibitory synapses in gastrointestinal tract, 38–9

ATPase, membrane-bound: and clearance of potassium from extracellular fluid, 192; digitoxigenin and, 186, 193

atropine: reduces time of opening of ion channel by ACh, 14; and response to phenyl diguanide in *Squalus*, 134; and salivary secretion, 35–6

autonomic nervous system: electrical transmission known at only two sites in, 23; responses in, to fastigial stimulation, activating, 351–5, and inhibiting, 356–7

axnos, fusimotor (γ): dynamic, supplying mainly bag$_1$ fibres, 49, and static, supplying both bag and chain fibres, 48, 49–50; motor endings of (p$_1$ plates, p$_2$ plates, trail), 46, 47–8; response patterns of, showing range from purely dynamic to purely static, 50–2, through mixtures obtained by simultaneous stimulation of pairs of single axons, one dynamic and one static, 52–3; unmyelinated terminals of, 87; *see also* muscle spindles

axons, skeleto-fusimotor (β), 53–4, 55; act through p$_1$ plates? 56–7; spindle fibres supplied by (mainly bag$_1$, some longchain), 54, 56

axons, skeleto-motor (α), 53; increased size of motor units of, in partly denervated muscle, 316–17, 319–21, 327–9; regeneration of, 310, 321; sprouting of collateral, after partial transection, 309, 316, 317, 325, 326; 'threshold' concentration of denervated fibres necessary to evoke sprouting of? 327, 329

baroreceptor reflexes, *see* carotid sinus

behaviour: development of cooperative, 384–5, 403, 408, 409; embryology of, 386–8; learning of, *see under* learning; modifiability of, by experience, 301

Bernard, Claude, on stability of internal environment, 413–16, 418; ideas of, still hold, with extension to individual intracellular environments, 420, 426

bicarbonate ions: in cells of extracellular fluids, 421–2, 426; in cerebrospinal fluid, 422–3; secretion of, into plasma by renal tubular epithelium, and into cerebrospinal fluid by choroid plexus, 425

blood, and maintenance of stable internal environment (Bernard), 414, 415, 416

blood pressure: decrease in, as response to increased CSTMP, 227–9, effected by fall in peripheral resistance, 228; increase

in, as response to decreased CSTMP, 227, effected by increase in cardiac output, and perhaps also splanchnic vasoconstriction, 228; increased by fastigial stimulation, 351–2; responses of, to change of CSTMP, in hypertensive subjects, 229, and in rabbits, 230, 231–2; in *Squalus*, affected by phenyl diguanide, and by section of branchial nerves, 133–5, 140

blood volume: anoxia and, 197, 198; increase of, in response to stimulation of dorsal root of spinal cord, 194, 195

botulinum toxin, sprouting from motor axons in muscle treated with, 326

bradykinin, excitant of nociceptors, 294

brain: anatomy of imagery and intention in, 388–90; angiotensin effects in, 333, 335–6, 338; angiotensin receptors in membranes of, 334, 339–44; angiotensin and renin produced in, 334, 339, 344; asymmetry of, 393–4, 402; connections of fastigial nuclei in, 358; control of environment of cells of, 416–17; development of, 386–7, 388, 401–2, (in human infants) 403–6; different kinds of mind in two halves of, 394, 397–401; learning only possible where circuitry already exists in, 306; memory not located in any particular part of, 302; memory as small modifications of intercellular connectivity in, 306; normal duplicity of, 401–3; split (with cerebral hemispheres disconnected), *see* cerebral commissurotomy

breathing, during exercise, 235–8; chemical drive in, 236, 240–1, 249–50; experiments on neural mechanisms for increase of frequency and tidal volume in, (in anaesthetized animals) 241–2, (in humans) 247–9, (in unanaesthetized animals) 242–7; neural inspiratory drive in, 238–40; neurogenic theories of control of, 236–7; ponto-medullary organization in control of, 236, 239–41; specific neural mechanisms in control of, 250; in *Squalus*, affected by phenyl diguanide or section of branchial nerves, 133–5, 140

bulbopontine oscillator, in control of breathing during exercise, 241–2, 250

calcium ions: can partially replace sodium in maintaining response of isolated muscle spindles, 66–7, 71; on inside of nerve membrane, in secretion of ACh at nerve terminal, 4–5; provide reactive attachment sites for ACh? 8; and responses of isolated carotid body, 121, 123

carbachol, depolarization of endplate by, 12–13